D0930183

# REPRODUCTION AND FITNESS IN BABOONS

DEVELOPMENTS IN PRIMATOLOGY: PROGRESS AND PROSPECTS

**Series Editor:**
Russell H. Tuttle
University of Chicago, Chicago, Illinois

---

This peer-reviewed book series will meld the facts of organic diversity with the continuity of the evolutionary process. The volumes in this series exemplify the diversity of theoretical perspectives and methodological approaches currently employed by primatologists and physical anthropologists. Specific coverage includes: primate behavior in natural habitats and captive settings: primate ecology and conservation; functional morphology and developmental biology of primates; primate systematics; genetic and phenotypic differences among living primates; and paleoprimatology.

**THE GUENOUS: DIVERSITY AND ADAPTATION IN AFRICAN MONKEYS**
Edited by Mary E. Glenn and Marina Cords

**ANIMAL MINDS, HUMAN BODIES**
Edited by W.A. Hillix and Duane Rumbaugh

**COMPARATIVE VERTEBRATE COGNITION**
Edited by Lesley J. Rogers and Gisela Kaplan

**ANTHROPOID ORIGINS: NEW VISIONS**
Edited by Callum F. Ross and Richard F. Kay

**MODERN MORPHOMETRICS IN PHYSICAL ANTHROPLOGY**
Edited by Dennis E. Slice

**BEHAVIORAL FLEXIBILITY IN PRIMATES: CAUSES AND CONSEQUENCES**
By Clara B. Jones

**NURSERY REARING OF NONHUMAN PRIMATES IN THE 21ST CENTURY**
Edited by Gene P. Sackett, Gerald C. Ruppenthal and Kate Elias

**NEW PERSPECTIVES IN THE STUDY OF MESOAMERICAN PRIMATES: DISTRIBUTION, ECOLOGY, BEHAVIOR, AND CONSERVATION**
Edited by Paul Garber, Alejandro Estrada, Mary Pavelka and LeAndra Luecke

**HUMAN ORIGINS AND ENVIRONMENTAL BACKGROUNDS**
Edited by Hidemi Ishida, Martin Pickford, Naomichi Ogihara and Masato Nakatsukasa

**PRIMATE BIOGEOGRAPHY**
Edited by Shawn M. Lehman and John Fleagle

**REPRODUCTION AND FITNESS IN BABOONS: BEHAVIORAL, ECOLOGICAL, AND LIFE HISTORY PERSPECTIVES**
Edited By Larissa Swedell and Steven R. Leigh

# REPRODUCTION AND FITNESS IN BABOONS: BEHAVIORAL, ECOLOGICAL, AND LIFE HISTORY PERSPECTIVES

Edited by

Larissa Swedell
*Department of Anthropology, Queens College-CUNY*
*New York Consortium in Evolutionary Primatology*

and

Steven R. Leigh
*Department of Anthropology, University of Illinois*

Larissa Swedell
Department of Anthropology
Queens College-CUNY
Flushing, NY 11367-1597
USA
larissa.swedell@qc.cuny.edu

Steven R. Leigh
Department of Anthropology
University of Illinois
109 Davenport Hall
Urbana, IL 61801
USA
sleigh@uiuc.edu

Library of Congress Control Number: 2005937077

ISSN 1574-3489
ISBN-10: 0-387-30688-9 e-ISBN 0-387-33674-5
ISBN-13: 978-0387-30688-9

Printed on acid-free paper.



Printed in the United States of America. (SPI/MVY)

9 8 7 6 5 4 3 2 1

springer.com

# CONTENTS

# CONTRIBUTORS

**Susan C. Alberts,** Duke University, Durham, NC, USA, alberts@duke.edu

**Jeanne Altmann,** Princeton University, Princeton, NJ, USA, altj@princeton.edu

**Louise Barrett,** University of Liverpool, Liverpool, UK and University of KwaZulu-Natal, Durban, South Africa, LBarrettl@uclan.ac.uk

**Jacinta C. Beehner,** Jacinta C. Beehner, University of Michigan, Ann Arbor, MI, USA, jbeehner@princeton.edu

**Thore J. Bergman,** Thore J. Bergman, University of Michigan, Ann Arbor, MI, USA, thore@sas.upenn.edu

**Robin M. Bernstein,** George Washington University, Washington, DC, USA, robinb@gwu.edu

**Dorothy L. Cheney,** University of Pennsylvania, Philadelphia, PA, USA, cheney@sas.upenn.edu

**Julia Fischer,** German Primate Center, Goettingen, Germany, jfischer@dpz.gwdg.de

**Gerard Galat,** Institut de Recherche pour le Développement (IRD), France, gerard.galat@wanadoo.fr

**Anh Galat-Luong,** Institut de Recherche pour le Développement (IRD), France, luong@orleans.ird.fr

**Suzanne Hagell,** City University of New York, New York Consortium in Evolutionary Primatology, New York, NY, USA, sehagell@gmail.com

**S. Peter Henzi,** University of Central Lancashire, Preston, UK and University of KwaZulu-Natal, Durban, South Africa, phenzi@uclan.ac.uk

**Sara E. Johnson,** California State University, Fullerton, CA, USA, sjohnson @fullerton.edu

**Clifford J. Jolly,** New York University, New York Consortium in Evolutionary Primatology, New York, NY, USA, clifford.jolly@nyu.edu

**Dawn M. Kitchen,** The Ohio State University, Columbus, OH, USA, kitchen.79@osu.edu

**Steven R. Leigh,** University of Illinois, Urbana, IL, USA, sleigh@uiuc.edu

**John E. Lycett,** University of Liverpool, Liverpool, UK and University of KwaZulu-Natal, Durban, South Africa, lycett@liv.ac.uk

**Ryne A. Palombit,** Rutgers University, New Brunswick, NJ, USA, rpalombit@anthropology.rutgers.edu

**Jane E. Phillips-Conroy,** Washington University School of Medicine, St. Louis, MO, USA, baboon@pcg.wustl.edu

**Drew Rendall,** University of Lethbridge, Lethbridge, Canada, d.rendall@uleth.ca

**Julian Saunders,** University of KwaZulu-Natal, Durban, South Africa, juls_saunders@yahoo.com

**Robert M. Seyfarth,** University of Pennsylvania, Philadelphia, PA, USA, seyfarth@psych.upenn.edu

**Joan B. Silk,** University of California, Los Angeles, CA, USA, jsilk@anthro.ucla.edu

**Larissa Swedell,** Queens College, City University of New York, New York Consortium in Evolutionary Primatology, New York, NY, USA, larissa.swedell@qc.cuny.edu

**Monica Uddin,** Wayne State University School of Medicine, Detroit, MI, USA, muddin@mac.com

CHAPTER ONE

# Perspectives on Reproduction and Life History in Baboons

*Larissa Swedell and Steven R. Leigh*

## INTRODUCTION

This volume explores reproductive behavior, social organization, and life history in baboons of the genera *Papio* and *Theropithecus*, contributing to a nascent discussion of the interrelations among these variables in recognition of their tremendous impacts on fitness (S. A. Altmann, 1998; Alberts and J. Altmann, 2002; Kappeler et al., 2002). Complicated sex- and age-specific strategies and tactics mediate ties among these variables, resulting in considerable diversity depending on ecological conditions, social variables, survivorship, population size, and age structure. The complexity of relations among these variables opens significant opportunities to enhance our understanding of primate adaptation and evolution. Our view is that processes of mating and ways of investing in offspring are related in extremely important, but often neglected, ways. This book aims to address ties between reproduction and life history variation in order to understand the evolution of social, behavioral, genetic, and morphological diversity. We direct our attention primarily to a single genus (*Papio*) that is characterized by remarkable variation in reproduction

---

**Larissa Swedell** • Department of Anthropology, Queens College, CUNY and the New York Consortium in Evolutionary Primatology, New York, NY, USA **Steven R. Leigh** • Department of Anthropology, University of Illinois, Urbana, IL, USA

and life history, providing exceptional and perhaps unparalleled opportunities to appraise issues about these aspects of life.

Our exploration of the links between reproductive behavior and life history centers on variables defined as fitness components, or factors directly related to reproductive success (Charlesworth, 1994; Hughes and Burleson, 2000). Fitness components include a wide array of traits, including, among others, mate competition and attraction, offspring growth rates, and age at maturation. While the relationship of fitness to specific fitness components can be straightforward, the relations among these variables and the ways in which they fit into overall courses of life histories remain largely unexamined (but see J. Altmann et al., 1988; Kappeler et al., 2002). This is unfortunate because the behaviors associated with both mating and offspring rearing occupy the most important life history phases in primates, with the greatest impacts on fitness. Moreover, the relationship between these two particular life phases—mating behavior and offspring rearing—remains poorly understood. Certain aspects of social organization, such as dominance rank and ways of maintaining rank or of acquiring mates, may be dynamically interrelated with the attributes of offspring. For example, from the adult perspective, long interbirth intervals (IBIs) may limit opportunities for mating and increase the risk of infanticide. From the offspring's viewpoint, maternal rank and condition affect growth trajectories, body condition, and age at maturation. This kind of complexity, particularly relationships among variables such as rank, morbidity, and age at maturation, requires exploration in various contexts, including analyses of both adults and offspring. Thus, at a broad theoretical level, this volume examines the relations of fitness components to one another at two especially important life history periods. Our major goal in this volume is to evaluate how patterns of behavior associated with rank attainment, mating, and reproduction interdigitate with ecology and life history attributes, particularly those involving allocation of reproductive effort and rearing of offspring.

## BABOONS IN PERSPECTIVE

This volume is largely restricted to baboons of the genus *Papio*, although Uddin et al. do consider data from *Theropithecus gelada* and Leigh and Bernstein include data from several papionin genera in their analyses. (In order to streamline the volume, the term "baboon" is used only in reference to *Papio*.) The genus *Papio* is widespread across Africa, being perhaps the most commonly observed African primate besides humans. *Papio* baboons are taxonomically diverse as well, occurring in at least five main forms recognized to date: hamadryas (*Papio hamadryas hamadryas*), olive (*P. h. anubis*), yellow (*P. h. cynocephalus*), chacma (*P. h. ursinus*), and Guinea (*P. h. papio*). The complexities of biogeography and phenotypic variability among baboons (Jolly, 1993, 2003; Frost et al., 2003), combined with a lack of reproductive isolation at most species/subspecies borders—in particular the well-documented hybrid zone between hamadryas and olive baboons in Ethiopia—suggest that a single-species classification for baboons may be most appropriate (Phillips-Conroy & Jolly, 1986; Williams-Blangero et al., 1990; Jolly, 1993, 2003; Frost et al., 2003; Disotell, 2000; Alberts and Altmann, 2001). Despite their classification as a single species, evolutionarily significant differences characterize baboon subspecies with respect to behavior, adult morphology (Jolly, 1993, 2003; Frost et al., 2003), and some aspects of development (Leigh, in press). As Jolly (1993, 2003) has perceptively recognized, this kind of patterned diversity provides opportunities to study evolutionary dynamics. While baboon taxonomy is still controversial and no one classification is universally accepted, we follow Groves (1993) and Jolly (1993, 2003) in adopting the single-species classification in this volume. Regardless of taxonomic preferences, phylogenetic relationships—both within the genus *Papio* (Newman et al., 2004) and among genera (Disotell, 1994)—are now relatively well understood, facilitating phylogenetically informed comparisons at a variety of taxonomic levels. In part because of their close phylogenetic relations, baboons provide exceptional opportunities to answer questions about life history periods, life history phases, social organization, reproductive behavior, fitness components, and the relations among these variables.

Beyond phylogenetic issues, several other desirable characteristics define baboons as excellent candidates for this kind of investigation. First, recent analyses have provided compelling evidence that the attributes and capabilities of extraordinarily young baboons significantly impact lifetime reproductive success (S. A. Altmann, 1998). The juvenile phase thus requires analytical weight equal to that of adult studies in understanding evolutionary dynamics. Second, the genus *Papio* shows a surprising array of variation in social structure (size and composition of groups) and social organization (patterns of social and sexual interactions within groups). For example, we see a range from the strict, male-driven multilevel social structure characteristic of hamadryas baboons (*P. h. hamadryas*) to the looser, multimale/

multifemale groups with female philopatry and matrilineal dominance hierarchies that typify olive and yellow baboons. Significant variability occurs even within subspecies, most notably among chacma baboons. Third, hybrid baboons—particularly between the two extremes of hamadryas on the one hand and olives and yellows on the other—commonly express intermediate characteristics. As several of our contributors discuss, the presence of these intermediates provides excellent opportunities to explore the genetics and evolution of social behavior, setting the foundation for investigations of how variables such as behavior, social organization, and life histories evolve at a genetic level. Fourth, the genus occupies an impressive range of habitats, providing ideal opportunities for "natural experiments" on the relations between ecological variables, reproduction, and life history. At the same time, the presence of hybrid zones and areas of geographic overlap among subspecies facilitates analyses that effectively control for large-scale habitat differences. Finally, understanding the relations between reproduction and life history mandates multigenerational, longitudinal data. Baboons are ideal subjects for such analyses because several field studies have spanned decades.

## REPRODUCTIVE BEHAVIOR AND LIFE HISTORIES

Fundamental tenets of parental investment and sexual selection, as defined in classic theoretical contributions (Darwin, 1871; Fisher, 1930; Trivers, 1972), anticipate sex differences in reproductive behavior, investment in offspring, and the course of life histories. These tenets predict that, when parental investment is asymmetrical, the sex that invests more in offspring is effectively a limiting resource. As a consequence, the sex that invests less should compete to gain access to members of the sex that invests more. Life history considerations have a prominent role to play in this framework. In mammals, females produce a tiny fraction of the number of gametes produced by males, release them at long intervals, expend extra energy on pregnancy and lactation, and usually invest more in offspring postpartum than males. Female mammals are thus limited to relatively few potential offspring and, from the outset, invest much more than males in each individual offspring. Factors that impact life histories, such as the length of the infant and juvenile periods, also influence female energy allocation (Altmann et al., 1978). Moreover, energy investment may vary with time, such that mothers experience peak periods of energy investment in offspring while males may expend variable amounts of energy on reproduction, varying their investments by season or with changes in group composition.

Given these principles, we generally expect males to focus on gaining access to females so as to increase offspring *quantity*, while females prioritize investments that maximize the *quality* of each offspring. At a more refined level, we expect sex differences in energy investments to covary with ecological conditions, group size, and composition. In terms of life histories, bimaturism should characterize baboons, with sexual selection favoring a long male developmental period culminating in large body size, significant canine weaponry, and perhaps the social skills needed to gain reproductive opportunities (Wiley, 1974; Jarman, 1983; Leigh, 1995; Leigh et al., 2005). Once adult, investment in mating opportunities should comprise the largest proportion of male reproductive energy allocation, because the greatest factor contributing to his fitness is his access to female mates and number of successful fertilizations. A male's fitness is also affected by the survival of his offspring, a function of both developmental rates and maternal investment.

Male baboons present a fascinating array of variation in terms of how these general goals are met. For example, a primary concern with access to mates may translate into an exclusion strategy, as seen in hamadryas baboons, whereby a male defends a group of females from all other males and gains exclusive reproductive access to those females for the length of his tenure. More commonly, though, male baboons cannot defend a group of females exclusively and instead tolerate other males in a group and compete for access to females only when they are in estrus. In any case, the allocation of male effort over the lifetime, shaped by immediate ecological and social considerations, is inherently a life history problem. This problem centers on classic tradeoffs between current and future reproduction (Fisher, 1930; Williams, 1966; Roff, 2002) as well as tradeoffs between reproduction and somatic maintenance (van Noordwijk and de Jong, 1986). Tradeoffs occur both in the direct process of insemination and in terms of social bonds to maintain access to females. Baboons are particularly interesting in this regard because of the sheer number of ways in which males seem to cope with such tradeoffs. Several contributors to this volume consider the implications of differences in male reproductive strategies for behavior, physiology, and the evolution of reproduction and life history in baboon males (notably, Bergman in Chapter 4 and Jolly & Phillips-Conroy in Chapter 11).

Female baboons also express a variety of reproductive options related to classic life history tradeoffs, so that no uniform pattern holds across all taxa. For females, life history theory has an especially vital role to play in defining these options by providing "an elaborate answer to the simple question of why having more offspring is not always selected for" (van Noordwijk and de Jong, 1986, p. 137; see also Williams, 1966). Kappeler et al. (2002) have identified a number of links between life histories and social behavior in primates, emphasizing life history variables that are likely to impact social organization. For example, female investment in the form of gestation and lactation, rates of infant development, and lifespan duration all influence how males and females allocate reproductive effort. As noted above, the general expectation is that the most important factor contributing to female fitness is the degree to which the survival and overall "quality" of each offspring can be maximized. This means ensuring that each infant is as healthy as possible (through adequate nutritional intake by the mother and/or the infant) and survives to reproductive age and beyond. The optimal allocation of reproductive effort for a female may include conceiving, giving birth, and/or weaning at the most appropriate time (with regard to maximizing food resources for herself or her offspring at critical periods); choosing the "best" mates (either to maximize offspring quality or to promote offspring survival); increasing access to high-quality mates by inciting male–male competition (either agonistic or sperm competition); and competing effectively against other females (so as to increase resources available for her own offspring). This array of reproductive considerations results in the expression of significant variation among baboon female reproductive and life history strategies and tactics. Several contributors to this volume illustrate variation in "optimal" reproductive strategies and offspring investment by female baboons. Notably, Barrett et al. offer an "elaborate answer" to a seemingly simple question about allocation of investment in offspring, defining significant correlates between life history and reproductive behavior. Leigh and Bernstein argue that baboon females make exceptional allocations to offspring during the early growth. Swedell and Saunders suggest that the mating strategies of female baboons are shaped primarily by the importance of ensuring the survival of their young infants, but that hamadryas and savanna baboons approach this problem in fundamentally different ways.

Infant mortality, through either predation or infanticide by males, has emerged recently as a major factor influencing the reproductive behavior and life history of baboons. As discussed by Palombit et al. (1997, 2000; see also Palombit, 2003) and Cheney and colleagues in Chapter 7, both infanticide and predation pose clear risks for infant baboons, with sexually selected infanticide by males impacting the evolution and maintenance of social and reproductive strategies of baboon females and infants as a result. For females, this may select for a motivation to mate with multiple males or to form associations with protective males. For infants, this may result in a life history pattern that reflects their greater vulnerability at certain stages of development. The links among infant mortality and morbidity, reproductive and social strategies of both females and males, and aspects of life history such as juvenile development are crucial to a full understanding of the evolution of behavior and social organization in baboons and other primates.

## CHAPTER OVERVIEWS

The scope of a project that considers multiple life history phases is broad, but, as noted, we have chosen to focus on only two major life history periods. Specifically, Part I examines what it takes for adults to reproduce, concentrating on mating behavior and general mating strategies and tactics. The chapters in this section investigate links between social organization, mating behavior, and various measures of fitness. Part II broadly considers what it takes for offspring to reach adulthood. Contributors to this section dissect the consequences of social interactions among adults on offspring-weaning behaviors, condition, and mortality. Still other chapters consider how morphologies relate to social variables, exploring the relationship between morphologies and the scheduling of reproduction. Coupling a focus on reproductive parameters and life history provides a more complete view of fitness in baboons (and primates more generally) than could be attained by concentrating on either in isolation. In effect, we evaluate how baboons go about the process of reproducing as a lifetime commitment. Our contributors ask how it is that male and female baboons go about finding mates, scheduling reproductive events, allocating reproductive investment, and successfully raising offspring.

In Chapter 2, Larissa Swedell and Julian Saunders use a comparative perspective to elucidate the relationship between female mating behavior and fitness in hamadryas baboons. Unique among *Papio* baboons, hamadryas have a rigid, multilayered social system in which mating occurs mainly within one-male units and female behavior is largely controlled by males. Female mating strategies and tactics are difficult to even detect in such a system. Swedell and Saunders argue that hamadryas female behavior, though tempered by highly structured relationships with males, is nevertheless similar to that of other baboon females in that it is closely tied to infanticide avoidance. In hamadryas society, the amount of protection females receive from their leader males—both for themselves and their offspring—appears to be a direct determinant of their own fitness. From this point of view, the one-male unit social structure characteristic of hamadryas baboons is advantageous with regard to female fitness.

Jacinta Beehner and Thore Bergman further clarify the role of female reproductive strategies in baboon social organization with their analysis of female social and mating strategies among hamadryas–olive hybrid baboons in the Awash hybrid zone. A comparison of females of varying phenotypes reveals that females exhibit mating strategies consistent with their phenotype, suggesting a correlation between genetics and patterns of association and mating behavior among female baboons. Beehner and Bergman's results complement Swedell and Saunders' contribution by providing more evidence supporting the notion that female baboons derive fitness benefits from a one-male unit social structure. Both contributions suggest that infants may be more likely to survive to adulthood in one-male units than in the looser, multimale multifemale aggregations typical of olive, yellow, and chacma baboons. It should be emphasized that these two studies effectively hold macroenvironmental variables constant by conducting their investigations in the same geographic region.

Thore Bergman's contribution helps complete this picture by providing a male perspective on baboon reproductive strategies. His study of hamadryas–olive hybrid baboons in the Awash hybrid zone capitalizes on behavioral variation among hybrid males to shed light on the evolutionary origins of the inflexible, stereotypical behavior of hamadryas males. Bergman proposes several evolutionary "precursors" for hamadryas male behavior and then tests for the presence of these precursors in the hybrid population. Bergman concludes that it is the temporary consortships of nonhamadryas baboons that are most likely to have led to the suite of male traits that shape hamadryas society today.

Guinea baboons may have a multilayered social structure similar to that of hamadryas, but this inference is based on sketchy data that derive mainly from captive populations (Boese, 1975). Anh Galat-Luong, Gerard Galat, and Suzanne Hagell address this issue with their contribution, but data on wild Guinea baboons remain frustratingly difficult to acquire. These authors suggest that Guinea social organization is only superficially similar to that of hamadryas: Subgroupings seem to be looser and less consistent in composition, males do not herd females in the same manner, and females do not appear to be as constrained in their behavior nor as monandrous. Galat-Luong et al. argue that the social flexibility of Guinea baboons provides adaptive benefits in that groups are able to adjust in size and composition as a response to what may be fairly significant swings in food availability. Selection in the highly seasonal and unpredictable West African environment may favor this kind of social flexibility, with important implications for understanding the social organization of both Guinea and hamadryas baboons.

The contribution by Monica Uddin, Clifford Jolly, and Jane Phillips-Conroy provides further insight into the relationship between behavior and fitness, but this time within the context of the evolution and maintenance of baboon endogenous virus (BaEV). Specifically, Uddin et al. test the hypothesis that differing patterns of reproductive behavior in various baboon populations influence BaEV diversity and patterning. Their results show that populations that reproduce within smaller, more closed breeding units—resulting in higher levels of inbreeding and relatedness among individuals—maintain higher copy numbers of BaEV than populations with more open reproductive units and lower levels of relatedness among individuals. Uddin et al. argue that, under the conditions of an increased BaEV copy number, inbreeding confers a selective advantage by decreasing the likelihood of ectopic exchange, which may lead to deleterious gene rearrangements. In this context, inbreeding in itself can be viewed as a reproductive tactic that leads to higher fitness under certain conditions. Uddin et al.'s analysis provides a clear starting point for future discussions of social and genetic evolution in primates, and shows how genetic data may be used to track social parameters.

Contributors to Part II explore a range of questions relating to life history in baboons. Life history adaptations condition opportunities for mating and the allocation of reproductive effort. The contribution by Dorothy Cheney and colleagues provides a fine-grained perspective on the dynamics of life history and reproduction in Botswana's Moremi chacma baboon population. Their study, when coupled with Johnson's analysis (Chapter 8), clearly reveals articulations among variables such as social behavior (notably rank acquisition and maintenance), demography, life history, and fitness. In so doing, they provide a necessary complement to Barrett et al.'s (Chapter 9) exposition of ecological dimensions of reproduction, life history, and fitness. More specifically, long-term research in Botswana's Okavango Delta offers extensive longitudinal data enabling studies of demography, life history, and behavior. Cheney et al. investigate an entire decade of demographic data, focusing on the reproductive consequences of mortality and rank in this population. Predominant sources of mortality include predation and infanticide, which operate on a strongly seasonal cycle. Slight reproductive advantages accrue from rank when sources of mortality that relate to resource acquisition are considered. On the other hand, "mediocrity" pays off because cessation of reproductive investment through infanticide affects both higher- and lower-ranking females more than middle-ranking individuals. However, predation affects reproductive success independent of rank.

Cheney et al. illustrate subtle but important relations between behavior, life history, and reproduction. For example, interactions among adults (rank maintenance and competitive interactions), maternal behaviors, and day-to-day decisions such as travel paths have important consequences for whether or not offspring can be brought to adulthood. One major result is that rank has comparatively small effects on offspring mortality in this population, raising important theoretical questions about the evolvability of social systems. This result should stimulate considerable discussion on the evolutionary significance of female dominance hierarchies.

Further analyses of the Moremi population by Sara Johnson nicely complement Cheney et al.'s study. Johnson moves beyond the "life or death" binary to explore relations between maternal attributes and offspring condition. At first pass, the effects of rank on offspring condition seem to be minimal. However, Johnson shows readily evident, but complicated, consequences of rank on offspring growth parameters. For example, female offspring of low-ranking females are much more likely to be smaller than comparably aged offspring of high-ranking females, but males appear to present a more complicated picture. Unfortunately, the effects of small size-for-age on offspring fitness are presently unknown, in part because of uncertainty regarding the consequences of small size-for-age at first reproduction. Similarly, maternal age independently affects offspring condition, with age regressive effects. Thus, while Cheney and colleagues illustrate few, if any, consistent effects of rank on mortality, Johnson shows that rank matters, at least in terms of the condition of infant and juvenile females. The consequences of offspring condition may be especially important during times of resource shortfalls.

One of the most important determinants of reproductive allocation and fitness is interbirth interval (IBI) length. Louise Barrett, Peter Henzi, and John Lycett address this issue with their intriguing analysis of factors that affect IBI in two baboon populations. Their study offers a strong and convincing critique of models based on direct links between habitat "quality" and reproductive or life history parameters (ideas generally compatible with a traditional r–K-selection continuum). Specifically, Barrett and coauthors compare reproductive parameters, particularly IBI, in a population occupying what might be considered a marginal habitat (the Drakensberg) with a population occupying what might be seen as a highly productive habitat (De Hoop). Drakensberg baboons have strongly seasonal births and a comparatively lengthy average IBI, while the De Hoop population distributes births more evenly across seasons. Paradoxically, infant mortality and other life history parameters fail to meet key predictions of traditional life history theories. In addition to addressing this interesting pattern, the chapter provides new insights into genetic conflicts of interests, a key issue in discussions of fitness. Thus, the scheduling of reproduction, patterns of infant care, behaviors surrounding weaning, and ultimately, reproductive success, can be seen as highly responsive to particular sources of mortality (intrinsic versus extrinsic). Traditional life history perspectives account poorly for different sources of mortality and thus do not adequately explain the key aspects of reproduction and life history in these populations. Barrett et al. note that sources of mortality result from complex interrelations among variables such as infant growth rates, conditions animals face at weaning, habitat quality and predictability, providing a complex picture of life history consequences of ecological variation.

Steven Leigh and Robin Bernstein position baboon life history within a larger context established by comparisons among several papionin primate species. Their comparisons suggest that baboons manifest an unusual and perhaps derived suite of life history characteristics in comparison to closely related species. Most notably, heavy investments in brain growth during pre- and early postnatal periods distinguish baboons from other papionins. These expenditures have important consequences for how papionins reach maturity and for the scheduling of reproductive events. In comparison to other papionins, *Papio* baboons invest heavily and early in each offspring, possibly reflecting a tradeoff between offspring quality and lifetime fecundity. Analyses of ontogenetic patterning in baboons offer strong critiques of traditional life history perspectives that rely on concepts of r- and K-selection. Leigh and Bernstein argue that the concept of a "life history mode" offers insight into questions about life histories that cannot be extracted from a traditional viewpoint. This critique aligns closely with that of Barrett et al. (Chapter 9), despite major differences in taxonomic levels of analysis.

Clifford Jolly and Jane Phillips-Conroy emphasize male life histories and reproductive attributes by analyzing relative testicular ontogeny across baboon subspecies. Their research reveals morphological and developmental dimensions of problems considered in Chapters 3 and 4 by Beehner and Bergman, bringing reproduction and life history together in very direct ways. More generally, males often receive short shrift in life history studies, so Jolly and Phillips-Conroy redress a palpable lack of literature on males. Testicular relative growth trajectories vary considerably, particularly in the phase immediately prior to attainment of adulthood. Of special interest are comparisons between hamadryas baboons and other subspecies, where differences in testicular developmental trajectories are interpreted in social terms. Specifically, the importance of sperm competition varies in tandem with social organization. For example, previous research by these authors comparing testicular growth trajectories between hamadryas and olive baboons indicates that sperm competition appears to be much less important for hamadryas than for olives. In the present contribution, they broaden this comparison to other subspecies, revealing unexpected patterns for yellow baboons. Importantly, Guinea baboons closely resemble hamadryas in their testicular proportions, a result that complements Galat-Luong et al.'s exposition of this understudied subspecies. In general, analyses of testicular growth trajectories reveal links among such diverse variables as male reproductive behavior, social organization, morphology, and life history.

The final "capstone" chapter, contributed by Susan Alberts and Jeanne Altmann, evaluates baboons in a broad evolutionary sense. Their investigation, tempered by the kinds of intimate details that can only be obtained from a commitment to long-term research, positions baboon adaptive flexibility in relation to climatic variation. Alberts and Altmann's analysis, couched in a theoretical context developed in paleoanthropology (Potts, 1996), defines and interprets responses of baboons to both short- and long-term climatic variability. Among their conclusions are that certain species, including baboons and humans, have evolved under circumstances of environmental variability and unpredictability and that key aspects of baboon life history and social organization were shaped by these processes.

## PROSPECTIVE

In our view, these contributions go far toward establishing goals for future studies of the ties between reproductive behaviors and life histories in primates. We have prioritized contributions from newly established scholars, partly in the hopes of encouraging further research into these areas. In any case, seeking to understand these links necessitates expertise in numerous fields, suggesting the potential for fruitful collaborations among behaviorists, geneticists, and morphologists. Intensive study of reproductive behavior, life history, and fitness in baboons provides a strong foundation for comparable studies at higher taxonomic levels. Addressing questions about reproduction and life history may yield especially valuable insights when posed in interspecific studies, particularly in cases in which social organization differs radically among taxa (see Garber and Leigh, 1997). This approach melds a number of specialties, offering unique insights into the evolution of social organization, morphology, and life history. We anticipate that such research will reveal a fundamentally important role for life histories and reproduction in driving variation in social organization among primates.

## REFERENCES

Alberts, S. C. and Altmann, J., 2001, Immigration and hybridization patterns of yellow and anubis baboons in and around Amboseli, Kenya, *Am. J. Primatol.* **53**:139–154.

Alberts, S. C., Altmann, J., 2002, Matrix models for primate life history analysis, in: *Primate Life Histories and Socioecology*, P. M. Kappeler, and M. E. Pereira, eds., University of Chicago Press, Chicago, pp. 66–102.

Altmann, S. A., 1998, *Foraging for Survival: Yearling Baboons in Africa*, University of Chicago Press, Chicago.

Altmann, J., Altmann, S. A., and Hausfater, G., 1978, Primate infant's effects on mother's future reproduction, *Science* **201**:1028–1030.

Altmann, J., Hausfater, G., and Altmann, S., 1988, Determinants of reproductive success in savannah baboons, *Papio cynocephalus*, in: *Reproductive Success: Studies of Individual Variation in Contrasting Breeding Systems*, T. H. Clutton-Brock, ed., University of Chicago Press, Chicago, pp. 403–418.

Boese, G. K., 1975, Social behavior and ecological considerations of west African baboons (*Papio papio*), in: *Socioecology and Psychology of Primates*, Tuttle R. H., ed., Mouton, The Hague, pp. 205–230.

Charlesworth, B., 1994, *Evolution in Age-Structured Populations*, Cambridge University Press, New York.

Darwin, C. R., 1871, *The Descent of Man, and Selection in Relation to Sex*, Princeton University Press, Princeton. Facsimile of First Edition.

Disotell, T. R., 1994, Generic level relationships of the Papionini (Cercopithecoidea), *Am. J. Phys. Anthropol.* **94**:47–57.

Disotell, T. R., 2000, Molecular systematics of the Cercopithecidae, in: *Old World Monkeys*, P. F. Whitehead, and C. J. Jolly, eds., Cambridge University Press, Cambridge, pp. 29–56.

Fisher, R. A., 1930, *The Genetical Theory of Natural Selection*, Clarendon Press, Oxford.

Frost, S. R., Marcus, L. F., Bookstein, F. L., Reddy, D. P., and Delson, E. (2003) Cranial allometry, phylogeography, and systematics of large-bodied papionins (Primates: Cercopithecinae) inferred from geometric morphometric analysis of landmark data, *Anat. Rec.* **275A**:1048–1072.

Garber, P. A., and Leigh, S. R., 1997, Ontogenetic variation in small-bodied New World primates: Implications for patterns of reproduction and infant care, *Folia Primatologica* **68**:1–22.

Groves, C. P., 1993, Order primates, in: *Mammal Species of the World: A Taxonomic and Geographic Reference*, D. E. Wilson, and D. M. Reeder, eds., Smithsonian Institution Press, Washington, DC, pp. 243–277.

Hughes, K. A., and Burleson, M. H., 2000, Evolutionary causes of genetic variation in fertility and other fitness components, in: *Genetic Influences on Human Fertility and Sexuality*, J. L. Rodgers, D. C. Rowe, and W. B. Miller, eds., Kluwer Academic Publishers, Boston, pp. 7–33.

Jarman, P. J., 1983, Mating systems and sexual dimorphism in large terrestrial mammalian herbivores, *Biol. Rev.* **58**:485–520.

Jolly, C. J., 1993, Species, subspecies, and baboon systematics, in: *Species, Species Concepts, and Primate Evolution*, W. H. Kimbel, and L. B., Martin, eds., Plenum Press, New York, pp. 67–108.

Jolly, C. J., 2003, Commentary: Cranial anatomy and baboon diversity, *Anat. Rec.* **275A**:1043–1047.

Kappeler, P. M., Pereira, M. E., and van Schaik, C. P., 2002, Primate life histories and socioecology, in: *Primate Life Histories and Socioecology*, P. M. Kappeler, and Pereira, M. E., eds., University of Chicago Press, Chicago, pp. 1–24.

Leigh, S. R., 1995, Socioecology and the ontogeny of sexual size dimorphism in anthropoid primates, *Am. J. Phys. Anthropol.* **97**:339–356.

Leigh, S. R., in press, Cranial ontogeny of *Papio* baboons (*Papio hamadryas*), *Am. J. Phys. Anthropol.* Published online 12 Dec 2005 DOI: 10.1002/ajpa.20319

Leigh, S. R., Setchell, J. M., and Buchanan, L. S., 2005, Ontogenetic bases of canine dimorphism in anthropoid primates, *Am. J. Phys. Anthropol.* **127**:296–311.

Newman, T. K., Jolly, C. J., and Rogers, J., 2004, Mitochondrial phylogeny and systematics of baboons (*Papio*), *Am. J. Phys. Anthropol.* **124**:17–27.

Palombit, R. A., 2003, Male infanticide in wild Savanna baboons: Adaptive significance and intraspecific variation, in: *Sexual Selection and Reproductive Competition in Primates: New Perspectives and Directions*, C. B., Jones, ed., The American Society of Primatologists, Norman, OK, pp. 3–47.

Palombit, R. A., Seyfarth, R. M., and Cheney, D. L., 1997, The adaptive value of "friendships" to female baboons: Experimental and observational evidence, *Anim. Behav.* **54**:599–614.

Palombit, R. A., Cheney, D. L., Fischer, J., Johnson, S., Rendall, D., Seyfarth, R. M., and Silk, J. B., 2000, Male infanticide and defense of infants in chacma baboons. In: *Infanticide by Males and Its Implications*, C. P. van Schaik, and C. H. Janson, eds., Cambridge University Press, Cambridge, pp. 123–152.

Phillips-Conroy, J. E. and Jolly, C. J., 1986, Changes in the structure of the baboon hybrid zone in the Awash National Park, Ethiopia, *Am. J. Phys. Anthropol.* **71**:337–350.

Potts, R., 1996, Evolution and climate variability, *Science* **273**:922–923.

Roff, D., 2002, *Life History Evolution*, Sinauer Associates, Inc., Sunderland, MA.

Trivers, R. L., 1972, Parental investment and sexual selection, in: *Sexual Selection and the Descent of Man, 1871–1971*, Campbell, B., ed., Aldine Press, Chicago, pp. 136–179.

van Noordwijk, A. J. and de Jong, G., 1986, Acquisition and allocation of resources: Their influence on variation in life history tactics, *Am. Naturalist* **128**:137–142.

Wiley, R. H., 1974, Evolution of social organization and life history patterns among grouse, *Q. Rev. Biol.* **49**:201–226.

Williams, G. C., 1966, *Adaptation and Natural Selection:A Critique of Some Current Biological Thought*, Princeton University Press, Princeton, NJ.

Williams-Blangero, S., Vandeberg, J. L., Blangero, J., Konigsberg, L., and Dyke, B., 1990, Genetic differentiation between baboon subspecies: Relevance for biomedical research, *Am. J. Primatol.* **20**:67–81.

PART I

# Reproductive Behavior and Mating Strategies

CHAPTER TWO

# Infant Mortality, Paternity Certainty, and Female Reproductive Strategies in Hamadryas Baboons

*Larissa Swedell and Julian Saunders*

## CHAPTER SUMMARY

Hamadryas differ from other *Papio* baboons in that their social organization centers around reproductively exclusive one-male units. Infanticide and aggression toward infants are risks for hamadryas and other baboons and, as has been suggested for other primates, these risks may have played a role in shaping female baboon reproductive strategies. One way that females may reduce aggression toward (and promote protection of) infants is by increasing paternity uncertainty through promiscuity and the incitement of male contest and sperm competition. Presentations to multiple males, postcopulation darts, and copulation calling in particular have been suggested as mechanisms whereby females may incite male competition at both the pre- and postcopulatory levels. Accordingly, a coupling of infanticide risk and multiple mating by females (and the associated male competition)

---

**Larissa Swedell** • Department of Anthropology, Queens College, CUNY and the New York Consortium in Evolutionary Primatology, New York, NY, USA **Julian Saunders** • Behavioural Ecology Research Group, University of KwaZulu-Natal, Durban, South Africa

characterizes many baboon societies. Another, alternate route to ensure protection against infanticide and other forms of infant mortality is association and exclusive copulation with a single protective male. Paternity certainty is probably quite high among hamadryas leader males, and protective behavior toward infants has likely been selected for. Correspondingly, compared to other baboons, female hamadryas are less promiscuous, do not frequently initiate copulation, and rarely behave in ways that might incite male–male competition. We suggest that, while all baboon females use a combination of paternity concentration and confusion to varying degrees, hamadryas baboon females in particular focus on paternity concentration rather than confusion and that this can be explained by changes in male and female reproductive strategies during the evolution of hamadryas social organization.

## 1. INTRODUCTION

While olive (*Papio hamadryas anubis*), yellow (*P. h. cynocephalus*), and most populations of chacma (*P. h. ursinus*) baboons are characterized by a multimale, multifemale social system in which there is little consistent substructuring, hamadryas baboon (*P. h. hamadryas*) social groups split regularly and consistently into progressively smaller subsets (Kummer, 1968; Swedell, 2006). The smallest stable social unit in hamadryas society is the one-male unit (OMU), consisting of a single "leader male" and several females. OMUs are often accompanied by *follower males*, which socialize with, but do not usually have sexual access to, the unit's females. Several OMUs comprise a *clan*, whose male members are thought to be related (Abegglen, 1984; Swedell, 2006), and two or more clans comprise large aggregations called *bands*, analogous to the "groups" or "troops" of other baboons. Finally, two or more bands may assemble at sleeping cliffs for the night, forming *troops*.

Hamadryas female behavior is different from that of other female baboons in that it is, on the surface at least, largely controlled by males. Male herding—through visual threats, chasing, and neckbiting—is the cohesive force holding OMUs together, and each female is conditioned by her leader male to remain near him, copulate only with him, and avoid interaction with individuals outside the unit. Within such a society, it is hard to imagine that females have social or reproductive strategies of their own, or that they are able to exert such strategies.

As with other females, however, we expect female hamadryas to act in ways that maximize individual reproductive success. From a female's point of view, enhancing the "quality" of each of her offspring (e.g., through better nutrition, socialization, or protection) is one of the most important ways in which she can do so. Ultimately, the fitness of a female baboon is determined by the survival and eventual reproductive success of her infants.

As shown by Cheney and colleagues for the Moremi chacma baboon population (Cheney et al., this volume), infant survival may be impacted by ecological factors such as seasonality and predation as well as social factors such as infanticide by males. In many primates and other mammals, immigrant or newly dominant males sometimes kill dependent infants that are present at the time of the immigration or takeover (Hrdy, 1974, 1977; Brooks, 1984; Packer and Pusey, 1984; Vogel and Loch, 1984; Sommer, 1994; Blumstein, 2000; van Schaik, 2000a,c). In most of these taxa, such behavior appears to be a male competitive strategy that has evolved via sexual selection (Hrdy, 1979; Hausfater and Hrdy, 1984; van Schaik, 2000a). Infanticide and attempted infanticide by males—either directly observed or strongly inferred—has been reported for most populations of baboons that have been studied to date (summarized in Palombit, 2003). While there is wide variation among baboon populations in its occurrence, the prevalence of infanticide in the genus *Papio* as a whole would suggest that it is a behavioral predisposition shared by all baboon males (Palombit, 2003). Palombit (2003) argues that variation in infanticide rate across baboon populations can be best explained by looking at specific demographic and reproductive characteristics of each population. He explains the high rate of infanticide among chacma baboons of the Drakensberg of South Africa as resulting from a combination of long interbirth intervals, low infant mortality (from sources other than infanticide), and high reproductive skew (see Barrett et al., this volume for further discussion of this population). At least the latter two of these factors—high reproductive skew and low infant mortality—are shared by hamadryas as well, suggesting that hamadryas females should, in theory, confront at least as high a risk of infanticide as females in other baboon populations.

But what evidence is there for infanticide in hamadryas baboons? Reports of infanticide in hamadryas derive mainly from captive populations, in some of which an exceptionally high rate of infanticide occurs (Angst and Thommen, 1977; Rijksen, 1981; Gomendio and Colmenares, 1989; Kaumanns et al., 1989; Chalyan and Meishvili, 1990; Zinner et al., 1993).

Infanticide also takes place in the wild (Kummer et al., 1974; Swedell, 2000, 2006; Swedell and Tesfaye, 2003), but at a far lower frequency. The first incidence of infanticide among wild hamadryas occurred during the field experiments reported by Kummer et al. (1974): Two mothers with infants were moved into new OMUs, after which one infant disappeared and the other was found dead with large canine-inflicted wounds on its skull and thighs. Although the evidence was only circumstantial, these infants may well have been killed by their mothers' new leader males. More recent evidence of infanticide in wild hamadryas derives from the Filoha population: After four takeovers of known females, the only black infant (aged less than 6 months) associated with each takeover either (a) disappeared (in two cases), (b) was the victim of prolonged kidnapping with no protective behavior on the part of its mother's new leader male (in one case), or (c) was attacked and killed by its mother's new leader male (in one case; Swedell, 2000; Swedell and Tesfaye, 2003). The first two cases were initially conservatively interpreted as accidental infant death by prolonged kidnapping resulting from a lack of protection by the females' new leader males (Swedell, 2000, 2006). Hamadryas leader males normally defend infants from harassment and kidnapping by extra-unit individuals, and the absence of such protection is unusual within the context of hamadryas society (Swedell, 2006). The more recent observation of direct infanticide in the same wild population, however, suggests the possibility that the first two infants may have been killed, rather than just neglected, by their mothers' new leader males (Swedell and Tesfaye, 2003; Swedell, 2006).

The relatively few observations of infanticide in hamadryas baboons compared to other taxa, including those on other baboons and other mammals living in one-male groups, might suggest that infanticide in hamadryas is a relatively rare occurrence and not much of a risk for females. This apparent rarity is misleading, however, for two reasons. The first is that the number of observation hours spent on individually identified wild hamadryas baboons is a tiny fraction of that spent on groups of other monkeys in which infanticide has been reported. For example, the Hanuman langurs of Ramnagar, Nepal (e.g., Borries et al., 1999) and the baboons of the Moremi Game Reserve in Botswana (e.g., Palombit et al., 1997, 2000; Cheney et al., this volume; Johnson, this volume) have each been observed for tens of thousands of hours over several decades, compared to less than 1,500 hr of observation for the hamadryas baboons at the Filoha site in Ethiopia. The second reason behind the apparent rarity of infanticide in hamadryas is that its occurrence appears to be closely tied to OMU takeovers, which are in themselves rare occurrences, having been observed only a handful of times (see Swedell 2000; Swedell and Tesfaye 2003). When takeovers do occur, typically only 1–4 females are involved, most of which may not have a black infant at the time. Thus, the circumstances under which infanticide would be expected to occur—male takeovers of females with young infants—do not arise very often. Overall, therefore, the actual rate of infanticide in hamadryas populations is probably quite low, but it is still undoubtedly a risk for females after takeovers. This can be described in terms of chronic versus acute risk: In savanna baboon populations such as that at Moremi (e.g., Palombit et al., 2000; Cheney et al., this volume; Johnson, this volume), where adult males are commonly in contact with infants they likely did not sire, there is a *chronic* risk of infanticide. In hamadryas populations, however, the chronic risk of infanticide is quite low but the *acute* risk after takeovers is high. In fact, the normally high rate of infant survival in hamadryas baboons (Sigg et al., 1982; Swedell, 2006) coupled with the observed and inferred infant mortality after takeovers (Swedell 2000; Swedell and Tesfaye 2003) suggests that infanticide may be the primary cause of death for hamadryas baboon infants.

Regardless of the actual number of successful infanticides that occur, infanticide is clearly a selective factor affecting hamadryas and other baboon females. Even if infanticide occurs, on average, only once in a female's lifetime, it reduces her lifetime reproductive success by negating a period of maternal investment and should therefore have an impact on the evolution of female behavior (van Schaik et al., 1999; van Schaik, 2000b). In female baboons, adaptive responses to male infanticide may include minimizing one's losses through abortion, premature birth or weaning, or an otherwise earlier return to reproductive condition following immigration or takeovers (Pereira, 1983; Colmenares and Gomendio, 1988; Alberts et al., 1992; Swedell, 2000, 2006); the manipulation of paternity assessment through "pseudoestrus" (Zinner and Deschner, 2000) or mating with multiple males (Hausfater, 1975; Smuts, 1985; Bercovitch, 1987b; Swedell, 2006); and social bonding with a protective male to obtain protection for one's infants (Smuts, 1985; Palombit et al., 1997; Weingrill, 2000; Swedell 2006). van Schaik et al. (1999) emphasize the duality of female counterstrategies to infanticide, hypothesizing that "female sexuality in species vulnerable to male infanticide has been molded by the dual need for paternity concentration and confusion: concentration in order to elicit infant protection from the likely father, confusion in order to prevent infanticide from non-likely fathers" (p. 207).

In this chapter, we compare four components of female sexual behavior across baboons as a preliminary examination of the qualitative and quantitative differences between the reproductive strategies of hamadryas females and females of other baboon subspecies. We regard our interpretations as hypotheses for further testing rather than empirically supported conclusions. We begin with the assumption that baboon infants are at a risk of mortality from infanticide or other sources and that female baboons may employ one of the two general strategies—paternity concentration and paternity confusion—to counteract this risk. We focus on four components of behavior in particular: female exclusivity of mating, female initiation of mating, postcopulatory darts, and copulation calls. We have chosen these behavioral elements because they are largely female initiated and thus indicative of sexual motivation and underlying strategies of females rather than behavioral compromises between females and males (which would be reflected by measures such as copulation frequency and grooming rates). Each of these variables is used for heuristic purposes only and is simply meant to give us an indication of whether females are using a general strategy of paternity confusion or paternity concentration (cf. van Schaik et al., 1999). We use female *exclusivity of mating* as a direct measure of the number of males that each female mates with during an estrus period. We acknowledge that the number of males with whom a female ultimately copulates is, in part, a result of male as well as female strategies. Nevertheless, we expect this number to increase with a general strategy of paternity confusion and decrease with a strategy of paternity concentration. We use the variable *female initiation of mating* as a second measure of female promiscuity. We assume that females that are using a paternity confusion strategy would be more likely to initiate copulations with multiple males than females using a paternity concentration strategy. On the other hand, females using a paternity concentration strategy have little need to expend energy in either initiating copulations or even maintaining this behavioral element in their repertoire. We acknowledge, however, that a behavioral pattern whereby a female repeatedly initiates copulations with only one male would not be indicative of a general strategy of paternity confusion. Finally, we use female *postcopulatory darts* (the postcopulation withdrawal response, in which females run away from a male at the end of a copulation) and *copulation calls* (loud vocalizations given by females during or just after copulation) as measures of a female's motivation to mate with multiple males. Both behavioral elements have been interpreted as means by which females attract attention of other males during copulation and thereby incite competition among males (O'Connell and Cowlishaw, 1994, 1995). While we recognize that copulation calls have conversely been argued to be mechanisms to assure paternity certainty and promote mate guarding (Henzi, 1996; Maestripieri et al., 2005), we view the male–male competition hypothesis to be better supported by the available evidence and use it as our working assumption for the purposes of this chapter (Hamilton and Arrowood, 1978; Oda and Masataka, 1995; Cowlishaw and O'Connell, 1996; but see Maestripieri et al., 2005).

We compare data from hamadryas baboons at the Filoha site in Ethiopia to reports of sexual behavior drawn from the literature in olive, yellow, and chacma baboons, all of which are generally characterized by a multimale, multifemale social structure (except mountain chacmas) and a female-bonded social organization. Where possible, we also include Guinea baboons, *P. h. papio*, in our comparisons.

## 2. METHODS

Behavioral data on hamadryas baboons, *P. h. hamadryas*, were collected from a population inhabiting the lowlands of the northern Rift Valley of East Africa. The study site is the Filoha outpost of the Awash National Park in Ethiopia (see Swedell, 2002a, b, 2006 for details). At least five different groups ("bands" cf. Kummer, 1968) of hamadryas baboons range throughout the Filoha area, each showing the characteristic hamadryas social structure (OMUs nested within clans and bands, as described above) and frequent male herding and neckbiting described by Kummer (1968). The main study group at Filoha consists of about 200 baboons, including about 25 one-male units and about 55 reproductively active females. This group has been under observation on and off over a 6-year period: 986 hr from October 1996 through September 1998 and over 250 hr from 2000 to 2004 (outlined in more detail in Swedell and Tesfaye, 2003; Swedell, 2006).

Comparative data from other baboon subspecies were drawn from the literature. Due to both the limitations of our hamadryas data set and differences in methods and presentation of results among studies, we restrict our comparison to the four behavioral elements listed below. Because we have ranges of values for each behavioral element from only one subspecies and cannot assume that the single values obtained from the literature are robust indicators of the distribution of those data in other subspecies, we cannot perform statistical comparisons with sufficient power. Instead, we have contrasted the values graphically for heuristic purposes.

We compared the following four behavioral elements:

1. *Exclusivity of mating:* average number of different males that a female typically copulates with during the sexually swollen phase of one monthly cycle.
2. *Female initiation of copulation:* percentage of copulations or consortships initiated by females. For most populations, we refer here to percentage of *copulations* initiated by either a female presentation of the hindquarters with the tail raised, commonly referred to as a "sexual present," or an approach. (This measure is somewhat inconsistent across studies because some authors consider an approach by a female to be an initiation of copulation while others limit their definition to a sexual present.) For one data point, we use Bercovitch's (1991) measure of initiation of *consortships* rather than copulations, because females at Gilgil apparently initiated consortships by sexually presenting to males but then did not initiate the actual copulations once in consort (Bercovitch, 1991). For hamadryas, we consider only sexual presents to be initiation of copulation because females are frequently herded and often approach males in response to a threat or a brief look, and thus an approach alone is not indicative of a female's motivation to copulate. In fact, female hamadryas also often present to males in response to threats, so our measure of female-initiated copulations for hamadryas is likely an overestimate. Unfortunately, for most of the copulations reported here we do not know if the female presentations were preceded by a male threat, so we cannot control for this factor in this preliminary analysis.
3. *Postcopulatory darts:* percentage of copulations followed by the postcopulation withdrawal response, or postcopulatory dart, described by Hall (1962) as a "short running-away by the female" during or after the male's dismount. It has been suggested that postcopulatory darts function to incite male–male competition in baboons by drawing attention to the copulating pair (O'Connell and Cowlishaw, 1995).

4. *Copulation calls:* percentage of copulations accompanied or followed by copulation calls, vocalizations by females that are largely, though not exclusively, given during copulation. These calls have been described as "intermittent roars" (Bolwig, 1959), "staccato grunts" (Saayman, 1970), "gurgling growls" (Hall, 1962), or "a series of grunts...accompanied by loud barks in longer calls" (O'Connell and Cowlishaw, 1994). Females have been reported to give these vocalizations while defecating as well (Hall, 1962; Boese, 1973; Bercovitch, 1985), and the calls may thus be an involuntary reaction to compression of the vaginal wall. Many authors suggest, however, that copulation calls are costly signals and therefore must serve a communicative function, one of which may be the incitation of competition among males (Hamilton and Arrowood, 1978; Dunbar, 1988; Dixson, 1998; O'Connell and Cowlishaw, 1994).

## 3. RESULTS

### 3.1. Exclusivity of Mating

Female exclusivity in mating, compared across the four subspecies, is shown in Figure 1. In olive baboons at Gilgil, Kenya (Smuts, 1985; Bercovitch, 1987b), and yellow baboons at Amboseli (Hausfater, 1975), females copulate

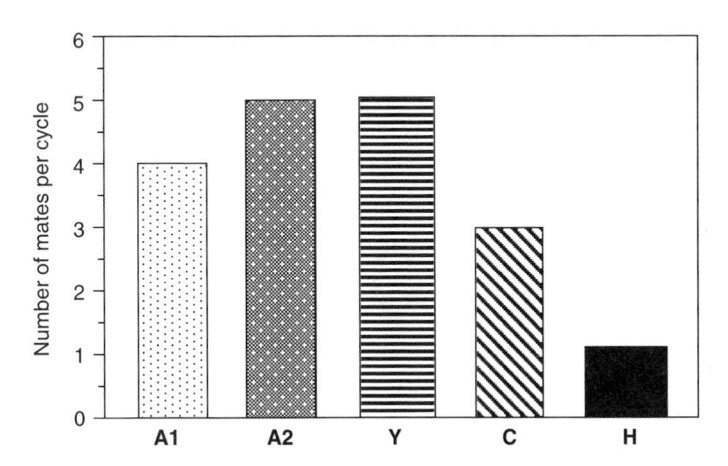


**Figure 1.** Female exclusivity of mating, defined as the average number of different males that a female typically copulates with during the sexually swollen phase of her monthly cycle. **A1** = olive baboons at Gilgil (Bercovitch, 1987); **A2** = olive baboons at Gilgil (Smuts, 1985); **Y** = yellow baboons at Amboseli (Hausfater, 1975); **C** = chacma baboons at Cape Point (Hall, 1962); and **H** = hamadryas baboons at Filoha (this study).

with an average of four to five different males during a single cycle. In chacma baboons, the alpha male usually has priority of access during peak estrus (Bulger, 1993), but Hall (1962) reported that females often copulated with all three adult males in the group on a given day (though the frequency of mating by each male was highly variable). In olive and yellow baboons, turnovers in sexual consortship are quite frequent (Hall and DeVore, 1965; Smuts, 1985; Bercovitch, 1988; Noë and Sluijter, 1990), and, unless a female is in a multiday consortship (as is common in chacmas), she will often copulate with more than one male each day (Hausfater, 1975; Noë and Sluijter, 1990). While olive and yellow baboon males occasionally monopolize a female for her entire period of probable conception, this is the exception rather than the rule (Hausfater, 1975; Bercovitch, 1987b). For example, of the 19 conceptive cycles reported by Bercovitch (1987b), only one of them included the monopolization of a female by a single male throughout her entire period of probable conception.

In hamadryas baboons, by contrast, most females copulate with only their leader male. Of the 76 copulations observed at Filoha for which the identity of the male could be determined, 15 were with nonleader males. Of these, 6 were with juvenile males (3–5 years of age using the age classes of Sigg et al., 1982; Swedell, 2006), 5 with adolescent males (5–6 years), and 4 with subadult males (6–9 years). No multiple mounts with nonleader males were observed, and only one of these copulations, with a young male (aged 5 or 6), included an ejaculatory pause. In captive Guinea baboons, copulation also occurs mainly between females and the one adult male in their subgroup (analogous to one-male units of hamadryas), though females do apparently copulate with other males as well on occasion (Boese, 1973; Maestripieri et al., 2005). Because these observations of Guinea baboons are from captivity, they are not included in our graphical comparisons.

### 3.2. Female Initiation of Copulation

Figure 2 shows the proportion of observed copulations (or consortships; see below) initiated by females in the four subspecies. In most baboons, females initiate at least 20 percent and often up to three-quarters of all copulations observed. Hausfater (1975) found that female yellow baboons initiate 44 percent of copulations, and Hall (1962) and Seyfarth (1978) found that

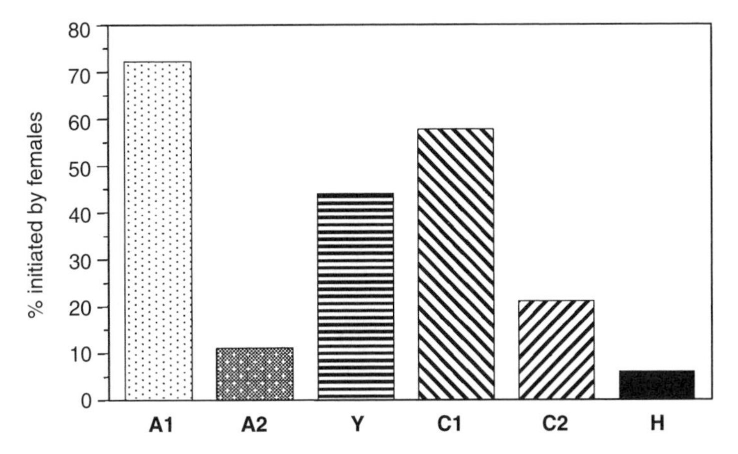


**Figure 2.** Percentage of copulations or consortships initiated by females. **A1** = percentage of consortships initiated by females in olive baboons at Gilgil (Bercovitch, 1991). **A2** = percentage of copulations initiated by females in olive baboons at Gombe (Ransom, 1981); **Y** = percentage of copulations initiated by females in yellow baboons at Amboseli (Hausfater, 1975); **C1** = percentage of copulations initiated by females in chacma baboons at Cape Point (Hall, 1962); **C2** = percentage of copulations initiated by females in chacma baboons at Mountain Zebra National Park, South Africa (Seyfarth, 1978); and **H** = percentage of copulations initiated by females in hamadryas baboons at Filoha (this study).

female chacmas initiate 58 percent and 21 percent of copulations, respectively. For olive baboons, Ransom (1981) found that females initiate only 11 percent of copulations, but Bercovitch (1991) pointed out that, while females do not often initiate copulation itself, they initiate 72 percent of *consortships* by presenting to males. Olive baboon females also regularly present to one male while in consort with another, often leading to consort turnovers (Smuts, 1985). In hamadryas, by contrast, only 4 of the 48 copulations (8 percent) for which the initiator of the copulation could be determined were initiated by females. In Guinea baboons, both Boese (1973) and Galat-Luong (pers. commun.) observed females presenting to males, and Galat-Luong et al. (this volume) report that females initiate slightly fewer copulations than do males. As these are only preliminary observations and the sample size is quite small, we did not include these data in our graphical representation.

### 3.3. Postcopulatory Darts

The postcopulation withdrawal response, or postcopulatory dart, is commonly seen in olive, yellow, and chacma baboons (Hall and DeVore, 1965; Ransom, 1981; Smuts, 1985; Bercovitch, 1995; O'Connell and Cowlishaw, 1995; K. Rasmussen, pers. commun.; Semple, pers. commun.). It varies in its occurrence from 25 percent (Ransom, 1981) to 92 percent (Bercovitch, 1985) in olive baboons, but occurs after at least 75 percent of copulations in chacmas (78 percent: Hall, 1962; 75 percent: Hall and DeVore, 1965; 86–89 percent: Saayman, 1970) (Figure 3). In olive baboons, females have been observed to run away from one male (with whom copulation had just occurred) and directly to another (Hall and DeVore, 1965), and such behavior often leads to consort turnovers (Smuts, 1985). We have not included yellow baboons in our graphical comparison because we could not find any published reports of quantitative data on darting, though we have been told that yellow baboon females dart after 80 percent (K. Rasmussen, pers. commun.) to virtually 100 percent (Semple, pers. commun.) of observed copulations. In hamadryas baboons, we have seen postcopulatory darts only by (a) adolescent females and (b) females that were mounted by males other

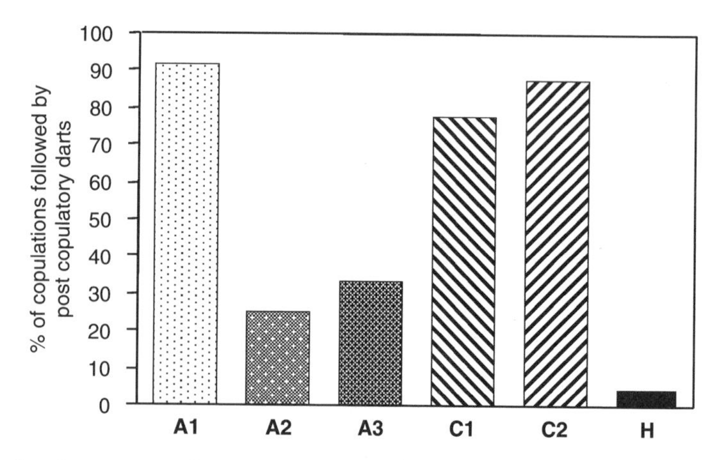


**Figure 3.** Percentage of copulations followed by postcopulatory darts. **A1** = olive baboons at Gilgil (Bercovitch, 1985); **A2** = olive baboons at Gombe (Ransom, 1981); **A3** = olive baboons at Nairobi Park (Hall and DeVore, 1965); **C1** = chacma baboons at Cape Point (Hall, 1962); **C2** = chacma baboons at Honnet (Saayman, 1970); and **H** = hamadryas (this study).

than their leader male, and postcopulatory darts occurred after only 6 percent of observed copulations for which the postcopulatory behavior was also observed (*N*=69). While Galat-Luong et al. (this volume) report that Guinea baboon females sometimes dart after copulations, their sample size was too small to include here.

### 3.4. Copulation Calls

In chacma, yellow, and some populations of olive baboons, females often give loud vocalizations, or copulation calls, during and/or just after copulation, and in most populations these calls are given in the majority of copulations that occur (Hamilton and Arrowood, 1978; O'Connell and Cowlishaw, 1994). Figure 4 shows the frequency of copulation calling in each subspecies. In Guinea baboons, copulation calls occur in 39 percent (Boese, 1973) of observed copulations; in olive baboons they occur in 19 percent (Ransom, 1981) to 68 percent (Bercovitch, 1985) of copulations (68 percent with subadult males and 62 percent with adult males, the latter of which is shown in Figure 4); in yellow baboons they occur in 80 percent (Collins, 1981) to 97 percent (Semple, 2001; Semple et al., 2002) of copulations; and in chacma

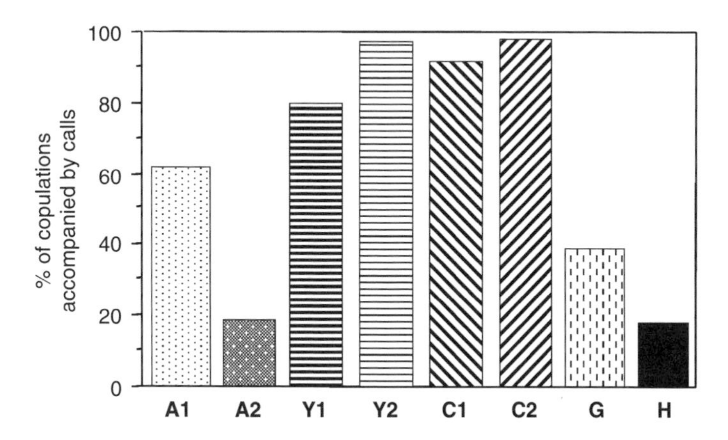


**Figure 4.** Percentage of copulations accompanied or followed by copulation calls. **A1** = olive baboons at Gilgil (Bercovitch, 1985); **A2** = olive baboons at Gombe (Ransom, 1981); **Y1** = yellow baboons in Tanzania (Collins, 1981); **Y2** = yellow baboons at Amboseli (Semple, 2001; Semple et al., 2002); **C1** = chacma baboons at Cape Point (Hall, 1962); **C2** = chacma baboons at Honnet (Saayman, 1970); **G** = Guinea baboons at Niokolo Koba (Boese, 1973); and **H** = hamadryas (this study).

baboons they occur in 92 percent (Hall, 1962) to 98 percent (Saayman, 1970) of copulations.

In hamadryas, only about 18 percent of observed copulations (*N*=86) included female calls, and copulation calls were given by only 30 percent (4 out of 13) of the females in the study group who were observed both sexually swollen and copulating. When calls were examined individually, we found that those of hamadryas were quieter and substantially reduced in both length and complexity compared to those of chacma and yellow baboons. For purposes of comparison, Figure 5 shows a representative call of (a) a female chacma baboon from the De Hoop Nature Reserve in South Africa and (b) a

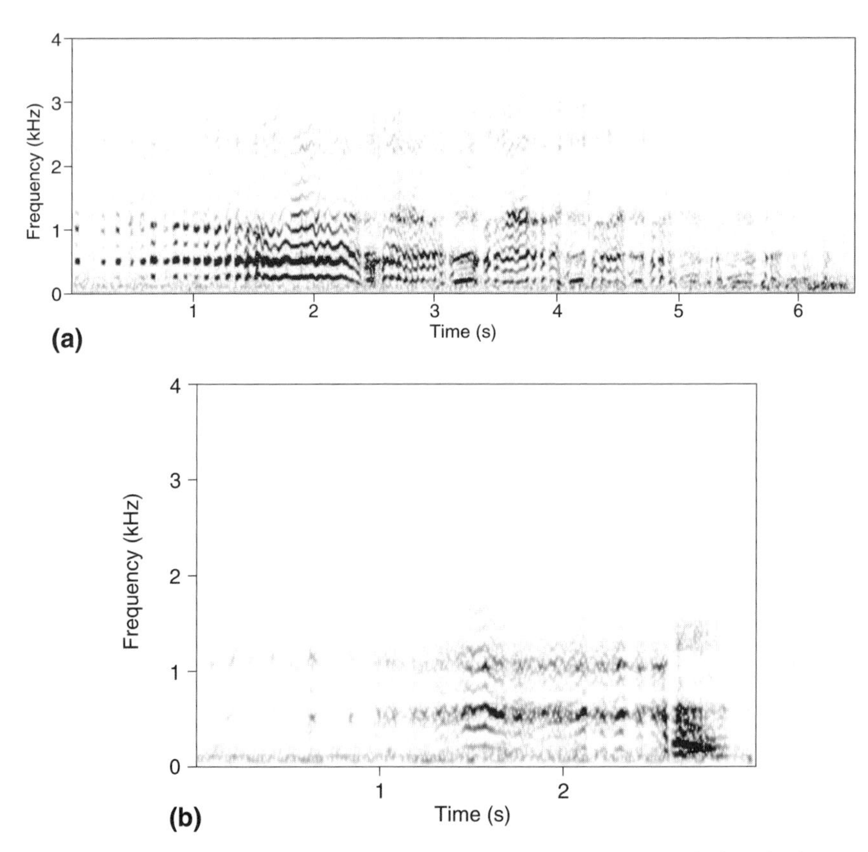


**Figure 5.** Spectrographic representations of (**a**) a copulation call of a female chacma baboon from the De Hoop Nature Reserve in South Africa and (**b**) a copulation call of a female hamadryas baboon from the Filoha field site in Ethiopia.

female hamadryas baboon from the Filoha field site in Ethiopia. A quantitative comparison of the copulation calls of chacma and hamadryas baboons will be reported elsewhere (Saunders, in preparation).

## 4. DISCUSSION

### 4.1. Comparisons Among Baboon Taxa

The most obvious difference between hamadryas and most other baboons is the number of males with whom each female copulates during the sexually swollen phase of her monthly cycle. In olive, yellow, and chacma baboons, this number typically approaches the number of males in the group as a whole (Hall and DeVore, 1965; Hausfater, 1975; Bercovitch, 1995). While the highest-ranking male in the group often manages to exclude other males during the peak of a female's sexual swelling, females do copulate with other males before and after maximal swelling and are generally characterized as "promiscuous" in their mating behavior (Hall and DeVore, 1965; Saayman, 1970; Hausfater, 1975; Seyfarth, 1978; Smuts, 1985; Bercovitch, 1987b). Even in chacmas, in which the alpha male usually consorts exclusively with females during peak estrus (Bulger, 1993; Weingrill et al., 2000), promiscuity may be the rule for females at other times: "The roving, appetitive behaviour of inflating females from one male to another was conspicuous. It was not uncommon for an inflating female to present to, and be mounted by, as many as three males within the space of two or three minutes" (Saayman, 1970, p. 86).

By contrast, in hamadryas baboons, most females have never been observed to sexually solicit or copulate with more than a single male in a given cycle, and that male is invariably the leader male of her OMU. The few copulations with nonleader males that have been observed in the Filoha population were with subadult males, and most did not appear to include ejaculation (see below). This relative exclusivity of hamadryas mating patterns confirms reports from previous observational field studies of wild hamadryas (Kummer, 1968; Abegglen, 1984) as well as a management "experiment" conducted by Biquand et al. (1994) in Saudi Arabia in which leader males were vasectomized and their females did not subsequently reproduce (though this study is not conclusive, as the only nonleader male in the group died shortly after the vasectomies of the leader males, so the females had few, if any, options other than their infertile leader male).

The proportion of copulations initiated by females also differs dramatically between hamadryas and other baboons. While most baboon females initiate either most copulations or the consortships themselves, hamadryas females initiate copulation only infrequently. Interestingly, female yellow and chacma baboons are less likely to initiate copulation when they are in consort with a male than when they are out of consort (Hall and DeVore, 1965; Seyfarth, 1978), and in olive baboons, females initiate consortships but do not usually initiate the actual copulations once they are in consort (Ransom, 1981; Bercovitch, 1995). If female baboons in general initiate copulations outside of consort but not while in consort, then female hamadryas are no different from other baboon females insofar as they can be viewed as being in a permanent consortship with their leader male.

The frequency of postcopulatory darts and copulation calls also differs between hamadryas and other baboon populations, though the patterns shown by these two behavioral elements are somewhat inconsistent. While olive baboon populations in general are quite variable in the percentage of copulations followed by darts, females in chacma and yellow populations as well as the olive baboon population at Gilgil dart away from males after the majority of copulations. Similarly, while olive baboon females vary in their tendency to give copulation calls, yellow and chacma baboons give calls during (or after) the majority of their copulations. Compared to olive, yellow, and chacma baboons as a whole, hamadryas females engage in both of these behavioral elements far less frequently.

Differences among baboons in their tendency to give copulation calls may be related to differences in their physiology. Because female baboons have been reported to give copulation calls while defecating as well (Hall, 1962; Boese, 1973; Bercovitch, 1985), it has been suggested that the calls are simply an involuntary reaction to compression of the vaginal wall. Whether during copulation or defecation, calls are almost invariably given when a female is sexually swollen, and Bercovitch (1985) suggested that they be called "sexual swelling vocalizations" rather than copulation calls for that reason. If variation in calling is tied to the size of a female's sexual swelling, then we would expect sexual swellings to be smaller in hamadryas than in other baboons. This does not appear to be the case, however. There may be other physiological differences among baboon subspecies that underlie differences in copulation calling, though it is not obvious what those differences might be other than the slightly smaller body size of hamadryas females compared to other female baboons.

## 4.2. Reproductive Strategies in Female Baboons: Overall Patterns

If we apply van Schaik et al.'s (1999) framework to baboons and assume that the risk of infanticide is present (either today or in ancestral populations), it appears that most populations of "savanna" baboons use a predominant strategy of paternity *confusion* to counteract male infanticide and other sources of infant mortality (Hall and DeVore, 1965; Hausfater, 1975; Bercovitch, 1991, 1995). Sexual presentations to multiple males, copulation with multiple males, postcopulatory darts, and copulation calls may all serve to elicit male–male contest and/or sperm competition and confuse paternity among males, which may in turn elicit protection of infants by multiple males and inhibit infanticide (O'Connell and Cowlishaw, 1994, 1995; Dixson, 1998; Soltis, 2002). Even in chacma baboons, in which the alpha or resident male has exclusive access to females during their periovulatory period, females expend what appears to be a substantial amount of energy soliciting and mating with *other* males earlier and later in their cycles. Such behavior would have the effect of confusing paternity from the perspective of these other males, even if paternity is *not* confused from the female's (or the alpha male's) perspective.

This pattern of behavior can be viewed as a high-energy strategy with fitness costs. Not only are such behavioral elements likely to be energetically demanding, but if they result in an increase in male–male competition then they are also likely to increase a female's risk of injury from male aggression (Manson, 1994). Moreover, these behavioral elements are likely to both render a female more conspicuous to predators and decrease her time spent engaged in predator and social vigilance, the combination of which may further reduce her fitness. Overall, the energetic demands on females that use a strategy of paternity confusion—as manifested in the above ways— are likely to be relatively high.

The general pattern shown by female hamadryas, on the other hand, appears to be one of lower energy expenditure, a high degree of concentration on a single male, and fewer behavioral elements that would promote male–male competition and increase a female's risk of injury. Further evidence of monandry in hamadryas females and the consequent reduced (or absent) sperm competition in hamadryas males can be drawn from the smaller testis size of hamadryas compared to olive baboons reported by Jolly and Phillips-Conroy (2003, this volume).

One might argue that the hamadryas social system in itself is an outcome of a high degree of male–male competition, leading to a constant high risk of injury for females. However, overt competition among males for sexual access to females is not a regular occurrence in hamadryas society and generally only occurs during takeovers and attempted takeovers. Similarly, one might argue that hamadryas females incur daily aggression from their leader males through neckbiting and are thus more prone to injury in general than nonhamadryas females. While this may be true, we do not view herding and neckbiting to be elements of the hamadryas *female* reproductive strategy, but rather of the hamadryas *male's*. Moreover, the behavioral elements comprising a paternity confusion strategy, if used in a hamadryas social system, would undoubtedly increase a hamadryas female's risk of injury. Thus, in the context of the hamadryas social system, if hamadryas females are to attempt to confuse paternity, they must do so surreptitiously.

### 4.3. Energetic Limitations on Hamadryas Baboon Females

Compared to other baboons, hamadryas inhabit drier, more resource-limited habitats in which female time budgets are likely constrained by foraging needs. Although found today in a wide range of ecosystems (Nagel, 1973; Zinner et al., 2001), the majority of hamadryas distribution coincides with the semidesert regions of the Horn of Africa and the southwestern Arabian peninsula. Hamadryas baboons likely spent most of their evolutionary history since divergence from other baboons in a dry, semidesert region (possibly the Arabian peninsula: Kummer et al., 1981; Kamal et al., 1994; Wildman, 2000), and it is this environment that may have provided the selective pressures leading to their rigid, male-dominated social organization (Kummer, 1968, 1971, 1990).

Food availability is closely tied to reproduction in baboons: Females with access to fewer food resources have been shown to either spend more time foraging during lactation, when their energetic needs are highest (Dunbar et al., 2002), or suffer from lowered fertility (Strum and Western, 1982; Bercovitch, 1987a; Bercovitch and Strum, 1993; Barton, 1990). Evidence from mountain chacmas (Lycett et al., 1998) as well as cross-subspecies comparative analyses (Hill et al., 2000; Barrett et al., this volume) suggest that severe environmental conditions lengthen interbirth intervals in female baboons, thereby reducing female reproductive output. Moreover, low-ranking females—i.e., those at a competitive disadvantage with regard to access to food—not only have longer interbirth intervals (see Cheney et al., this volume) but also give birth to infants with lower growth rates than high-ranking females (Johnson, this volume). Current evidence does not point to shorter interbirth intervals in hamadryas than in other baboons (Sigg and Stolba, 1981; Hill et al., 2000; Swedell, 2006), but available reproductive data on wild hamadryas females derive from relatively mild habitats compared to those in most of the hamadryas range. If interbirth intervals are longer in hamadryas baboons as a whole compared to other baboons, then hamadryas females would incur an increased cost of infanticide, as an already high cost of reproduction would be exacerbated by a short-term loss in maternal investment.

In addition to yielding lower-quality food resources, hamadryas habitats are also hotter year round than those of other baboons, which may impose an additional cost on hamadryas females. The mean annual afternoon temperature at the Filoha site is about 34°C with no more than four degrees of variation (32.7–36.0°C) throughout the year (Swedell, 2006). This is higher than, for example, both the average (23°C) and the maximum (33°C) temperatures reported for Amboseli, Kenya (Bronikowski and Altmann, 1996), and is generally expected to be higher than most other habitats in which baboons are found (with the exception of desert chacmas, e.g., in Namibia: Brain and Mitchell, 1999). Hill et al. (2003) demonstrated that the mean annual temperature affects the amount of time that baboons spend in different activities: Higher temperatures are associated with less time spent feeding and moving as well as more time spent resting, presumably to compensate for the high heat load. Likewise, Brain (1991) found that desert chacmas were far less active during intergroup encounters after several days of water deprivation, compared to high rates of activity during such encounters when they had drunk water more recently. Hamadryas are arguably heat stressed year round and appear to thermoregulate by seeking shade throughout the day regardless of the season. Most other baboons, by contrast, live in more seasonally variable habitats in which females are probably environmentally stressed during only parts of the year. The costs of an energetically demanding paternity confusion strategy may therefore be especially high for hamadryas females compared to other female baboons.

### 4.4. Paternity Concentration in Hamadryas Baboons

A strategy of paternity concentration by hamadryas females is intrinsically tied to, and likely evolved in association with, the very specific behavioral strategy of hamadryas males. Instead of competing for access to any estrous female like other baboons, male hamadryas mate exclusively with a small subgroup of females, their "possession" of which is "respected" by other leader males (Kummer et al., 1974). Assuming that leader males sire most, if not all, of the infants born into their units, paternity certainty is probably quite high and protective behavior by leader males of infants born into their units (and their mothers) is likely to be selected for. That protection of infants by leader males is important in hamadryas society is suggested by two lines of evidence from this population: (1) leader males often threaten and ultimately retrieve infants from individuals outside the unit who handle those infants and (2) females tend to remain closer to their leader male during the first month after the birth of an infant, when the infant is most vulnerable, compared to subsequent months or when pregnant (Swedell, 2006; unpublished data). Rohrhuber (1987 (in German), cited in Kaumanns et al., 1989) found that, in a group of captive hamadryas with an unusually high rate of infanticide, a high rate of grooming as well as close proximity between a female and her leader male lowered the probability that her offspring would be killed. This suggests that hamadryas leader males play an important role in infant survival and that females benefit reproductively from associating with and copulating exclusively with a protective leader male. Sigg et al. (1982) pointed out that infant survival is higher in hamadryas than in other baboons and suggested that the OMU social structure might provide a safer environment for infants and juveniles than the multimale, multifemale social structure of yellow or olive baboons. Such a conclusion is supported by both the findings of Beehner and Bergman (this volume), who found a higher survival rate of infants born to hybrid females in OMUs compared to those born to non-OMU females, and the lower infant mortality found in chacma baboons living in one-male groups compared to those living in multimale groups (Lycett et al., 1998; Barrett et al., this volume). A hamadryas female's leader male is her main protector from aggression, whether it be toward herself or her offspring, and, consequently, she probably benefits from ensuring his paternity.

Such an extreme version of a paternity concentration strategy can be explained by the coevolution of male and female strategies in hamadryas baboons. It is probable that the evolution of hamadryas social organization began with a larger group splintering into foraging parties (consisting of female kin groups and a few associated males) due to the low food availability in their arid, semidesert habitat (Kummer, 1990). Males would presumably have found it advantageous to remain with small groups of females on a semipermanent basis so as to be nearby when females came into estrus, and those that managed to keep these small groups of females with them at all times via herding and neckbiting would have been favored by natural selection. Once this began, a powerful evolutionary feedback loop would have led rather quickly to the system we see today because of the dual advantages of such a system to both males and females (Henzi and Barrett, 2003). For males, it would have become even more advantageous to maintain exclusive access to a small group of females due to the high degree of paternity certainty that would result and the corresponding reduction in fitness incurred by males that did not adopt this strategy. For females, it would have become advantageous to participate in such a system because of the protection they would receive for themselves and their infants. Such protection may be even more important for hamadryas than other baboon females due to the high energetic costs of reproduction in a semidesert habitat. Ironically, a system would have evolved that, while protecting females against infanticide, would have also inherently increased the potential risk of infanticide by extra-unit males (due to their far lower chance of paternity), thereby increasing the power of the feedback loop.

## 4.5. Promiscuity in Hamadryas Baboons

Despite their near exclusivity in mating, hamadryas females also occasionally copulate with nonleader males. In the Filoha population, females copulated with nonleader males in 15 of the 76 copulations for which individual identities were known, and one of these copulations included an ejaculatory pause. Kummer (1968) also reported copulations between females and nonleader males. Noting that these males were almost always subadult, Kummer suggested that they were probably not sexually mature and that such copulations were probably unlikely to ever result in pregnancy. Jolly and Phillips-Conroy (2003, this volume), however, found that hamadryas males undergo testicular enlargement at an earlier age than other baboon subspecies and suggested that copulations between hamadryas females and subadult nonleader males may, in fact, result in fertilization. Whether or not they ever do is not yet known.

Most copulations with nonleader males, both at Filoha and Erer Gota, appear to be surreptitious (Kummer, 1968; Swedell, 2006). At Filoha, most took place while the OMU was traveling and the leader male was ahead of the copulating pair and unable to see them. Likewise, Kummer (1968) reported such copulations to take place mainly "behind the backs of" the females' leader males (pp. 41–42). In these cases, such copulations may still be interpreted as part of a general strategy of paternity concentration. Provided that (a) the leader male provides the protective benefits consistent with such a strategy, as outlined above, and (b) the majority of a female's offspring are sired by her leader male, a strategy of paternity concentration with occasional inseminations by nonleader males could still be selected for. This assumes, of course, that leader males cannot distinguish their own offspring from those of other males, an assumption that does not appear to be met in yellow baboons (Alberts, 1999; Buchan et al., 2003).

If, as has been suggested by Kummer (1968) and Abegglen (1984), leader males and their followers are usually closely related, then copulations between females and follower males may actually provide inclusive fitness benefits to the leader male. On three occasions, copulations with nonleader males took place in full view of the females' leader males. In two of these cases, the nonleader male was a follower of that unit and may have been a close relative of the leader male.

It is possible that some hamadryas females, in effect, combine the two strategies of paternity concentration and confusion. Females may "concentrate" paternity by engaging in multiple-mount copulations with their leader males around the time of ovulation, but they might also "confuse" paternity via occasional, surreptitious single-mount copulations with nonleader males. Although such copulations probably rarely result in conception, they would nevertheless have the outcome of confusing paternity from the perspective of the nonleader male. A nonleader male should be selected to protect, tolerate, or at least avoid killing infants of any female with whom he has copulated recently as long as the chances of inseminating her are greater than zero. Whether or not the female gains—or the leader male loses—fitness benefits from these extra-unit copulations, the fact that they occur at all suggests that, at the very least, hamadryas females retain the behavioral motivation to mate promiscuously that presumably characterized the ancestral female baboon.

Our observations suggest that for most females, this motivation is unexpressed, but for a few females, it is strong enough to act upon despite the associated risk of aggression from their leader males.

### 4.6. Paternity Concentration and Confusion in Baboons

Such a system whereby males and females engage in mutualistic strategies of close association, protection, and (relatively) exclusive mating has evolved to an extreme only in hamadryas baboons, but it appears to exist to varying degrees in other baboon subspecies as well (Boese, 1973; Smuts, 1985; Anderson, 1989, 1990; Palombit et al., 1997). For example, many researchers have described strong social bonds between baboon females and specific adult males, sometimes called "friendships," and have suggested that these relationships benefit females via protection from infanticide or aggression by males (Smuts, 1985; Palombit et al., 1997; Weingrill, 2000; also see Beehner and Bergman, this volume). In chacma baboons in particular, females may be in effect pursuing a strategy of paternity concentration by focusing their periovulatory mating on the alpha male (or the resident male in one-male groups), while, at the same time, confusing paternity by mating with other males at other times. As with hamadryas, such a strategy would have the outcome of confusing paternity from the perspective of these other males and could thus provide a female the protective benefits gained via both strategies. As long as the alpha (or resident) male fathered most of a female's offspring while the chances of other males' paternity was greater than zero, such a pattern could be selected for in both males and females.

Obviously, we must bear in mind that female reproductive strategies do not exist outside of the context of those of males. Neither chacma nor hamadryas females may be "choosing" to mate exclusively with one male. Rather, one male is able to exclude other males from a female for a longer period than is typical in other baboons. The ability of chacma and hamadryas males to do this may be related to the absence in these taxa of the coalitionary behavior seen in olive and yellow baboons. The concentration of paternity in a single male for hamadryas and chacma baboon females, therefore, may be as much of a consequence of male mating strategies as those of females. Females in these two subspecies can thus be viewed as "making the best out of a bad situation" in that they likely garner benefits from close association with a single male even though such relationships may in effect be imposed on them by males.

In chacma baboons living at high altitudes, the similarities to hamadryas are more striking: Compared to lowland chacmas, mountain chacmas are more commonly found in one-male groups, cross-sex bonds are stronger, female–female bonds are weaker, and herding behavior by males occurs more frequently (Anderson, 1981, Anderson, 1990; Byrne et al., 1987, 1989; Whiten et al., 1987; Henzi et al., 1990; Hamilton and Bulger, 1992; Henzi et al., 1999). These features, however, are not seen to the extent that they are in hamadryas, and chacma one-male groups do not coalesce to form the larger bands seen in hamadryas. That chacma one-male groups with exclusive mating are more prevalent at high altitudes than at low altitudes suggests that, as in hamadryas, one-male groups and strong cross-sex relationships are adaptive responses to food scarcity and reduced predator pressure (Whiten et al., 1987; Byrne et al., 1987, 1989; Henzi et al., 1990, 1999; Anderson, 1990).

Some authors have suggested that the evolution and maintenance of strong cross-sex relationships in *all* baboons is related primarily to protection from infanticide or aggression by males rather than to ecological factors (Busse and Hamilton, 1981; Smuts, 1985; Palombit et al., 1997; Weingrill, 2000). Palombit (1999), for example, discusses infanticide avoidance as a primary reason for close bonds between female chacma baboons and adult males, and attributes a significant portion of infant mortality in this species to sexually selected infanticide. He points out that in both chacma baboons and gorillas, infanticide accounts for a large portion of infant mortality *and* females develop and maintain bonds with males. In both taxa it is the females, not the males, which are most responsible for proximity maintenance and do most of the grooming. Weingrill (2000) points out the prevalence of close relationships between chacma baboon females and the likely sires of their offspring specifically during the periods of pregnancy and lactation. As infanticide by males has been observed in the same population, Weingrill suggests that the driving force behind such associations may be infanticide avoidance. Many authors have suggested that infanticide and aggression toward infants has been a primary selective force leading to strong intersexual bonds in primates as a whole (Wrangham, 1979, 1982; Fossey, 1984; Watts, 1989; van Schaik and Dunbar, 1990; Smuts and Smuts, 1993; Clutton-Brock and Parker, 1995; Sterck et al., 1997; Palombit, 1999; van Schaik et al., 1999). Treves (1998) proposes the conspecific-threat hypothesis for the evolution of primate social systems, an extension of Brereton's (1995) coercion–defense hypothesis, in which females are always at risk of aggression (to themselves or their infants) from unrelated males, and so must adopt one or more defensive strategies, one of which is the association with a male for protection.

## 4.7. CONCLUSIONS

As a means of ensuring protection for their infants—whether it be from infanticide or other sources—most primate females, and baboons in particular, seem to show some combination of paternity concentration and paternity confusion (Bercovitch, 1991, 1995; Palombit et al., 1997; Henzi and Barrett, 2003). Olive and yellow baboon females can be described as focusing largely on paternity confusion, as suggested by the prevalence of multiple mating and behavioral elements that incite male–male competition. Hamadryas females, by contrast, focus mainly, though not exclusively, on the concentration of paternity in a single male. Chacma baboon females appear to combine these two general strategies by focusing on a single male around ovulation and mating promiscuously at other times. The predominance of one strategy over the other may depend on both demographic factors such as sex ratio and ecological factors such as seasonality, food availability, and predator pressure (Wrangham, 1980; van Schaik, 1989; Dunbar, 1992; Barrett et al., submitted). The collection of additional data from other populations of wild baboons—and in particular from Guinea baboons, for which we know very little—promises to shed further light on both the variation among and flexibility of baboon females in their responses to infanticide and other threats to the survival of their offspring.

## ACKNOWLEDGMENTS

The research contributing to this chapter was funded by the Wenner-Gren Foundation, the U.S. National Science Foundation, the L.S.B. Leakey Foundation, the National Geographic Society, the PSC-CUNY Research Award Program, the WISC Program of the American Association for the Advancement of Science, and the South African National Research Foundation. We thank the Ethiopian Wildlife Conservation Department and the Awash National Park Baboon Research Project for permission to conduct this research and logistical support, and Peter Henzi for financial support and research facilitation. Finally, we are grateful to Steve Leigh, Fred Bercovitch,

Ryne Palombit, Richard Wrangham, and four anonymous reviewers for valuable discussion and comments on previous versions of this manuscript.

## REFERENCES

Abegglen, J.-J., 1984, *On Socialization in Hamadryas Baboons*, Associated University Presses, London.

Alberts, S. C., 1999, Paternal kin discrimination in wild baboons, *Proc. R. Soc. Lond. B* **266**:1501–1506.

Alberts, S. C., Sapolsky, R. M., and Altmann, J., 1992, Behavioral, endocrine, and immunological correlates of immigration by an aggressive male into a Natural Primate Group, *Horm. Behav.* **26**:167–178.

Anderson, C. M., 1981, Subtrooping in a chacma baboon (*Papio ursinus*) population, *Primates* **22**:445–458.

Anderson, C. M., 1989, The spread of exclusive mating in a chacma baboon population, *Am. J. Phys. Anthropol.* **78**:355–360.

Anderson, C. M., 1990, Desert, mountain and savanna baboons: A comparison with special reference to the Suikerbosrand population, In: *Baboons: Behaviour and Ecology. Use and Care.* M. Thiago de Mello, A. Whiten, and R. W., Byrne, eds., *Selected Proceedings of the XIIth Congress of the International Primatological Society*, Brasil, Brasilia, pp. 89–103.

Angst, W. and Thommen, D., 1977, New data and a discussion of infant killing in old world monkeys and apes, *Folia Primatol.* **27**:198–229.

Barrett, L., Echeverria-Lozano, E., Weingrill, T., Gaynor, D., Rendall, D., and Henzi, S. P., submitted, Operational sex ratio determines male social tactics in a baboon mating market. *Behav. Ecol. Sociobiol.*

Barton, R., 1990, Feeding, reproduction and social organisation in female olive baboons (*Papio anubis*), In *Baboons: Behaviour and Ecology. Use and Care.* M. Thiago de Mello, A. Whiten, and R. W. Byrne, eds., *Selected Proceedings of the XIIth Congress of the International Primatological Society*, Brasil, Brasilia, pp. 29–37.

Bercovitch, F. B., 1985, Reproductive tactics in adult female and adult male olive baboons. Ph.D. Dissertation, University of California, Los Angeles.

Bercovitch, F. B., 1987a, Female weight and reproductive condition in a population of olive baboons (*Papio anubis*), *Am. J. Primatol.* **12**:189–195.

Bercovitch, F. B., 1987b, Reproductive success in male savanna baboons, *Behav. Ecol. Sociobiol.* **21**:163–172.

Bercovitch, F. B., 1988, Coalitions, cooperation and reproductive tactics among adult male baboons, *Anim. Behav.* **36**:1198–1209.

Bercovitch, F. B., 1991, Mate selection, consortship formation, and reproductive tactics in adult female savanna baboons, *Primates* **32**:437–452.

Bercovitch, F. B., 1995, Female cooperation, consortship maintenance, and male mating success in savanna baboons, *Anim. Behav.* **50**:137–149.

Bercovitch, F. B. and Strum, S. C., 1993, Dominance rank, resource availability, and reproductive maturation in female savanna baboons, *Behav. Ecol. Sociobiol.* **33**:313–318.

Biquand, S., Boug, A., Biquand-Guyot, V., and Gautier, J.-P., 1994, Management of commensal baboons in Saudi Arabia. *Rev. Ecol. (Terre Vie)* **49**:213–222.

Blumstein, D. T., 2000, The evolution of infanticide in rodents: A comparative analysis, in: *Infanticide by Males and Its Implications*, van Schaik, C. P. and Janson, C. H., eds., Cambridge University Press, Cambridge, pp. 178–197.

Boese, G. K., 1973, Behavior and social organization of the Guinea baboon *(Papio papio)*. Ph.D. Dissertation, Johns Hopkins University.

Bolwig, N., 1959, A study of the behaviour of the chacma baboon, *Behaviour* **14**:136–163.

Borries, C., Launhardt, K., Epplen, C., Epplen, J. T., and Winkler, P., 1999, DNA analyses support the hypothesis that infanticide is adaptive in langur monkeys, *Proc. R. Soc. Lond. B* **266**:901–904.

Brain, C., 1991, Activity of water-deprived baboons (*Papio ursinus*) during intertroop encounters, *S. Afr. J. Sci.* **87**:221–222.

Brain, C. and Mitchell, D., 1999, Body temperature changes in free-ranging baboons (*Papio hamadryas ursinus*) in the Namib Desert, Namibia. *Int. J. Primatol.* **20**:585–598.

Brereton, A. R., 1995, Coercion–defence hypothesis: The evolution of primate sociality, *Folia Primatol.* **64**:207–214.

Bronikowski, A. M. and Altmann, J., 1996, Foraging in a variable environment: Weather patterns and the behavioral ecology of baboons, *Behav. Ecol. Sociobiol.* **39**:11–25.

Brooks, R. J., 1984, Causes and consequences on infanticide in populations of rodents, in: *Infanticide: Comparative and Evolutionary Perspectives*, G. Hausfater, and S. B. Hrdy, eds., Aldine Publishing Company, New York, pp. 331–348.

Buchan, J. C., Alberts, S. C., Silk, J. B., and Altmann, J., 2003, True paternal care in a multi-male primate society, *Nature* **425**:179–181.

Bulger, J. G., 1993, Dominance rank and access to estrous females in male savanna baboons, *Behaviour* **127**:67–103.

Busse, C. and Hamilton, W. J. I., 1981, Infant carrying by male chacma baboons, *Science* **212**:1281–1283.

Byrne, R. W., Whiten, A., and Henzi, S. P., 1987, One-male groups and intergroup interactions of mountain baboons, *Int. J. Primatol.* **8**:615–633.

Byrne, R. W., Whiten, A., and Henzi, S. P., 1989, Social relationships of mountain baboons: Leadership and affiliation in a non-female-bonded monkey, *Am. J. Primatol.* **18**:191–207.

Chalyan, V. G. and Meishvili, N. V., 1990, Infanticide in hamadryas baboons, *Biologischeskie Nauki* **3**:99–106.

Clutton-Brock, T. H. and Parker, G. A., 1995, Sexual coercion in animal societies, *Anim. Behav.* **49**:1345–1365.

Collins, D. A., 1981, Social behaviour and patterns of mating among adult yellow baboons (*Papio c. cynocephalus. L.* 1766). Ph.D. Dissertation, University of Edinburgh.

Colmenares, F. and Gomendio, M., 1988, Changes in female reproductive condition following male take-overs in a colony of hamadryas and hybrid baboons, *Folia Primatol.* **50**:157–174.

Cowlishaw, G. and O'Connell, S. M., 1996, Male–male competition, paternity certainty and copulation calls in female baboons, *Anim. Behav.* **51**:235–238.

Dixson, A. F., 1998, *Primate Sexuality: Comparative Studies of the Prosimians, Monkeys, Apes, and Human Beings*, Oxford University Press, Oxford.

Dunbar, R. I. M., 1988, *Primate Social Systems*, Cornell University Press, Ithaca, NY.

Dunbar, R. I. M., 1992, Time: A hidden constraint on the behavioural ecology of baboons, *Behav. Ecol. Sociobiol.* **31**:35–49.

Dunbar, R. I. M., Hannah-Stewart, L., and Dunbar, P., 2002, Forage quality and the costs of lactation for female gelada baboons, *Anim. Behav.* **64**:801–805.

Fossey, D., 1984, Infanticide in mountain gorillas (*Gorilla gorilla beringei*) with comparative notes on chimpanzees, in: *Infanticide: Comparative and Evolutionary Perspectives*, G. Hausfater, and S. B. Hrdy, eds., Aldine, New York, pp. 217–235.

Gomendio, M. and Colmenares, F., 1989, Infant killing and infant adoption following the introduction of new males to an all-female colony of baboons, *Ethol.* **80**:223–244.

Hall, K. R. L., 1962, The sexual, agonistic and derived social behaviour patterns of the wild chacma baboon, *Papio ursinus, Proc. Zool. Soc. Lond.* **139**:283–327.

Hall, K. R. L., and DeVore, I., 1965, Baboon social behavior, In: *Primate Behavior: Field Studies of Monkeys and Apes*, I. DeVore, ed., Holt, Rinehart and Winston, Inc., New York, pp. 53–110.

Hamilton, W. J., III and Arrowood, P. C., 1978, Copulatory vocalizations of chacma baboons (*Papio ursinus*), gibbons (*Hylobates hoolock*), and humans. *Science* **200**:1405–1409.

Hamilton, W. J., III and Bulger, J., 1992, Facultative expression of behavioral differences between one-male and multimale savanna baboon groups, *Am. J. Primatol.* **28**:61–71.

Hausfater, G., 1975, *Dominance and Reproduction in Baboons (Papio cynocephalus)*, Karger, Basel.

Hausfater, G. and Hrdy, S. B., eds., 1984, *Infanticide: Comparative and Evolutionary Perspectives*, Aldine Publishing Company, New York.

Henzi, S. P., 1996, Copulation calls and paternity in chacma baboons, *Anim. Behav.* **51**:233–234.

Henzi, P. and Barrett, L., 2003, Evolutionary ecology, sexual conflict, and behavioral differentiation among baboon populations, *Evol. Anthr.* **12**:217–230.

Henzi, S. P., Dyson, M. L., and Deenik, A., 1990, The relationship between altitude and group size in mountain baboons (*Papio cynocephalus ursinus*), *Int. J. Primatol.* **11**:319–325.

Henzi, S. P., Weingrill, T., and Barrett, L., 1999, Male behaviour and the evolutionary ecology of chacma baboons, *S. Afr. J. Sci.* **95**:240–242.

Hill, R. A., Lycett, J. E., and Dunbar, R. I. M., 2000, Ecological and social determinants of birth intervals in baboons, *Behav. Ecol.* **11**:560–564.

Hill, R. A., Barrett, L., Gaynor, D., Weingrill, T., Dixon, P., Payne, H., and Henzi, S. P., 2003, Day length, latitude and behavioural (in)flexibility in baboons (*Papio cynocephalus ursinus*), *Behav. Ecol. Sociobiol.* **53**:278–286.

Hrdy, S. B., 1974, Male–male competition and infanticide among the langurs (*Presbytis entellus*) of Abu, Rajasthan, *Folia Primatol.* **22**:19–58.

Hrdy, S. B., 1977, *The Langurs of Abu: Female and Male Strategies of Reproduction*, Harvard University Press, Cambridge, MA.

Hrdy, S. B., 1979, Infanticide among animals: A review, classification, and examination of the implications for the reproductive strategies of females, *Ethol. Sociobiol.* **1**:13–40.

Jolly, C. J. and Phillips-Conroy, J. E., 2003, Testicular size, mating system, and maturation schedules in wild anubis and hamadryas baboons, *Int. J. Primatol.* **24**:125–142.

Kamal, K. B., Ghandour, A. M., and Brain, P. F., 1994, Studies on new geographical distribution of hamadryas baboons *Papio hamadryas* in the western region of Saudi Arabia, J. *Egypt. Vet. Med. Ass.* **54**:81–89.

Kaumanns, W., Rohrhuber, B., and Zinner, D., 1989, Reproductive parameters in a newly established colony of hamadryas baboons (*Papio hamadryas*), *Primate Rep.* **24**:25–33.

Kummer, H., 1968, *Social Organization of Hamadryas Baboons: A Field Study*, University of Chicago Press, Chicago.

Kummer, H., 1971, *Primate Societies: Group Techniques of Ecological Adaptation*, Aldine, Chicago.

Kummer, H., 1990, The social system of hamadryas baboons and its presumable evolution, in: *Baboons: Behaviour and Ecology. Use and Care. M. Thiago de Mello, A. Whiten, and R. W. Byrne, eds., Selected Proceedings of the XIIth Congress of the International Primatological Society*, Brasil, Brasilia, pp. 43–60.

Kummer, H., Götz, W., and Angst, W., 1974, Triadic differentiation: An inhibitory process protecting pair bonds in baboons, *Behaviour* **49**:62–87.

Kummer, H., Banaja, A. A., Abo-Khatwa, A. N., and Ghandour, A. M., 1981, Mammals of Saudi Arabia: Primates: A survey of hamadryas baboons in Saudi Arabia, *Fauna Saud. Arab.* **3**:441–471.

Lycett, J. E., Henzi, S. P., and Barrett, L., 1998, Maternal investment in mountain baboons and the hypothesis of reduced care, *Behav. Ecol. Sociobiol.* **42**:49–56.

Maestripieri, D., Leoni, M., Raza, S. S., Hirsch, E. J., and Whitham, J. C., 2005, Female copulation calls in Guinea baboons: Evidence for post-copulatory female choice? *Int. J. Primatol.* **26**:737–758.

Manson, J. H., 1994, Male aggression: A cost of female mate choice in Cayo Santiago Rhesus Macaques, *Anim. Behav.* **48**:473–475.

Nagel, U., 1973, A comparison of anubis baboons, hamadryas baboons and their hybrids at a species border in Ethiopia, *Folia Primatol.* **19**:104–165.

Noë, R. and Sluijter, A. A., 1990, Reproductive tactics of male savanna baboons, *Behaviour* **113**:117–170.

O'Connell, S. M. and Cowlishaw, G., 1994, Infanticide avoidance, sperm competition and mate choice: The function of copulation calls in female baboons, *Anim. Behav.* **48**:687–694.

O'Connell, S. M. and Cowlishaw, G., 1995, The post-copulation withdrawal response in female baboons: A functional analysis, *Primates* **36**:441–446.

Oda, R. and Masataka, N., 1995, Function of copulatory vocalizations in mate choice by females of Japanese Macaques (*Macaca fuscata*), *Folia Primatol.* **64**: 132–139.

Packer, C. and Pusey, A. E., 1984, Infanticide in carnivores, in: *Infanticide: Comparative and Evolutionary Perspectives*, G. Hausfater, and S. B. Hrdy, eds., Aldine Publishing Company, New York, pp. 31–42.

Palombit, R. A., 1999, Infanticide and the evolution of pair bonds in nonhuman primates, *Evol. Anthr.* 7:117–129.

Palombit, R. A., 2003, Male infanticide in wild savanna baboons: Adaptive significance and intraspecific variation, In: *Sexual Selection and Reproductive Competition in Primates: New Perspectives and Directions*, C. B. Jones, ed., The American Society of Primatologists, Norman, OK, pp. 3–47.

Palombit, R. A., Seyfarth, R. M., and Cheney, D. L., 1997, The adaptive value of "friendships" to female baboons: Experimental and observational evidence, *Anim. Behav.* **54**:599–614.

Palombit, R. A., Cheney, D. L., Fischer, J., Johnson, S., Rendall, D., Seyfarth, R. M., and Silk, J. B., 2000, Male infanticide and defense of infants in chacma baboons, in: *Infanticide by Males and Its Implications*, C. P. van Schaik, and C. H. Janson, eds., Cambridge University Press, Cambridge, pp. 123–152.

Pereira, M. E., 1983, Abortion following the immigration of an adult male baboon (*Papio cynocephalus*), *Am. J. Primatol.* **4**:93–98.

Ransom, T. W., 1981, *Beach Troop of the Gombe*, Bucknell University Press, Lewisburg.

Rijksen, H. D., 1981, Infant killing: A possible consequence of a disputed leader role, *Behaviour* **78**:138–168.

Saayman, G. S., 1970, The menstrual cycle and sexual behaviour in a troop of free ranging chacma baboons (*Papio ursinus*), *Folia Primatol.* **12**:81–110.

Semple, S., 2001, Individuality and male discrimination of female copulation calls in the yellow baboon, *Anim. Behav.* **61**:1023–1028.

Semple, S., McComb, K., Alberts, S. C., and Altmann, J., 2002, Information content of female copulation calls in yellow baboons, *Am. J. Primatol.* **56**:43–56.

Seyfarth, R. M., 1978, Social relationships among adult male and female baboons. I. Behaviour during sexual consortship, *Behaviour* **64**:204–226.

Sigg, H. and Stolba, A., 1981, Home range and daily march in a hamadryas baboon troop, *Folia Primatol.* **36**:40–75.

Sigg, H., Stolba, A., Abegglen, J.-J., and Dasser, V., 1982, Life history of hamadryas baboons: Physical development, infant mortality, reproductive parameters and family relationships, *Primates* **23**:473–487.

Smuts, B. B. (1985) *Sex and Friendship in Baboons*, Aldine Publishing Company, New York.

Smuts, B. B. and Smuts, R. W., 1993, Male aggression and sexual coercion of females in nonhuman primates and other mammals: Evidence and theoretical implications, *Adv. Stud. Behav.* **22**:1–63.

Soltis, J., 2002. Do primate females gain nonprocreative benefits by mating with multiple males? Theoretical and empirical considerations, *Evol. Anthr.* **11**:185–197.

Sommer, V., 1994, Infanticide among the langurs of Jodhpur: Testing the sexual selection hypothesis with a long-term record, in: *Infanticide and Parental Care*, S. Parmigiani, and F. S., vom Saal, eds., Harwood Academic Publishers, New York, pp. 155–187.

Sterck, E. H. M., Watts, D. P., and van Schaik, C. P., 1997, The evolution of female social relationships in nonhuman primates, *Behav. Ecol. Sociobiol.* **41**:291–309.

Strum, S. C. and Western, J. D., 1982. Variations in fecundity with age and environment in olive baboons (*Papio anubis*), *Am. J. Primatol.* **3**:61–76.

Swedell, L., 2000, Two takeovers in wild hamadryas baboons, *Folia Primatol.* **71**:169–172.

Swedell, L., 2002a, Affiliation among females in wild hamadryas baboons (*Papio hamadryas hamadryas*), *Int. J. Primatol.* **23**:1205–1226.

Swedell, L., 2002b, Ranging behavior, group size and behavioral flexibility in Ethiopian hamadryas baboons (*Papio hamadryas hamadryas*), *Folia Primatol.* **73**:95–103.

Swedell, L., 2006, *Strategies of Sex and Survival in Hamadryas Baboons: Through A Female Lens*, Pearson Prentice Hall, Upper Saddle River, NJ.

Swedell, L. and Tesfaye, T., 2003, Infant mortality after takeovers in wild Ethiopian hamadryas baboons, *Am. J. Primatol.* **60**:113–118.

Treves, A., 1998, Primate social systems: Conspecific threat and coercion–defense hypotheses, *Folia Primatol.* **69**:81–88.

van Schaik, C. P., 1989, The ecology of social relationships amongst female primates, in: *Comparative Socioecology: The Behavioural Ecology of Humans and Other Mammals*, V. Standen, and R. A. Foley, eds., Blackwell Scientific Publications, Oxford, pp. 195–218.

van Schaik, C. P., 2000a, Infanticide by male primates: The sexual selection hypothesis revisited, in: *Infanticide by Males and Its Implications*, C. P. van Schaik, and C. H. Janson, eds., Cambridge University Press, Cambridge, pp. 27–60.

van Schaik, C. P., 2000b, Social counterstrategies against infanticide by males in primates and other mammals, in: *Primate Males: Causes and Consequences of Variation in Group Composition*, P. M. Kappeler, ed., Cambridge University Press, Cambridge, pp. 34–52.

van Schaik, C. P., 2000c, Vulnerability to infanticide by males: Patterns among mammals, in: *Infanticide by Males and Its Implications*, C. P. van Schaik, and C. H. Janson, eds., Cambridge University Press, Cambridge, pp. 61–71.

van Schaik, C. P. and Dunbar, R. I. M., 1990, The evolution of monogamy in large primates: A new hypothesis and some crucial tests, *Behaviour* **115**:30–62.

van Schaik, C. P., van Noordwijk, M. A., and Nunn, C. L., 1999, Sex and social evolution in primates, in: *Comparative Primate Socioecology*, P. C. Lee, ed., Cambridge University Press, Cambridge, pp. 204–231.

Vogel, C. and Loch, H., 1984, Reproductive parameters, adult-male replacements, and infanticide among free-ranging langurs (*Presbytis entellus*) at Jodhpur (Rajasthan), India, in: *Infanticide: Comparative and Evolutionary Perspectives*, G. Hausfater, and S. B. Hrdy, eds., Aldine, New York, pp. 237–255.

Watts, D. P., 1989, Infanticide in mountain gorillas: New cases and a reconsideration of the evidence, *Ethology* **81**:1–18.

Weingrill, T., 2000, Infanticide and the value of male–female relationships in mountain chacma baboons, *Behaviour* **137**:337–359.

Weingrill, T., Lycett, J. E., and Henzi, S. P., 2000, Consortship and mating success in chacma baboons (*Papio cynocephalus ursinus*), *Ethology* **106**:1033–1044.

Whiten, A., Byrne, R. W., and Henzi, S. P., 1987, The behavioral ecology of mountain baboons, *Int. J. Primatol.* **8**:367–388.

Wildman, D., 2000, Mammalian zoogeography of the Arabian peninsula and horn of Africa with a focus on the cladistic phylogeography of hamadryas baboons (*Primates: Papio hamadryas*). *PhD Dissertation, New York University.*

Wrangham, R. W., 1979, On the evolution of ape social systems, *Soc. Sci. Inf.* **18**:335–368.

Wrangham, R. W., 1980, An ecological model of female-bonded primate groups, *Behaviour* **75**:262–300.

Wrangham, R. W., 1982, Mutualism, kinship and social evolution, in: *Current Problems in Sociobiology*, K. C. S. Group, ed., Cambridge University Press, Cambridge, pp. 269–289.

Zinner, D. and Deschner, T., 2000, Sexual swellings in female hamadryas baboons after male take-overs: "Deceptive" swellings as a possible female counter-strategy against infanticide, *Am. J. Primatol.* **52**:157–168.

Zinner, D., Kaumanns, W., and Rohrhuber, B., 1993, Infant mortality in captive hamadryas baboons (*Papio hamadryas*), *Primate Rep.* **36**:97–113.

Zinner, D., Peláez, F., and Torkler, F., 2001, Distribution and habitat associations of baboons (*Papio hamadryas*) in Central Eritrea, *Int. J. Primatol.* **22**:397–413.

CHAPTER THREE

# Female Behavioral Strategies of Hybrid Baboons in the Awash National Park, Ethiopia

*Jacinta C. Beehner and Thore J. Bergman*


## CHAPTER SUMMARY

The harem-based social organization of hamadryas baboons has been attributed primarily to the predisposition of hamadryas males to herd females into one-male units (OMUs). Hamadryas females, by contrast, are thought to be behaviorally flexible. In this chapter, we describe and analyze female behavior in a hybrid group located in the Awash hybrid zone of Ethiopia. All individuals in the group are hamadryas–olive hybrids, but individual phenotypes range from mostly hamadryas to mostly olive. We use data across a 40-month period to assess whether or not females of different ancestry exhibit different behavioral strategies within the same mixed group. We found that females



**Jacinta C. Beehner** • Departments of Psychology and Anthropology, University of Michigan, Ann Arbor, MI, USA **Thore J. Bergman** • Department of Psychology, University of Michigan, Ann Arbor, MI, USA

follow three distinct behavioral strategies: strict OMU, loose OMU, and non-OMU. Behaviors that suggested a hamadryas-like social organization (e.g., strong intersexual bonds) were associated with the strict-OMU females, and behaviors that suggested an olive-like social organization (e.g., strong intra-sexual bonds) were associated with the non-OMU females. Furthermore, there was a significant relationship between morphological phenotype and behavior: Females that had more hamadryas-like morphological phenotypes exhibited behaviors characteristic of hamadryas baboons, while females with more olive-like morphological phenotypes exhibited behaviors like that of typical olive baboons. Loose-OMU females ranged across the spectrum both morphologically and behaviorally. Finally, strict-OMU females enjoyed higher reproductive success during this study than females in other kinds of groups. In sum, females under identical ecological pressures still exhibit particular behavioral strategies consistent with their morphological phenotype, suggesting that some aspects of female grouping behavior have a genetic basis.

## 1. INTRODUCTION

Members of the genus *Papio* generally demonstrate remarkable behavioral flexibility in response to local conditions, resulting in many studies evaluating the socioecological factors that affect the social organization of various *Papio* groups (e.g., Kummer, 1968b; Aldrich-Blake et al., 1971; Henzi et al., 1990; Barton et al., 1992; Biquand et al., 1992; Hamilton and Bulger, 1992; Byrne et al., 1993; Cowlishaw, 1997; but see Henzi and Barrett, 2003). In contrast to most of the genus, however, the social organization of the hamadryas baboon (*Papio hamadryas hamadryas*) appears to be relatively inflexible. Hamadryas form one-male units (OMUs), consisting of a single leader male and one or more females and their offspring, whether they are in captivity (Kummer and Kurt, 1965; Sigg, 1980; Caljan et al., 1987), in the wild (Kummer, 1968a; Abegglen, 1984; Swedell, 2006), or under varying conditions of food availability and predation pressure (Kummer et al., 1985; Biquand et al., 1992).

Most research on hamadryas groups reveals that females spend more time interacting with their leader male than with other OMU females (Kummer, 1968a, 1971; Sigg, 1980; Abegglen, 1984). The term "cross-sex bonding" has been used to describe this hamadryas-like social organization in which *inter*sexual bonds are stronger than *intra*sexual bonds (Byrne et al., 1989). In contrast, the females of nonhamadryas baboon populations form a "female-bonded" social organization (Wrangham, 1980). In these groups, females are philopatric and maintain closer bonds with females, especially maternal relatives, than with males (Seyfarth, 1976; Altmann, 1980; Melnick and Pearl, 1987; Barton et al., 1996).

The obvious difference in the hamadryas social organization with respect to most of the genus certainly requires an explanation. It is generally accepted that the OMU behavior of hamadryas baboons is derived with respect to the social organization of other baboon populations and probably arose as an adaptation to a semidesert climate and dispersed food resources (Jolly, 1963; Kummer, 1968a; Dunbar, 1988; see also Bergman, this volume and Swedell and Saunders, this volume). Particularly for hamadryas females, cross-sex bonding may be favored as a foraging strategy suited for a group that lacks cohesion. Where resources are scarce, foraging in small groups is most efficient. Therefore, choosing to associate with a single male creates both a reproductive unit and a foraging unit, simultaneously enhancing protection from predators and other males. However, most of the stereotypical behavioral traits that accompany the hamadryas social organization are thought to be "male driven." For example, the cohesiveness of OMUs is maintained primarily by the herding behavior of hamadryas males, while the following behavior of hamadryas females has been reported as only a response to the aggressive gestures of leader males nearly twice her size (Kummer, 1968a,b; Kummer et al., 1970; Abegglen, 1984; Swedell, 2006).

In a series of behavioral experiments widely cited as demonstrating the behavioral flexibility of female baboon behavior, Kummer et al. (1970) released olive females into a wild hamadryas group and hamadryas females into a wild olive group. The hamadryas males immediately "claimed" the unattached olive females and began herding them into their respective units. Initially, the olive females fled the aggressive advances of the hamadryas males, but within a few hours, the olive females learned to follow their leader males just like hamadryas females. Likewise, in the olive group the hamadryas females roamed independently within their new groups, interacting with several males, just as olive females would. The authors concluded that because both types of females were able to modify their behavior to fit a different social organization, "the main cause of hamadryas polygyny must be the hamadryas male" (Kummer et al., 1970, p. 357).

In the first study of female hamadryas behavior in the wild, Swedell (2002) found that hamadryas females did not necessarily focus all their social time on the leader male, but rather spent much of their time socializing with other females, in some cases even crossing unit boundaries to interact with one another. Swedell (2002) concluded that female bonding may have been constrained during the evolutionary history of the hamadryas female by the development of male herding behavior and the male-driven hamadryas social organization but that hamadryas females may still retain the flexibility in social behavior and motivation to form affiliative relationships with other females seen in other baboons.

In contrast with these studies in the wild, behavioral observations in captivity provide some evidence that female hamadryas baboons do preferentially bond to a central individual. In one case, a dominant hamadryas female assumed the role of leader male in the absence of a male, and all other females in the enclosure focused the majority of their social interactions on the dominant female rather than any other females (Stammbach, 1978; Coelho et al., 1983). In a similar study, a hamadryas female assumed the leader male position when the only male in the enclosure did not herd females (Pfeiffer et al., 1985). Because a hamadryas-like social organization persisted despite the absence of a male, these authors concluded that hamadryas females must have an innate disposition to orient themselves around a central individual.

Although the social behavior of hamadryas females is certainly more flexible than that of their male counterparts, there is the possibility that, when faced with limited opportunities for behavioral expression, a female's *realized* behavior may be different from her *preferred* behavior (for clarity, we refer to a female's preferred behavior as a *behavioral strategy*). For instance, once released into a hamadryas group, an olive female has little choice but to be herded into an OMU or suffer continuous attacks from one or more leader males. The best way to distinguish between different female behavioral strategies, in this case OMU versus non-OMU behavior, is if females are in a situation where they have opportunities for both. In other words, it is difficult to evaluate female OMU behavior in a fully hamadryas group where the choice to associate with multiple males is virtually nonexistent. Likewise, in an olive baboon society where male–male competition can be fierce and maternal kin alliances are readily available, a female has little opportunity or incentive to bond with a single male (Smuts, 1985; Strum, 1994). The hybrid zone between hamadryas and olive baboons in the Awash National Park of Ethiopia offers an ideal opportunity to address whether females have different behavioral strategies, as many of the baboon groups with mixed ancestry have both hamadryas and olive behavioral patterns coexisting in the same group (Sugawara, 1982; Nystrom, 1992; Beyene, 1993). Just as some socioecological studies look at the same baboons under different ecological pressures to examine the flexibility in baboon social organization, we analyze animals along a morphological spectrum (hamadryas to olive) under the *same* ecological circumstances to examine the innate preferences for one behavioral strategy over another.

We base our analyses on a study of a hybrid group of baboons to determine whether or not morphological phenotypes correlate with behavioral phenotypes. In effect, we expect females with more hamadryas ancestry to "act" more hamadryas-like while females with more olive ancestry "act" more olive-like. We chose an Awash hybrid group that, since its first published description (Nagel, 1973; Sugawara, 1979, 1982, 1988), has consisted of individuals that range phenotypically and behaviorally from hamadryas-like to olive-like.

Our study addresses the following questions: First, do differences in behavioral strategies exist among females? While we knew that there were OMUs embedded within the larger group structure based on earlier studies (Sugawara, 1979, 1982, 1988), we had no information on how females were distributed within this structure. Therefore, we begin by describing the current composition of the group and changes since earlier published accounts. Second, if female behavioral differences do exist, can morphological phenotype (as a proxy for ancestry; see Phillips-Conroy and Jolly, 1986; Phillips-Conroy et al., 1991) predict a female's behavioral strategy? Using this proxy for ancestry, we looked at the relationship between ancestry and OMU behavior as well as between ancestry and several behaviors that mark females as cross-sex bonded or female bonded. Third, do females play an active role in these different behavioral strategies? Particularly in the case of OMU behavior, do females contribute to the cohesion of the OMU? Finally, can the reproductive consequences of choosing one behavioral strategy over another be defined?

## 2. METHODS

### 2.1. Study Site and Subjects

We collected data across a 40-month period from September 1997 to November 1998 (T.J.B.) and from December 1999 to December 2000 (J.C.B.) on a group of hybrid baboons (group H) in the Awash National Park of Ethiopia. The group is located at the phenotypic center of the Awash hybrid zone. Details on the layout of the hybrid groups along the Awash River and the ecology of the park can be found in Nagel (1973) and Phillips-Conroy and Jolly (1986).

The habitat of group H is more typical of hamadryas baboons than olive baboons. During our study, group H spent the majority of time inside the Awash River canyon, which consists mainly of rocky soil, semidesert *Acacia* thorn scrub, and densely packed bushes (*Acacia* and *Grewia* spp.). Faced with limited visibility and the dispersed resources of a semidesert environment, H frequently formed small foraging groups during the day (which we refer to as *subgroups*). During the dry season when food was scarce, H extended their foraging range outside the canyon to the grasslands, where they fed on grasses, corms, and seeds from a variety of *Acacia* tree species. Some of these subgroups reassembled back at the sleeping cliffs at night, but because there were several sleeping sites, subgroups often remained separate for several days.

All individuals in group H were individually recognizable and habituated to observers on foot. We collected behavioral data on 30 adult females in group H. Females were considered adults following their first full-term pregnancy. Four females disappeared and four females reached adulthood during the course of the study, and there were never more than twenty-six adult females in the group at any one time. The demographic composition of group H during this study period is described in detail elsewhere (Bergman and Beehner, 2003, 2004).

### 2.2. Behavioral Data Collection

Focal animal sampling could not be used in this study due to the extremely rough terrain, poor visibility, and limited habituation of this group; it was nearly impossible to observe the same individual for even short intervals. Instead, social behaviors of interest (grooming, approaches, follows, copulations, and other sexual behaviors) were recorded using all-occurrence sampling and then standardized (for time spent observing each individual) using 5-min scan samples (Altmann, 1974). During our scan samples we also recorded the identity of all nearest neighbors (within 10 m). We made every attempt to observe individuals for the same amount of time (i.e., by choosing to follow one subgroup over another), but as we could not control the exact individuals in each subgroup, some individuals necessarily had more observation time than others. On a daily basis, we recorded the identity of all individuals in subgroups and single-male units and all estrous females and their consort partners. We define female estrus based on visible signs of swelling in the perineal region (sexual swellings).

### *2.2.1. Phenotypic Hybrid Index Scores*

We assigned phenotypic hybrid index (PHI) scores to each female when the study began in 1997. The total PHI score consists of the summed scores across four traits: face color, hair color, hair length, and tail shape. The PHI ranges from 0 to 16, with each trait receiving a score between 0 and 4 (see Bergman and Beehner (2003) for a detailed explanation of the PHI). A lower PHI score denotes a typically hamadryas morphological phenotype, while a higher PHI score denotes a typically olive baboon morphological phenotype (Table 1).

### *2.2.2. Female Maternal Relationships*

We determined maternal relationships for adult females based on analysis of genetic material obtained during trapping seasons carried out by the Awash National Park Baboon Research Project (ANPBRP, codirected by C. Jolly and J. Phillips-Conroy; for trapping protocol see Brett et al., 1982). Bergman (2000) based a maternity analysis for these samples on 10 population-specific microsatellite loci. Sister relationships were assigned if two females were assigned to the same mother, cousin relationships were assigned if their mothers had the same mother, and so on. Female relatives included all adult female maternal relatives to the level of first cousin. Because genetic data from group H were available as far back as the 1973 trapping season, 27 out of 30 females (90 percent) were confidently assigned a mother. The remaining 3 females could not be assigned to a mother, most likely because we lacked genetic data from their mothers.

**Table 1.** All adult females in group H, OMU type, and phenotype

| Females | OMU type | PHI score |
|---|---|---|
| MX | Strict | 6.5 |
| IM | Strict | 7.5 |
| HA | Strict | 8.5 |
| GI | Strict | 9.0 |
| DO | Strict | 9.5 |
| CH | Strict | 13.0 |
| KO | Loose | 4.0 |
| PO | Loose | 7.0 |
| HE | Loose | 7.5 |
| LO | Loose | 8.5 |
| RA | Loose | 11.0 |
| SQ | Loose | 13.5 |
| TB | Loose | 14.0 |
| GX | Loose | 16.0 |
| KI | None | 10.0 |
| GA | None | 10.0 |
| WU | None | 11.5 |
| ST | None | 11.5 |
| TG | None | 12.0 |
| LA | None | 12.0 |
| AR | None | 13.0 |
| OW | None | 14.0 |
| NI | None | 14.0 |
| RI | None | 14.0 |
| SP | None | 15.5 |
| AT | None | 15.5 |
| CE | None/strict/none[a] | 15.0 |
| JE | Loose/none[a] | 4.5 |
| TR | None/loose[a] | 11.0 |
| QU | None/loose[a] | 14.0 |

[a]Indicates a change in OMU type during the study.

### *2.2.3. Female Behavioral Strategies*

In the analysis of the composition of group H, we use a categorical variable (*OMU type*, see results for details) to indicate an overall behavioral strategy for each female. However, individual female behaviors within these categories operated along a continuum from hamadryas to olive behavior. Therefore, to further analyze the relationship between female ancestry and behavior, we conducted a factor analysis (Sokal and Rohlf, 1981) on six behavioral measures of "hamadryas-ness" or "olive-ness."

We defined the six variables as follows: (1) *Leading males*—when the female initiated movement and the male immediately maintained proximity.

(2) *Following males*—when the male initiated movement and the female immediately maintained proximity. Leading and following were calculated as frequencies (occurrences/observation hour). (3) *Association with one male*—all days spent in association with one male (as determined by subgroup patterns and proximity throughout the day) standardized by total number of observation days. (4) *Grooming one male*—time spent grooming one male standardized by total grooming time. (5) *Grooming female relatives*—time spent grooming female relatives standardized by total grooming time. This variable was corrected for total number of female maternal relatives in the group. Females with either no relatives or unknown relatives (i.e., no DNA available) were excluded from analyses of this variable ($n$ = 5). (6) *Nearest neighbor the same male*—the proportion of scan samples in which the same male was a female's nearest neighbor. Grooming duration was calculated when the starting and ending times were known for female grooming bouts. Female behaviors scoring high with respect to cross-sex bonding were considered *hamadryas* behaviors, and behaviors that scored high with respect to female bonding, particularly among related females, were considered *olive* behaviors (Table 2).

To determine the active role that females played in maintaining a relationship with a single male, we analyzed a combined variable representing the rate of approaches, presents, and follows directed toward a single male (occurrences/hr). *Approaches* are defined as any approach to within 2 m of a male. *Presents* are defined as the presentation of one's hindquarters to a male. *Follows* are defined above. These three behaviors were combined into a single variable, *proximity maintenance behaviors.*

Because olive males interact frequently with estrous females during consortships, differences in behaviors that indicate cross-sex bonding versus female bonding were most salient when females were outside of their estrus

**Table 2.** Female social behaviors indicating hamadryas or olive behavior

| Female behavior | Hamadryas prediction | Olive prediction |
|---|---|---|
| Leading males | Low | High |
| Following males | High | Low |
| Association with one male | High | Low |
| Grooming one male | High | Low |
| Grooming female relatives | Low | High |
| Nearest neighbor the same male | High | Low |

period (i.e., no sexual swelling). Indeed, nearest male neighbor data and rates of interaction with a male were indistinguishable across estrous females. Therefore, in calculating values for variables related to cross-sex bonding, we included only data from anestrous periods.

### 2.3. Data Analysis

We tested all behavioral variables, including PHI scores, for deviations from normality using the Kolmogorov–Smirnov test. Tests showed no significant deviations from normal, thus permitting parametric statistical tests. When variables were separated into discrete categories, single-factor ANOVA was used to analyze differences in means. ANOVAs indicating a significant difference across categories were supplemented by a Tukey's multiple comparisons' test to determine which groups were significantly different from the others. Following a factor analysis, we used linear regression to determine the relationship between PHI score and values along one of the resulting factors. We used a binomial test to determine whether the OMU type for each female followed that of her mother or was random in distribution and a Kaplan–Meier survival analysis to examine reproductive success of OMU types. The statistical threshold was set at $p < 0.05$, and all tests were two tailed.

## 3. RESULTS

### 3.1. The Social Organization of a Hybrid Group

Overall, group H had an unusual social organization that reflected elements of both hamadryas and olive societies. Three distinct social components emerged in this study that will be used to characterize each adult female: strict OMU, loose OMU, and non-OMU. Out of 30 females, 6 (20.0 percent) were members of OMUs very much like those found in hamadryas society, termed here *strict OMUs*. There were four strict OMUs (i.e., four strict OMU leader males) in group H. These units shared several characteristics with traditional hamadryas OMUs. First, strict OMUs maintained consistent membership across the study period (with a few abrupt transitions, noted below). Second, strict OMUs were discrete units, visible even when observing group H as a whole. Nearest neighbor data indicate that the leader male and unit females were found in closer proximity with each other than with the rest of the group. Specifically, 83.1 percent of the time the nearest neighbor to an OMU female was another member of her unit, and in 64.1 percent of these observations, the nearest neighbor was the leader male. Conversely, females outside of strict OMUs (i.e., loose- and non-OMU females) had consistent neighbors only 39.2 percent of the time (using summed data on nearest male and nearest two females). Third, when the group separated into subgroups, OMU members, without exception, remained in the same subgroup or formed a subgroup of their own. Finally, the OMU leader male always consorted with his unit's females when they were in estrus.

Despite these attributes, we noticed three differences between the strict OMUs of group H and the OMUs of traditional hamadryas societies (Kummer, 1984). First, group H OMUs showed very low rates of male herding or aggression directed at establishing and maintaining unit cohesion. Second, males often followed females (instead of the reverse). Third, in the context of the entire group, strict-OMU females sometimes interacted with other group members, which is typically not seen in hamadryas societies (but see Swedell, 2002). Therefore, the term "strict OMU" is used only in the context of this hybrid group to distinguish between these OMUs and the next hybrid grouping (the loose OMUs) and is not meant to imply that these OMUs are any "stricter" than (or even as strict as) traditional hamadryas OMUs.

Eight females (26.7 percent) were members of a less cohesive type of OMU, the *loose OMU*. We recorded three loose OMUs in group H. Similarities between loose OMUs and strict OMUs included consistency in membership, cohesiveness among individuals during subgrouping, and exclusive consorting by the leader male with female members during estrus. However, while strict OMUs remained visually discrete units to an observer, loose OMU members often mixed throughout group H, and a loose OMU only became obvious when group H broke into subgroups. Less than half the time (45.8 percent), a loose-OMU female's nearest neighbor was another member of the unit, and only 19.4 percent of these neighbors were the leader male.

Thirteen females (43.3 percent) avoided OMUs entirely, exhibiting characteristics of olive females (termed here *non-OMU females*). In general, these females consorted with various males during estrus and did not associate with any one male at the conclusion of estrus. Five of these females associated with a male during lactation. These five females had one consistent male neighbor

26.2 percent of the time—higher than females in loose OMUs but considerably less than those in strict OMUs. However, outside of lactation, the amount of time spent with this male was halved (13.6 percent). These observations seem to approximate the male–female relationship found in olive baboon *friendships* (Smuts, 1985) rather than cross-sex bonding that extends throughout all reproductive stages (Kummer and Kurt, 1963). Like olive females, the non-OMU females of group H also formed a linear dominance hierarchy (Beehner, 2003) that, with some exceptions, mirrored the nested matrilineal hierarchies of olive and other savanna baboons whereby the daughters assume the rank just below that of their mother (Hausfater et al., 1982). This hierarchy showed stability across the 40 months of this study.

Only four females joined more than one type of social subunit (Table 1). Three females occupied one type during the first observation period, changing to a different type by the beginning of the second observation period. Two of these females switched from no OMU membership to loose OMUs, and one female switched from a loose OMU to no OMU membership. The fourth female (CE) spent most of the first observation period as a non-OMU female, then consorted with a strict-OMU male (the only strict-OMU male that aggressively herded his unit females) and subsequently remained in his OMU. Following the disappearance of this OMU male, she reverted to a non-OMU female for the entire second observation period. We thus analyzed CE as a non-OMU female.

Initially, to place females in one of these three categories, we observed the smallest subgroups formed during group fissions in combination with nearest neighbor data. However, because strict OMUs were spatially cohesive, discrete entities even in the context of the entire group (as in hamadryas societies), it quickly became apparent that this method of group classification was not necessary. Strict OMUs were obvious groupings of a single male and unit females set apart from the rest of the group. Loose OMUs were more difficult to score in the beginning; after observing consistent single-male subgroups and recording exclusive consortships between the same male and females, however, it became clear that these groups represented a third type of organization in the substructure of group H. Furthermore, the membership of these units remained consistent throughout the study period. Loose-OMU males consorted with extra-unit females on several occasions, but the females of the unit never changed and were always observed in consort with the leader male during their estrus periods.

### 3.2. Changes in Group Structure

The unique substructure of group H dates at least as far back as Sugawara's descriptions of the group in the 1970s (Sugawara, 1982). Sugawara used different definitions to describe the various single-male groups, thereby making it difficult to make direct comparisons of OMU females between Sugawara's study and ours. However, we can compare data on the females outside of OMUs. From 1975 to 1978, Sugawara reported that 48.7 percent of group H females were not involved in an OMU, compared to 46.6 percent during our study. Therefore, in both studies just over half of group females were involved in some type of OMU.

During our study period, we observed female movements into and out of strict OMUs, recording 17 transfers either *into* or *out of* a strict OMU. We defined a "transfer" as an abrupt change in association with an OMU leader male where a female was observed with one male on one day and another male the next. Interestingly, all females transferring into or out of strict OMUs were previous members of strict OMUs, with one exception (CE, see previous section). The leader male and two females in this unit disappeared at the same time, but all of the other unit females ($n$ = 3) were subsequently incorporated into new strict OMUs, suggesting that CE was only an OMU female as a result of the aggressive herding by the leader male. In all other transfers, females moved *among* strict OMUs. Despite many transfers among strict OMUs, no strict-OMU female ever changed to a loose- or non-OMU female. Furthermore, neither loose-nor non-OMU females ever became strict-OMU females (except CE). Female transfers between loose- and non-OMU groups, however, were more frequent (Table 1).

We compared our transfer data with those of Sugawara to assess the overall amount of movement into and out of strict OMUs. Sugawara observed the group for two different study periods (4–5 months each) spanning 41 months between 1975 and 1979, noting any transfers since the first observation period at the start of the second period. He estimated 21 movements into and out of OMUs, for a transfer rate of approximately once every 2 months. This closely matches our estimates of transfer rates (17 such movements in 40 months). Assuming a constant transfer rate since 1975, this suggests that over 100 OMU transfers have occurred since Sugawara's study, thus providing ample opportunity for reshuffling of females within group H. Evidence suggests that, just prior to Sugawara's study, group H fused with its downstream neighbor, group I, a more hamadryas-like group (Bergman and Beehner, unpublished data), which may account for much of the group's organization reported by Sugawara (1979,1982,1988). However, because there have been no further fusions since Sugawara's study, any current deviations from random with respect to phenotype and group membership are not the result of larger-scale historical events.

## 3.3. Social Behavior of Hybrid Females

### *3.3.1. Across-Unit Social Behavior*

We used factor analysis to determine the extent to which the variables listed in Table 2 could be accounted for by a smaller number of underlying variables. The descriptive statistics for these six variables are listed in Table 3. We retained a factor if its eigenvalue exceeded 1.0, supplying two factors, which together accounted for 72.4 percent of the total variance. Table 4 shows the loadings of each of the six original variables on the two factors. Factor 1 loaded positively on four variables, including *following males*, *association with one male*, *grooming one male*, and *nearest neighbor the same male*. Furthermore, factor 1 loaded negatively on *grooming female relatives*. Thus, factor 1 appears to approximate female social behavior by separating females into cross-sex bonded (higher scores) or female-bonded individuals (lower scores). Factor 2 only loaded strongly on *leading males*.

**Table 3.** Descriptives for female social behaviors

| Female behavior | Mean ± SD | Range | *n* |
|---|---|---|---|
| Leading males (occurrence/hr) | 0.25 ± 0.13 | 0.01–0.53 | 26 |
| Following males (occurrences/hr) | 0.07 ± 0.07 | 0.00–0.35 | 26 |
| Association with one male (days/total days observed) | 0.28 ± 0.25 | 0.00–0.86 | 26 |
| Grooming one male (min/total grooming time) | 0.27 ± 0.17 | 0.04–0.65 | 26 |
| Grooming female relatives (min/total grooming time) | 0.24 ± 0.04 | 0.00–0.56 | 22 |
| Nearest neighbor the same male (scans/total scans) | 0.48 ± 0.01 | 0.35–0.64 | 26 |

**Table 4.** Loadings of behavioral variables on first two factors

| Variable | Factor 1 | Factor 2 |
|---|---|---|
| Leading males (occurrences/hr) | 0.17 | 0.83 |
| Following males (occurrence/hr) | 0.75 | 0.01 |
| Association with one male (days/total days observed) | 0.92 | 0.08 |
| Grooming one male (min/total grooming time) | 0.71 | −0.53 |
| Grooming female relatives (min/total grooming time) | −0.73 | −0.53 |
| Nearest neighbor the same male (scans/total scans) | 0.80 | 0.44 |

As shown in Figure 1, OMU types had significantly different means for factor 1 ($F_{2,21} = 12.26$, $p < 0.001$). Strict-OMU females had significantly higher scores on factor 1 than either loose-OMU ($p < 0.01$) or non-OMU females ($p < 0.001$). There was no difference between loose- and non-OMU females' factor 1 scores ($p = 0.623$).

### 3.3.2. *Within-Unit Social Behavior*

In strict OMUs, most females restricted grooming to within the unit (75 percent). Of within-unit grooming, females directed 71 percent of this grooming toward the leader male and 29 percent toward other females in the unit. Among loose OMUs, females directed approximately half of their grooming

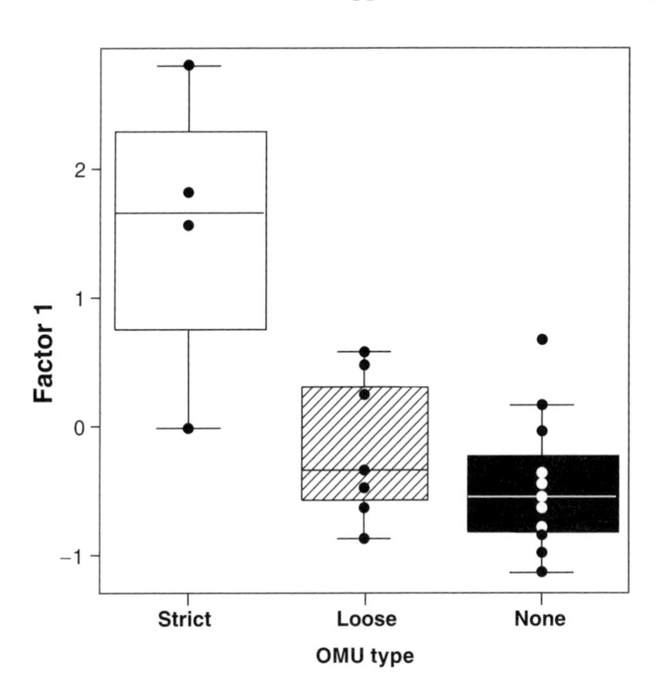


**Figure 1.** Female factor 1 scores (mean ± SEM) for each OMU. High factor 1 scores indicate cross-sex bonding (more hamadryas-like behavior) and low factor 1 scores indicate female bonding (more olive-like behavior). Circles represent individual females. The factor 1 means were significantly different across OMU types ($F_{2,21} = 12.26$, $p < 0.001$). Strict-OMU females had significantly higher scores on factor 1 than either loose-OMU ($p < 0.01$) or non-OMU females ($p < 0.001$). There was no difference in scores between loose- and non-OMU females ($p = 0.623$).

towards other unit members (55 percent). Of this within-unit grooming, 56 percent was directed toward the leader male and 44 percent was directed toward other unit females.

We analyzed a composite variable representing the rate of approaches, presents, and/or follows (*proximity maintenance behaviors*) toward a single male to determine whether or not females played an active role in maintaining an exclusive relationship with one male. As Figure 2 illustrates, strict-OMU females had higher rates of proximity maintenance behaviors than either loose-OMU or non-OMU females ($F_{2,25} = 8.70$, $p < 0.01$). Although loose-OMU females had higher rates of proximity maintenance behaviors than non-OMU females, differences between loose- and non-OMU females failed to reach statistical significance ($p = 0.068$).

### 3.4. Social Behavior and Ancestry

After removing the four females that switched between OMU types, we analyzed PHI scores for all females across the three OMU types. PHI differed

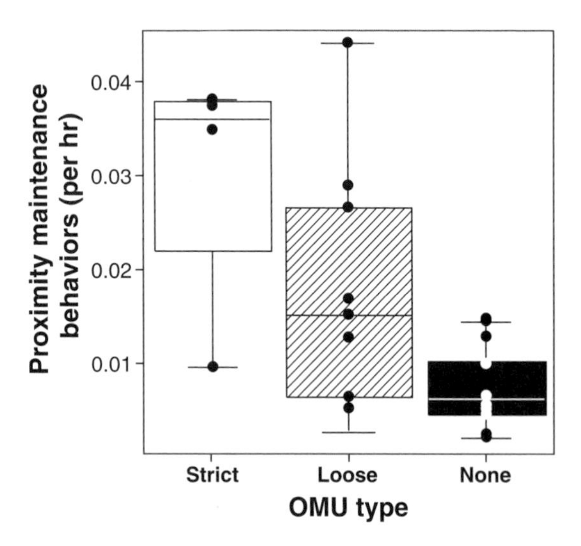


**Figure 2.** Female proximity maintenance scores (mean ± SEM) for each OMU type. Circles represent individual females. Strict-OMU females had higher rates of proximity maintenance behaviors than either loose-OMU or non-OMU females ($F_{2,25} = 8.70$, $p < 0.01$). Differences between loose- and non-OMU females were not statistically significant ($p = 0.068$).

significantly across types of social groups ($F_{2,25}$ = 4.14, $p < 0.05$; Figure 3). A multiple comparisons' test indicated significantly lower PHI scores in strict OMU females (i.e., more hamadryas-like) than in non-OMU females ($p$ = 0.036). However, loose-OMU females did not differ significantly from strict ($p$ = 0.719) or non-OMU females ($p$ = 0.137).

Factor 1 held a negative relationship with PHI scores ($R^2$ = 0.24, $p < 0.05$). Females with high scores on factor 1 (indicating cross-sex bonding) tended to have lower PHI scores (Figure 4). In other words, females with more hamadryas-like behavior had more hamadryas-like phenotypes, and females with more olive-like behavior had more olive-like phenotypes. There was no relationship between factor 2 and PHI score ($R^2$ = 0.00, $p$ = 0.869).

We then examined whether females were more likely to be in an OMU if they had fewer female relatives in the group. An analysis of variance revealed no significant differences across females ($F_{2,25}$ = 0.09, $p$ = 0.92). Females in

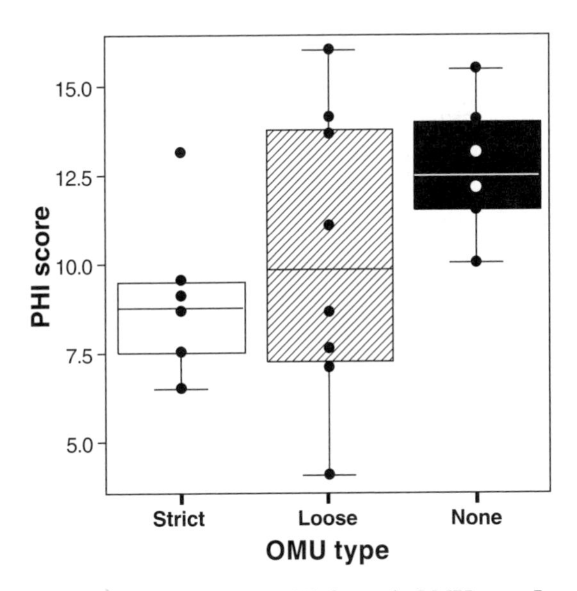


**Figure 3.** Female PHI scores (mean ± SEM) for each OMU type. Lower PHI scores indicate more hamadryas-like phenotypes. Circles represent individual females. PHI differed significantly across types of social groups ($F_{2,25}$ = 4.14, $p < 0.05$). Strict-OMU females had significantly lower PHI scores than non-OMU females ($p$ = 0.036). The differences between loose-OMU females and strict-OMU ($p$ = 0.719) or non-OMU females ($p$ = 0.137) were not statistically significant.

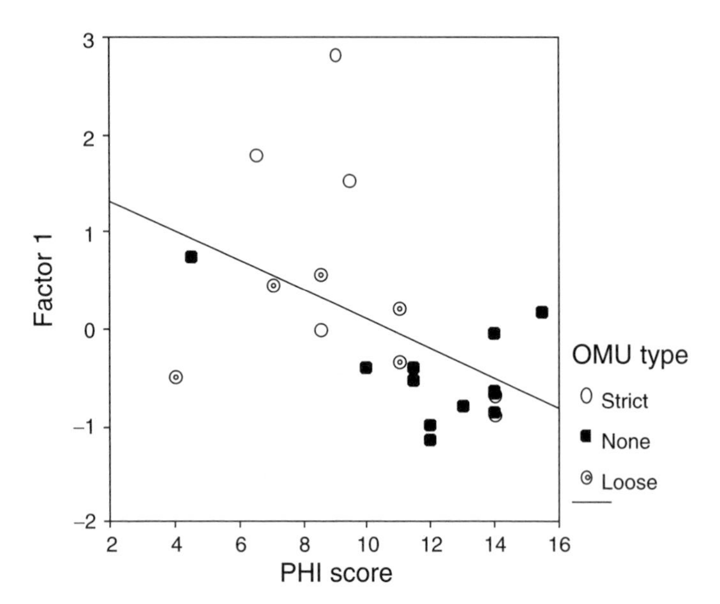


**Figure 4.** Least squares linear regression between factor 1 and PHI score ($R^2 = 0.24$, $p < 0.05$). High factor 1 scores indicate cross-sex bonding (more hamadryas-like behavior) and low factor 1 scores indicate female bonding (more olive-like behavior). Lower PHI scores indicate more hamadryas-like phenotypes. Circles represent individual females and are separated into OMU types.

strict OMUs had just as many maternal relatives in group H as did non- and loose-OMU females.

Next, we addressed the possibility that OMU membership might be socially inherited (i.e., that daughters learn OMU behavior from their mother). This hypothesis predicts that a daughter's OMU type will be the same as that of her mother. Out of 14 mother–daughter pairs, 9 mothers and daughters were found in *different* types of social groups, a result that is significantly different from the prediction (binomial test: $p < 0.001$) but not from what would be expected if the daughter's social group was random (binomial test: $p = 0.14$). When we compared the PHI scores of mother and daughter for the differing pairs, seven out of nine daughters' change in social group corresponded with the directional difference in PHI score. For example, a mother (PHI score: 6.0) in a strict OMU might have a daughter (PHI score: 11.5) in a loose OMU or no OMU.

### 3.5. Social Behavior and Reproductive Success

A survival analysis on the outcome of 38 pregnancies across 40 months from strict-OMU, loose-OMU, and non-OMU females (Figure 5) indicated differences in reproductive success across OMU types. Despite the small sample size, analyses revealed the highest reproductive success among strict-OMU females. The average strict-OMU female experienced 1.52 live births compared to 1.05 births for loose-OMU females and 1.31 births for non-OMU females. Pronounced differences in infant survival among groups exacerbated differences in total fertility. When infant survival was set at 3 months (resulting in no censored data), strict-OMU females had 100 percent survivorship (8 out of 8), loose-OMU females experienced 53.9 percent survivorship (7 out of 13), and survivorship of non-OMU female infants reached 76.5 percent (13 out of 17). Right censoring precluded survival estimates for strict-OMU females (as all infants were still alive at the end of the study), but we estimated survival of loose- and non-OMU females at 53.9 percent and 72.2 percent, respectively.

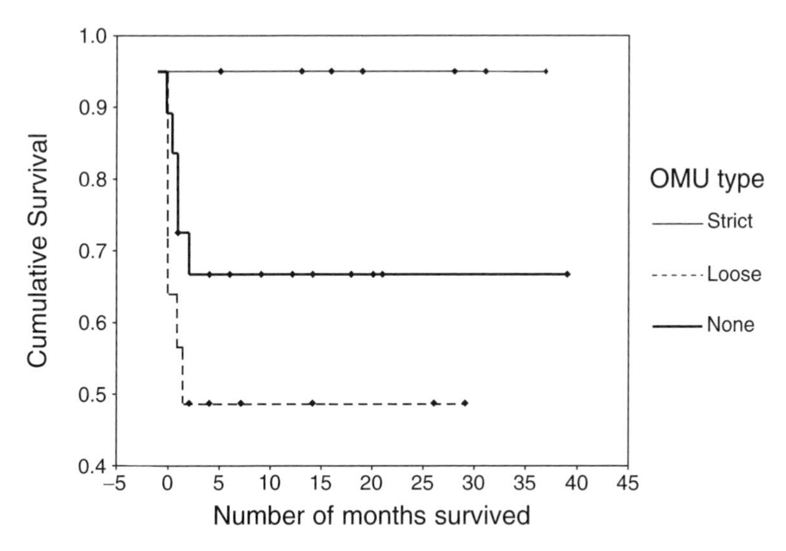


**Figure 5.** Survival analysis on the outcome of 38 pregnancies. Dots represent censored cases.

## 4. DISCUSSION

In a previous study, we found that there were several males in group H that clearly were "OMU males," i.e., males that consistently pursued a strict OMU strategy (Bergman, 2000; Bergman and Beehner, 2003). In this study we wanted to determine if, like hybrid males, there were similar differences in the behavioral strategies of hybrid females. Mainly, were there consistent females that preferentially bonded with a single male? Indeed, we found that there were different female affiliative tendencies within group H corresponding to hamadryas (strict-OMU females) and olive (non-OMU females) equivalents, with an intermediate group (loose-OMU females).

The high transfer rate during both Sugawara's and our study suggests that there was ample opportunity for a reshuffling of females into (and out of) various subunits. From these results, it appears that females in group H *do* have opportunities to determine their social affiliations in ways that females in hamadryas or olive societies do not. Nevertheless, analysis of this reorganization during the current study period indicates that the *same* females transferred among strict OMUs. With one exception, all females that joined a new strict OMU derived from strict OMUs. Additionally, strict-OMU females had significantly more hamadryas-like phenotypes than loose- or non-OMU females. Taken together, these data suggest that both males and females express innate tendencies to form OMUs.

As a general rule, the most hamadryas-like females exhibited a hamadryas-behavioral strategy, forming stronger bonds with one particular male; and the most olive-like females had behaviors that corresponded to the olive behavioral strategy, forming strong bonds with female maternal kin. However, group H included many females with intermediate behaviors, and these females covered the entire phenotypic spectrum (Figure 4). As illustrated in Figure 1, intermediate scores on factor 1 are primarily associated with a new social group altogether: the loose OMU. Loose-OMU females manage to maintain a bond with a single male while simultaneously maintaining relationships with other females.

In contrast with strict-OMU females, loose- and non-OMU females overlapped considerably along the middle and lower end of factor 1. We suggest two reasons for this. First, on the negative end of factor 1 (i.e., the behaviorally olive end), levels of bonding with males at any one time were similar among both loose- and non-OMU females, but while non-OMU females associated with different males across cycles, loose-OMU females stayed with the same male. In other words, a female with "serial" male associations had a similar factor 1 score to a female with a single male association due to the limited duration of this study (on average, females only cycled 2–3 times). Second, females in both groups had similar levels of bonding with female relatives. In fact, some loose-OMU females had stronger ties with female relatives than did non-OMU females, particularly when the female's unit included relatives. Consequently, while on average we found higher degrees of cross-sex bonding in loose-OMU than in non-OMU females, behavioral measures alone were not sufficient to separate these two groups.

Not surprisingly, females with intermediate ancestry showed high levels of both cross-sex and female bonding. Loose-OMU females spent a little more than half of their grooming bouts on within-unit partners, and their within-unit grooming was approximately evenly split among the leader male and the unit females. Strict-OMU females, however, spent three-quarters of their grooming bouts on within-unit partners, and most of these were with the leader male. For many of the other variables used in this analysis, loose-OMU females were intermediates along the cross-sex/female-bonded continuum. The idea that female behavior operates along such a continuum rather than a dichotomy between cross-sex bonding and female bonding has grown more popular with the emergence of more data on wild populations. Many female-bonded primates, including olive baboons, maintain high association rates between anestrous females and certain adult males (e.g., Seyfarth, 1978; Altmann, 1980; Smuts, 1985). There are several hypotheses for the adaptive advantages conferred upon the female participants of these long-term bonds outside of consortship (Smuts, 1985). Conversely, several studies of both captive and wild hamadryas baboons (a "non-female-bonded" primate) report that interfemale interactions occur more often than expected (Stammbach, 1978; Sigg, 1980; Stammbach and Kummer, 1982; Coelho et al., 1983; Chalyan et al., 1991; Vervaecke et al., 1992; Colmenares et al., 1994; Leinfelder et al., 2001; Swedell, 2002). Thus, variation in intra- and intersexual bonding among pure olive and hamadryas females may also account for much of the intermediacy found among hybrid females.

Despite the high degree of behavioral overlap between loose-OMU females and the other two groups, the fact remains that strict-OMU and non-OMU females are phenotypically and behaviorally very different. OMU females consistently remain OMU females despite the fact that they have female relatives outside of the unit. Individuals in these two social categories situate themselves very differently in the context of the group, with OMU females nearly always found immediately adjacent to their leader male. Their affiliative preferences can be accurately described as cross-sex bonded for strict-OMU females and female bonded for non-OMU females. Furthermore, OMU females play an active role in maintaining OMU cohesion. So, if females have different behavioral strategies under the same ecological pressures, the next question is *why?*

Barton et al. (1996) argued that the strength of bonds among females marks the critical difference between the social organization of hamadryas and "savanna" baboons. In other words, where females form extended alliances and grooming networks, males cannot partition groups into independent OMUs (Barton et al., 1996). This opens the possibility that group H females joined OMUs because they lacked female relatives in the group. With no female relatives, it might be advantageous to form an affiliative relationship with a male. However, on average, strict-OMU females had the same number of maternal relatives as loose- and non-OMU females. Therefore, the potential to form bonds with female relatives existed for nearly all group females. That some females were male bonded despite the opportunity to form strong female bonds provides further evidence that OMU females were not merely following the social agenda of OMU males.

If the number of female relatives in the group had no bearing on whether a female was in an OMU, what about the possibility that OMU behavior was the result of socially acquired knowledge? This hypothesis maintains that female affiliative tendencies may be learned from a female's mother. However, this does not appear to be the case. In an analysis of mother–daughter pairs, most daughters were involved in a type of social group that was different from that of their mothers. Furthermore, daughters were more likely to be in a social group that coincided with their own phenotype and not that of their mother.

An alternative explanation may be that OMU males had a preference for "already trained" females, i.e., females that had previously (and successfully) been in an OMU. However, OMU females (both strict and loose) were more likely to exhibit proximity maintenance behaviors toward a single male than were non-OMU females. This suggests that OMU females played active roles in maintaining bonds with a particular male. It is also important to point out that if, indeed, males preferred easily trained females, this would not preclude an active role on the part of the females: "Trainability" from the male perspective corresponds to an interest in cross-sex bonding from the female perspective.

Although merely suggestive at this stage, there is some evidence that strict-OMU females benefited from greater reproductive success than other females, particularly in terms of offspring survival. This matches the pattern found by Sigg et al. (1982), who reported higher infant survival in hamadryas than in other baboons. It has been widely reported that hamadryas females and their infants receive protection from their leader male (Sigg et al., 1982; Swedell, 2006). In the structured society of hamadryas baboons and presumably in any situation where a larger group breaks up into smaller groups, it would likely be adaptive for females to bond to a single male. Bonds with a particular male not only provide the "seams" for group division, but they also ensure that females are not left in a subgroup without a male. It has already been noted that when troop segments regularly split and join again, males who permanently maintain access to a few females may have an advantage over males who compete for all females who are in estrus (Kummer, 1968a; Anderson, 1983). Our results suggest that, in such situations, bonding to a single male carries reproductive benefits for females as well (see also Swedell and Saunders, this volume).

The main prediction of this study was that ancestry would be the best predictor of female behavioral strategies. While there was no relationship between the loose-OMU and any particular female phenotype, there was a strong relationship between female phenotype and strict- and non-OMU membership. Furthermore, we detected a significant relationship between female phenotype and the presence of hamadryas-like versus olive-like behaviors. In effect, more hamadryas-like females behaved in a more hamadryas-like fashion, and more olive-like females behaved in a more olive-like fashion. It is important to point out that these females encounter virtually identical ecological pressures, and they occupy a larger group consisting of both female relatives and males with and without a herding instinct. Yet each female exhibits a particular behavioral strategy that can be predicted on the basis of her morphological phenotype. These data thus offer strong support for a genetic basis for some of the attributes that underlie baboon social organization, particularly for female hamadryas baboons.

## ACKNOWLEDGMENTS

First, we would like to thank the organizers of the symposium and the editors of this volume, L. Swedell and S. Leigh, for inviting us to contribute this chapter. Second, we would like to thank our counterparts in Ethiopia, the Ethiopian Wildlife Conservation Department and the Biology Department of Addis Ababa University for granting us permission to conduct this research. As any project that is part of a larger one, we are greatly indebted to the codirectors, C. Jolly and J. Phillips-Conroy, and the many participants of the Awash National Park Baboon Research Project for all of their help in trapping the members of our study group. We would like to thank J. Phillips-Conroy, D. Kitchen, and two anonymous reviewers for their insightful comments on an earlier draft of this chapter. This work was supported by Boise Fund, A Fulbright Student Grant, the National Geographic Society, the National Science Foundation, Sigma Xi, Washington University, and the Wenner-Gren Foundation.

## REFERENCES

Abegglen, J. J., 1984, *On Socialization in Hamadryas Baboons: A Field Study*, Associated University Press, London.

Aldrich-Blake, F. P. G., Dunn, T. K., Dunbar, R. I. M., and Headley, P. M., 1971, Observations on baboons, *Papio anubis*, in an arid region in Ethiopia, *Folia Primatol.* **15**:1–35.

Altmann, J., 1974, Observational study of behavior: Sampling methods, *Behaviour* **49**:229–267.

Altmann, J. A., 1980, *Baboon Mothers and Infants*, Harvard University Press, Cambridge, MA.

Anderson, C. M., 1983, Levels of social organization and male–female bonding in the genus *Papio*, *Am. J. Phys. Anthropol.* **60**:15–22.

Barton, R. A., Whiten, A., Strum, S. C., Byrne, R. W., and Simpson, A. J., 1992, Habitat use and resource availability in baboons, *Anim. Behav.* **43**:831–844.

Barton, R. A., Byrne, R. W., and Whiten, A., 1996, Ecology, feeding competition and social structure in baboons, *Behav. Ecol. Sociobiol.* **38**:321–329.

Beehner, J., 2003, Female behavior and reproductive success in a hybrid baboon group (*Papio hamadryas hamadryas* × *Papio hamadryas anubis*). Ph.D. Dissertation, Washington University, St. Louis, MO.

Bergman, T. J., 2000, Mating behavior and reproductive success of hybrid male baboons (*Papio hamadryas hamadryas* × *Papio hamadryas anubis*). Ph.D. Dissertation, Washington University, St. Louis, MO.

Bergman, T. J. and Beehner, J. C., 2003, Hybrid zones and sexual selection: Insights from the Awash baboon hybrid zone (*Papio hamadryas anubis* × *P. h. hamadryas*), in: *Sexual Selection and Reproductive Competition in Primates: New Insights and Directions*, C. B. Jones, ed., American Society of Primatologists, Norman, OK, pp. 503–537.

Bergman, T. J. and Beehner, J. C., 2004, Social system of a hybrid baboon group (*Papio anubis* × *P. hamadryas*), *Int. J. Primatol.* **25**:1313–1330.

Beyene, S., 1993, Group-fusion and hybridization between anubis and hamadryas baboons at Gola, Ethiopia. *SINET: Ethiop. J. Sci.* **16**:61–70.

Biquand, S., Biquand-Guyot, V., Boug, A., and Gautier, J. P., 1992, Group composition in wild and commensal hamadryas baboons: A comparative study in Saudi Arabia, *Int. J. Primatol.* **13**:533–543.

Brett, F., Turner, T., Jolly, C. J., and Cauble, R., 1982, Trapping baboons and vervet monkeys from wild, free-ranging populations, *J. Wildl. Manage.* **46**:164–174.

Byrne, R. W., Whiten, A., and Henzi, S. P., 1989, Social relationships of mountain baboons: Leadership and affiliation in a non-female bonded monkey, *Am. J. Primatol.* **18**:191–207.

Byrne, R. W., Whiten, A., Henzi, S. P., and McCulloch, F. M., 1993, Nutritional constraints on mountain baboons (*Papio ursinus*): Implications for baboon socioecology, *Behav. Ecol. Sociobiol.* **33**:233–246.

Caljan, V. G., Meishvili, N. V., and Vancatova, M. A., 1987, Sexual behaviour of hamadryas baboons. *Anthropologie* **25**:183–187.

Chalyan, V. G., Meishvili, V., and Dathe, R., 1991, Dominance rank and reproduction in female hamadryas baboons. *Primate Rep.* **29**:35–40.

Coelho, A., Turner, S., and Bramblett, C., 1983, Allogrooming and social status: An assessment of the contributions of female behavior to the social organization of hamadryas baboons (*Papio hamadryas*). *Primates* **24**:184–197.

Colmenares, F., Lozano, M. G., and Torres, P., 1994, Harem social structure in a multiharem colony of baboons (*Papio* spp.): A test of the hypothesis of the "star-shaped" sociogram, in: *Current Primatology, Vol. II: Social Development, Learning & Behavior*, J. J. Roeder, ed., Université Louis Pasteur, Strasbourg, pp. 93–101.

Cowlishaw, G., 1997, Trade-offs between foraging and predation risk determine habitat use in a desert baboon population, *Anim. Behav.* **53**:667–686.

Dunbar, R. I. M., 1988, *Primate Social Systems*, Cornell University Press, New York.

Hamilton, W. J. and Bulger, J. B., 1992, Facultative expression of behavioral differences between one-male and multimale savanna baboon groups, *Am. J. Primatol.* **28**:61–71.

Hausfater, G., Altmann, J., and Altmann, S. A., 1982, Long-term consistency of dominance relations among female baboons (*Papio cynocephalus*), *Science* **217**:752–755.

Henzi, P. and Barrett, L., 2003, Evolutionary ecology, sexual conflict, and behavioral differentiation among baboon populations, *Evol. Anthropol.* **12**:217–230.

Henzi, S. P., Dyson, M. L., and Deenik, A., 1990, The relationship between group size and altitude in mountain baboons (*Papio cynocephalus ursinus*), *Int. J. Primatol.* **11**:319–325.

Jolly, C. J., 1963, A suggested case of evolution by sexual selection in primates, *Man* **63**:177–178.

Kummer, H., 1968a, *Social Organization of Hamadryas Baboons: A Field Study*, University of Chicago Press, Chicago.

Kummer, H., 1968b, Two variations in the social organization of baboons, in: *Primates: Studies in Adaptation and Variability*, P. C. Jay, ed., Holt, Rinehart, and Winston, New York, pp. 293–312.

Kummer, H., 1971, *Primate Societies: Group Techniques of Ecological Adaptation*, Harlan Davidson, Arlington Heights, IL.

Kummer, H., 1984, From laboratory to desert and back: A social system of hamadryas baboons, *Anim. Behav.* **32**:965–971.

Kummer, H. and Kurt, F., 1963, Social units of a free-living population of hamadryas baboons, *Folia Primatol.* **1**:4–19.

Kummer, H. and Kurt, F., 1965, A comparison of social behavior in captive and wild hamadryas baboons, in: *The Baboon in Medical Research*, H. Vagtborg, ed., University of Texas Press, Austin, pp. 65–80.

Kummer, H., Banaja, A. A., Abo-Khatwa, A. N., and Ghandour, A. M., 1985, Differences in the social behavior between Ethiopian and Arabian hamadryas baboons, *Folia Primatol.* **45**:1–8.

Kummer, H., Götz, W., and Angst, W., 1970, Cross-species modifications of social behavior in baboons, in: *Old World Monkeys: Evolution, Systematics, and Behavior*, J. R. Napier, and P. H. Napier, eds., Academic Press, New York, pp. 351–363.

Leinfelder, I., de Vreis, H., Deleu, R., and Nelissen, M., 2001, Rank and grooming reciprocity among females in a mixed-sex group of captive hamadryas baboons, *Am. J. Primatol.* **55**:25–42.

Melnick, D. J. and Pearl, M. C., 1987, Cercopithecines in multimale groups: Genetic diversity and population structure, in: *Primate Societies*, B. B. Smuts, D. L. Cheney, R. M. Seyfarth, R. W. Wrangham, and T. T. Struhsaker, eds., University of Chicago Press, Chicago, pp. 121–134.

Nagel, U., 1973, A comparison of anubis baboons, hamadryas baboons and their hybrids at a species border in Ethiopia, *Folia Primatol.* **19**:104–165.

Nystrom, P., 1992, Mating success of hamadryas, anubis and hybrid male baboons in a "mixed" social group in the Awash National Park, Ethiopia. Ph.D. Dissertation, Washington University, St. Louis, MO.

Pfeiffer, G., Kaumanns, W., Schwibbe, and M. H., 1985, A female "defeated leader" in hamadryas baboons: A case study, *Primate Rep.* **12**:18–26.

Phillips-Conroy, J. E. and Jolly, C. J., 1986, Changes in the structure of the baboon hybrid zone in the Awash National Park, Ethiopia, *Am. J. Phys. Anthropol.* **71**:337–350.

Phillips-Conroy, J. E., Jolly, C. J., and Brett, F. L., 1991, Characteristics of hamadryas-like male baboons living in anubis baboon troops in the Awash hybrid zone, Ethiopia, *Am. J. Phys. Anthropol.* **86**:353–368.

Seyfarth, R. M., 1976, Social relationships among adult female baboons, *Anim. Behav.* **24**:917–938.

Seyfarth, R. M., 1978, Social relationships among adult male and female baboons. II. Behaviour throughout the female reproductive cycle, *Behaviour* **64**:227–247.

Sigg, H., 1980, Differentiation of female positions in hamadryas one-male units, *Z. Tierpsychol.* **53**:265–302.

Sigg, H., Stolba, A., Abegglen, J. J., and Dasser, V., 1982, Life history of hamadryas baboons: Physical development, infant mortality, reproductive parameters, and family relationships, *Primates* **23**:473–487.

Smuts, B. B., 1985, *Sex and Friendship in Baboons*, Aldine, New York.

Sokal, R. R. and Rohlf, F. J., 1981, *Biometry*, W.H. Freeman, New York.

Stammbach, E., 1978, On social differentiation in groups of captive female hamadryas baboons, *Behaviour* **67**:322–338.

Stammbach, E. and Kummer, H., 1982, Individual contributions to a dyadic interaction: An analysis of baboon grooming, *Anim. Behav.* **30**:964–972.

Strum, S. C., 1994, Reconciling aggression and social manipulation as means of competition. I. Life-history perspective, *Int. J. Primatol.* **15**:739–765.

Sugawara, K., 1979, Sociological study of a wild group of hybrid baboons between *Papio anubis* and *Papio hamadryas* in the Awash Valley, Ethiopia, *Primates* **20**:21–56.

Sugawara, K., 1982, Sociological comparison between two wild groups of anubis–hamadryas hybrid baboons. *Afr. Study Monogr.* **2**:73–131.

Sugawara, K., 1988, Ethological study of the social behavior of hybrid baboons between *Papio anubis* and *P. hamadryas* in free-ranging groups, *Primates* **29**:429–448.

Swedell, L., 2002, Affiliation among females in wild hamadryas baboons (*Papio hamadryas hamadryas*), *Int. J. Primatol.* **23**:1205–1226.

Swedell, L., 2006, *Strategies of Sex and Survival in Hamadryas Baboons: Through a Female Lens*, Pearson Prentice Hall, Upper Saddle River, NJ.

Vervaecke, H., Dunbar, R. I. M., van Elsacker, L., and Verheyen, R., 1992, Interactions with and spatial proximity to the males in relation to rank of captive female adult hamadryas baboons (*Papio hamadryas*), *Acta Zoologica et Pathologica Anterverpiensia* **82**:61–77.

Wrangham, R. W., 1980, An ecological model of female-bonded primate groups, *Behaviour* **75**:262–299.

CHAPTER FOUR

# Hybrid Baboons and the Origins of the Hamadryas Male Reproductive Strategy

*Thore J. Bergman*


## CHAPTER SUMMARY

Hamadryas male baboons concentrate their reproductive effort on a small subset of females (i.e., females in their one-male unit) rather than competing with other males for estrous females throughout the group. Hamadryas males also exhibit a sustained, intense interest in females, regardless of estrus condition, an interest manifested by stereotypical herding behaviors. In this chapter, I use the behavioral variation expressed by hybrid (olive × hamadryas) baboon males to test predictions regarding behavioral precursors of the unique hamadryas male reproductive strategy. I analyze three potential precursors for hamadryas male behavior: (1) the herding of females during intergroup encounters, (2) the long-term bonds involving paternal care (i.e., "friendships"), and (3) temporary consortships. Because the male motivation behind each potential precursor is different (intolerance of other males, paternal care, and attraction to females, respectively), it is possible to make different predictions for the behavior of hybrid males. Consort behavior of hybrid males continued over a longer portion of female estrus than is usual for



**Thore J. Bergman** • Department of Psychology, University of Michigan, Ann Arbor, MI, USA

nonhamadryas baboons, and nonconsort behavior of hybrid males resembled the consort behavior of nonhamadryas males. Herding females and paternal care were rare, while all hybrid males expressed some interest in nonestrous females. Results support hypothesis 3, that consortships represent the behavioral homologue to the hamadryas male reproductive strategy.

## 1. INTRODUCTION

Hamadryas baboons (*Papio hamadryas hamadryas*) live in a unique, multilevel society based on long-term relationships among a male, one or more females, and their dependent offspring (Kummer, 1968a,b). These one-male units (OMUs) aggregate hierarchically into clans, bands, and troops. Clans consist of two or more OMUs and are thought to function as the foraging unit (Abegglen, 1984), while bands are aggregations of several clans that are probably the equivalent of the troop (or group) in other baboons. Troops are associations of bands at sleeping sites that often consist of several hundred baboons (Kummer, 1968a; Abegglen, 1984). This hierarchical structure facilitates troop division into smaller foraging parties, suggesting that hamadryas social organization serves as an adaptation to foraging in resource-poor habitats with limited sleeping sites (Kummer, 1968a).

The reproductive strategy of hamadryas males is correspondingly distinct. Hamadryas males concentrate their reproductive effort on a small subset of the females that they encounter on a regular basis. Rather than compete for females as they come into estrus, as do nonhamadryas males (Packer, 1979a; Bulger, 1993; Alberts et al., 2003), hamadryas males form long-term bonds with only a few females, and these bonds remain strong even when the females are not in estrus (Kummer, 1968a). For the leader male, the OMU represents both mating investment (mating occurs almost exclusively within OMUs (Kummer, 1968a; see also Swedell and Saunders, this volume)) and paternal investment (hamadryas males provide protection for the offspring in their OMUs (Swedell, 2006)). Within the OMU, males aggressively herd females that stray and, consequently, females learn to maintain proximity to their male by following him (Kummer, 1968a). Thus, the long-term, intersexual bonds in hamadryas societies are associated with two attributes unique to hamadryas male reproductive behavior: (1) an interest in OMU females regardless of their estrus condition and (2) the aggressive herding of females.

These two aspects of hamadryas male behavior depart from the social and behavioral flexibility that characterizes most nonhamadryas baboon males (but see Henzi and Barrett, 2003). While group size and higher-order group structure (i.e., clans, bands, and troops) of hamadryas societies may vary with predation risk, habitat quality, and the availability of sleeping sites (Kummer et al., 1985; Biquand et al., 1992; Swedell, 2002b), hamadryas form single-male units regardless of ecological setting, even continuing to form OMUs when in captivity (e.g., Kummer and Kurt, 1965; Caljan et al., 1987; Colmenares, 1992) or after migrating into an olive baboon troop (Nystrom, 1992). The consistency of hamadryas behavior across a wide spectrum of conditions suggests that the hamadryas male (and possibly female (see Beehner and Bergman, this volume)) propensity to form OMUs almost certainly has a genetic component (Kummer and Kurt, 1965).

Despite this consistency in hamadryas male behavior, hamadryas societies exhibit several characteristics generally associated with multimale, multifemale baboon groups (e.g., female receptivity advertised by sexual swellings), and thus hamadryas are thought to have evolved from an olive baboon-like ancestor that lived in multimale, multifemale groups (Kummer, 1968b). Recent genetic work supports this view and also indicates that hamadryas and olive baboons share a more recent common ancestor with each other than either does with chacma baboons (Newman et al., 2004). How, then, did the transition occur from more *generalized* male reproductive behavior (interest in and competition for only estrous females) to the highly specialized hamadryas OMU behavior? There has been much discussion of the causal connection between the two major transitions that distinguish the hamadryas lineage, a move to drier habitat and the appearance of the unique social organization and reproductive strategies just described. This chapter takes a different approach and focuses on exploring the proximate mechanisms of behavioral transitions. While the hamadryas male reproductive strategy is unique among *Papio* baboons, other (nonhamadryas) baboons exhibit elements of hamadryas-like behavior. These behaviors may represent evolutionary precursors that have evolved into the intense, consistent behavior exhibited by hamadryas males. The goal of this chapter is to use the behavioral variation found in hybrid (hamadryas × olive) male baboons to evaluate three potential behavioral precursors to hamadryas behavior: herding during intergroup encounters, male–female “friendships,” and consortships.

The first of these possibilities, herding behavior, is observed in nonhamadryas males in specific contexts, most notably during intergroup encounters. Herding during intergroup encounters targets both estrous and nonestrous females and functions primarily to increase the distance between the herded female and the extra-group males (Henzi et al., 1998). While this behavior has been observed in all *Papio* species (Packer, 1979b), it appears to be most pronounced in chacma baboons and is particularly common in high-altitude populations where groups are small and often contain only a single male (Hamilton et al., 1975; Cheney and Seyfarth, 1977; Byrne et al., 1987; Whiten et al., 1987). It is in these single-male chacma groups that herding during intergroup encounters appears the most similar to hamadryas herding. This may be, in part, why several authors have argued that these small groups are convergent with the hamadryas OMU (Byrne et al., 1987; Whiten et al., 1987) despite the conspicuous absence of male interest in nonestrous females outside the context of intergroup encounters. Byrne et al. (1989) suggest that an increase in predation pressure at a location with single-male chacma groups might cause the single-male groups to congregate into larger groups for predator defense, resulting in a hamadryas-like multilevel society. Thus, it is possible that hamadryas reproductive behavior evolved from the previously sporadic herding of females away from extra-group males during intergroup encounters.

The second possibility is suggested by the long-term relationships that nonhamadryas baboon males often form with one or more females outside of sexual consortships. These consistent associations between males and nonestrous females, termed "friendships," have been studied mainly in olive and chacma baboons (Smuts, 1985; Palombit et al., 1997). In chacma baboons, friendships occur almost exclusively between a lactating female and a potential father of her infant. Consequently, it has been argued that the primary adaptive function for male friendships is paternal investment (Palombit et al., 1997, 2000; Palombit, 2001). While most work on friendships and paternal investment has been done with chacma baboons, recent genetic analysis has indicated that paternal care also occurs in yellow baboons (Buchan et al., 2003). Furthermore, it has been hypothesized that olive baboon friendships primarily represent paternal investment as well (Bercovitch, 1988). Investment in infants may thus be the primary function of male–female friendships in baboons, but this does not preclude the possibility that males may also be investing in future mating possibilities, particularly in olive baboon societies where dominant males are unable to monopolize mating to the extent that dominant chacma males do (Smuts, 1983, 1985; Bulger, 1993). Consequently, it has been proposed that friendships for olive baboon males may represent an alternative mating strategy (Strum, 1982; Smuts, 1985; Bercovitch, 1986). Proximity between male and female friends is generally closer than between nonfriends, with females typically maintaining this proximity (Palombit et al., 1997). Thus, hamadryas male reproductive behavior could have derived from an intensification of male–female friendships by adding (or strengthening) mating investment to a relationship that was initially based on paternal investment alone.

Finally, all baboons engage in intensive consortships, that is, exclusive associations between a male and an estrous female. These consortships resemble temporary versions of the long-term bonds between hamadryas males and females, in that consorting males and estrous females maintain close proximity. In contrast to hamadryas OMUs, however, consorting males typically follow females rather than the hamadryas pattern whereby females follow males (Seyfarth, 1978; Ransom, 1981). Olive males frequently herd females when in consort, and this mate guarding behavior is particularly common when males are being harassed by "follower" males (Ransom, 1981). These attributes suggest that hamadryas male reproductive behavior may have arisen as an extension of typical nonhamadryas male consort behavior. In other words, while nonhamadryas males are attracted only to females with sexual swellings ("swollen" females), this attraction may have been lengthened for hamadryas males beyond the swollen period to all reproductive periods.

To gain insight into the evolutionary origins of hamadryas male behavior, I test several predictions for the behavior of hybrid baboon males based on these three potential evolutionary precursors: herding during intergroup encounters, friendships, and consortships. In conducting these analyses, I assume that the behavior of hybrids can shed light on the behavior of an evolutionary intermediate, which may not necessarily be the case. If, for example, a hybrid male were found with a hamadryas-like pink face and an olive-like bend at the base of the tail, this would not mean that a hamadryas ancestor went through a stage with this particular combination of traits. However, in this case, I am making a comparison between distinct behavioral motivations (intolerance of other males, paternal care, and attraction to females) that represent a single complex of behavioral traits, rather than a suite of unrelated traits (as in the face and tail example). Therefore, there is little need to speculate about the sequence of multiple evolutionary transitions. It seems reasonable to expect that an evolutionary intermediate would have had an intermediate level of motivation, and the candidate precursors, while not mutually exclusive, entail distinct predictions about the behavior of such an intermediate. If there is a genetic component to hamadryas reproductive behavior, hybrids will also have intermediate and polymorphic behavior and, thus, *should* display behaviors similar to evolutionary intermediates. Consequently, the behavior of hybrids may, indeed, be informative about the evolution of hamadryas behavior.

### 1.1. Hypotheses and Predictions

Several predictions about the behavior of hybrids can be made based on three evolutionary hypotheses.

**Hypothesis 1.** *Hamadryas male behavior represents an extension of the herding behavior observed in nonhamadryas males during intergroup encounters.* This hypothesis states that herding behavior evolved first in the evolutionary shift from nonhamadryas to hamadryas behavior. *Prediction 1.* Herding nonestrous females will be common even for males that do not exhibit other signs of interest in nonestrous females. For example, the herding of nonestrous females will occur at higher frequencies than either following or leading nonestrous females. *Prediction 2.* Herding will be correlated with increased distance to other males. The most behaviorally hamadryas-like males should exhibit the greatest intermale distance, comparable to chacma male herding behavior during intergroup encounters, which serves to increase the distance from group females to extra-group males (Henzi et al., 1998). *Prediction 3.* Males will frequently separate from the group to form their own subgroup.

**Hypothesis 2.** *Hamadryas male behavior represents an intensification of the intersexual bonds observed in friendship behavior.* Under this hypothesis, paternal care represents the first step toward the extended bonds between hamadryas males and females. *Prediction 1.* In the absence of herding behavior, females will frequently follow males (as observed in friendships). *Prediction 2.* Infant-directed behaviors will be common among potential fathers of those infants. *Prediction 3.* Males with the most hamadryas-like behavior will have the highest rates of infant-directed behavior.

**Hypothesis 3.** *Hamadryas male behavior represents an extension of consortships.* This hypothesis states that consortships between males and females represent the evolutionary precursor to the extended bonds between hamadryas males and females. *Prediction 1.* Males will extend their following behavior outside of consortship, much like nonhamadryas males follow females during consortship. Thus, the following of nonestrous females will be common. *Prediction 2.* The behavioral patterns for OMU males and consorting males will be similar. *Prediction 3.* All females will be in consort for a greater proportion of their cycle than is found among nonhamadryas baboons. Even more olive-like hybrids will exhibit some degree of attraction to females with very small sexual swellings, resulting in greater consort activity per female cycle.

To test these predictions, I analyze data from the males of a group comprised entirely of hybrid baboons. Reflecting its mixed ancestry, the group has an unusual mixed social system with several hamadryas-like OMUs embedded in a larger multimale group (Bergman and Beehner, 2004). The group frequently splits into subgroups that often remain separated for days. Hybrid males in this group exhibit different combinations of hamadryas-like and olive-like behavior with varying degrees of interest in nonestrous females. On the most hamadryas end of the behavioral spectrum several hybrid males maintain OMUs, while at the olive baboon end of the spectrum several males show interest primarily in estrous females. I use variation in male behavior (herding, leading, and following females; consortship activity; and infant-directed behaviors) to test the above hypotheses about the origins of hamadryas behavior.

## 2. METHODS

### 2.1. Study Group

The study group, "group H," consists of hybrids between hamadryas (*P. h. hamadryas*) and olive (*P. h. anubis*) baboons located in the hybrid zone along the southern border of the Awash National Park in Ethiopia. The layout of hybrid groups along the Awash River has been described previously (Nagel, 1973; Phillips-Conroy and Jolly, 1986), and a detailed description of the study group is given in Bergman and Beehner (2003, 2004) and Beehner and Bergman (this volume). Group H occupies the center of the Awash hybrid zone and includes a range of male and female hybrid individuals that span the phenotypic spectrum from olive-like to hamadryas-like (Bergman, 2000). Phenotypes for the 14 group H males in this study are described in Bergman and Beehner (2003) and all group H females are described in Chapter 3

(Beehner and Bergman, this volume). The behavior of group H males exhibited considerable variation, specifically with respect to their interest in nonestrous females. During this study, 4 males formed OMUs that were much like OMUs found in hamadryas societies (i.e., reproductive and social units), while the other 10 males had varying degrees of interest in nonestrous females. Some of these males frequently formed temporary associations with a series of nonestrous females, while others associated with females primarily when they were in estrus. There were 25–30 adult females in the group, and approximately half (14 females) were associated with an OMU.

Group H utilized multiple sleeping sites on the cliffs of the Awash canyon and frequently divided into subgroups. The term "subgroups" refers to foraging parties that consist of any assortment of group H individuals. Members of a single OMU were never split among subgroups, but beyond this there was no predictable pattern to the subgroup association of individuals (Bergman and Beehner, 2004). Group H subgroups often foraged and slept separately for several days at a time.

## 2.2. Behavior

Behavioral data were collected on the 14 adult males of group H between September 1997 and November 1999. Instances of *herding*, *leading*, *following*, and *infant-directed behavior* were recorded using all-occurrence sampling and then standardized (for time spent observing each individual) using 5-min scan samples (Altmann, 1974). I recorded 519 occurrences of these behaviors (range: 9–71 per male). Scan samples were collected whenever data were being recorded (providing a record of observation time for each male) and included nearest neighbor data (distance and identity of nearest neighbor, nearest adult male, and nearest adult female).

*Herding* behaviors included any of a suite of stereotypical hamadryas herding behaviors directed toward a female (i.e., neck/ear-biting, dragging, holding, eyebrow and yawn threats). Visual threats directed toward females were only observed within an OMU and frequently preceded an approach by the female. Instances of *leading* and *following* were scored when one individual moved along a route that no other baboons were using at the time, and another followed less than 5 m behind, maintaining proximity to the leader. Because I was interested in male behavior outside of sexual consortship, I analyzed these behaviors only when directed toward nonestrous females.

I define "nonestrous females" as any female outside of her sexually swollen period. In other words, nonestrous females include cycling females without sexual swellings, pregnant females, and lactating females. "Estrous females," on the other hand, specifically refers to females with any degree of sexual swelling.

I recorded two types of *infant-directed behavior*: infant handling (physical contact between a male and an infant) and infant-directed grunts (male grunts directed toward an adult female carrying an infant).

Intergroup encounters were rare. I observed auditory and/or visual contact between group H and another group approximately twice a month. On all but three occasions, more than 500 m separated the two groups and one of the groups moved off in a different direction following auditory or visual contact. On the other three occasions, the other group overlapped members of group H for a brief period (usually the result of one group ascending the steep canyon where the other group happened to be resting) before moving away. On all occasions, the presence of neighboring groups was associated with increased vigilance. Members of group H clearly desired to avoid extra-group animals, yet males conspicuously failed to either make vocal displays or herd females, behaviors characteristic of chacma intergroup encounters (Hamilton et al., 1975). Therefore, no data from intergroup encounters are included here. Rather, I look at intragroup male spacing to address the hypothesis that herding functions to increase the distance between one male's females and another male.

Analyses of male neighbor data exclude days where a male was in a subgroup with no other males. This was only a frequent occurrence for one group H male, also an OMU male.

Only one OMU male acquired a new female that resisted incorporation into his OMU. The male aggressively herded this female for more than two months and, consequently, the male had a rate of herding that was several times higher than that of other males. This has also been observed in pure hamadryas groups, in which males herd newly acquired females more intensively than other unit females (Sigg, 1980; Swedell, 2006). Therefore, I did not include this male's behavior from these 2 months in order to assess his "baseline" herding rate (which was the highest in the group).

For the analysis of the proportion of female cycle days spent in consort, I restricted the dataset to (1) only non-OMU females (i.e., females outside of an OMU) and (2) only cycles for which the consort status of the female (i.e., associated with a male or not) was known for at least 80% of her estrus period

($N = 17$). I included only non-OMU females because my predictions are based on the consort activity of the males that are behaviorally *less* hamadryas-like.

### 2.2. Analysis

Data were tested for deviations from normality using the Kolmogorov–Smirnov test and found not to deviate significantly from normal (at $p < 0.05$) and, thus, parametric statistics were used in all analyses. Data were analyzed using SPSS (version 11.0) statistical software.

## 3. RESULTS

### 3.1. Female-directed behavior

Figure 1 illustrates the rates of *following*, *leading*, and *herding* behavior for each male in group H. All males *followed* nonestrous females ($N = 14$) to some degree. A subset of these males ($N = 11$) *herded* nonestrous females, and only a subset of these herding males ($N = 5$) *led* nonestrous females. Males followed females significantly more than they herded females (pairwise $t$-test: $t = 3.32$, $N = 14$, $p = 0.005$), and they herded females significantly more than they led females (pairwise $t$-test: $t = 3.20$, $N = 14$, $p = 0.006$).

The four males that maintained OMUs for the duration of the study are clustered on the right side of the $x$-axis in Figure 1. After separating OMU males from non-OMU males, I compared mean rates of nonconsort *following*, *leading*, and *herding* between the two categories (Figure 2). OMU males had higher rates of all three behaviors, although the difference for *following* was not significant (ANOVA: *following*, $F = 2.48$, $p = 0.14$; *herding*, $F = 18.44$, $p = 0.001$; *leading*, $F = 24.34$, $p < 0.001$). OMU and non-OMU males had similar relative rates of these behaviors (i.e., they both followed females the most and led females the least).

OMU males resembled hamadryas males because they had relatively high rates of both herding and leading (Table 1, Figure 2). However, group H OMU males also had high rates of following nonestrous females. Hamadryas males do not typically follow females; this is a behavior more often found among nonhamadryas males in consort. Therefore, the most hamadryas-like males in group H resembled consorting olive males. To make a similar comparison within group H, I divided behavior into consort and nonconsort behavior, and I compared OMU and non-OMU males. OMU and non-

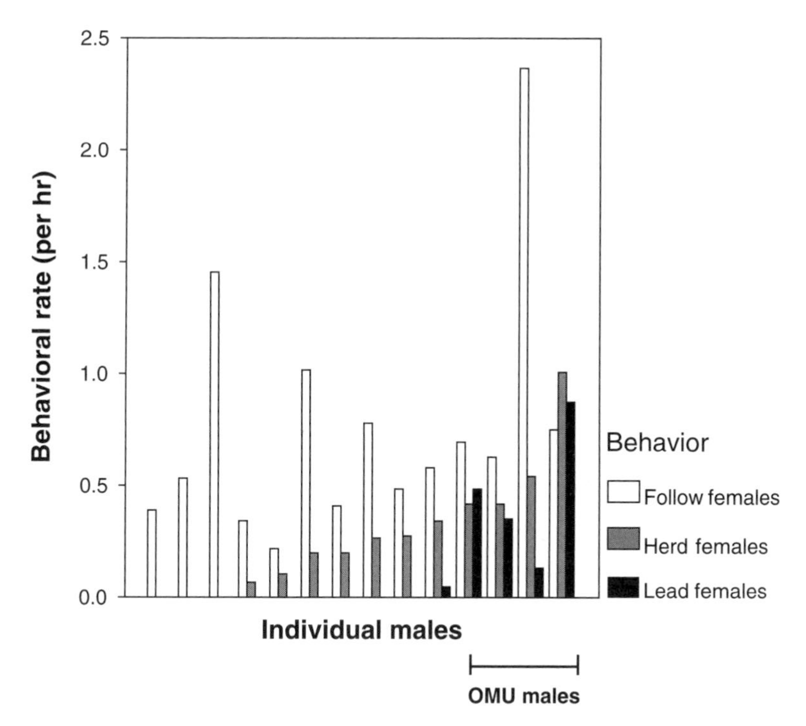


**Figure 1.** Behavioral rates (per hour) for each adult male in group H. Only male behavior directed at nonestrous females was included. Males followed females significantly more than they herded females (pair-wise $t$-test: $t = 3.32$, $N = 14$, $p = 0.005$), and males herded females significantly more than they led females (pair-wise $t$-test: $t = 3.20$, $N = 14$, $p = 0.006$).

OMU males did not differ in the rates of these behaviors when they were in consort (*following*, $F = 0.14$, $p = 0.71$; *herding*, $F = 0.79$, $p = 0.39$; *leading*, $F = 1.47$, $p = 0.25$), and the category "consorting males" (Figure 2) thus includes both OMU and non-OMU males. Within group H, OMU males behaved more like consorting males (Figure 2).

Males observed *leading* nonestrous females ($N = 5$) were separated from other males ($N = 9$) and *leading* was associated with higher rates of *herding* (ANOVA; $F = 18.44$, $p = 0.001$). Furthermore, *leading* and *herding* were positively correlated ($r = 0.87$, $p < 0.001$). *Herding* was not related to proximity to other males; there was no correlation between *herding* and the proportion of observations in which the nearest adult male neighbor was within 10 m ($r = -0.11$, $p = 0.70$). OMU males did not differ from other

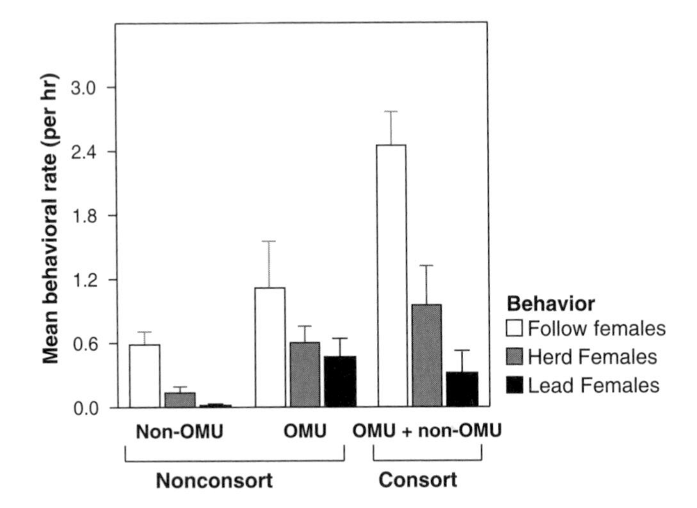


**Figure 2.** Mean behavior rates (+SEM, per hour) for non-OMU and OMU males when not in consort and for all males when in consort. Nonconsort OMU males had higher rates of all three behaviors than nonconsort non-OMU males (ANOVA: *following*: $F = 2.48$, $p = 0.14$; *herding*: $F = 18.44$, $p = 0.001$; *leading*: $F = 24.34$, $p < 0.001$). OMU and non-OMU males did not differ in the rates of these behaviors when they were in consort (*following*: $F = 0.14$, $p = 0.71$; *herding*: $F = 0.79$, $p = 0.39$; *leading*: $F = 1.47$, $p = 0.25$) and the category "consorting males" thus includes both OMU and non-OMU males.

males in mean distance to an adult male ($F = 0.02$, $p = 0.88$, Figure 3), but OMU males did have a lower mean distance to an adult female ($F = 6.26$, $p = 0.03$, Figure 3).

### 3.2. Infant-Directed Behavior

*Infant-directed behavior* among group H males was rare. Rates of infant–male interactions for this group were much less common, for example, than in the Moremi population of chacma baboons (Cheney and Seyfarth, unpublished data). Only eight males in group H had any interactions with infants. Infant-directed behavior was not correlated with *herding* nonestrous females ($r = 0.27$, $p = 0.18$) nor did OMU and non-OMU males differ in infant-directed behavior ($F = 0.29$, $p = 0.60$).

Because not all males fathered offspring during this study (Bergman, 2000), I controlled for the differences in opportunities for paternal behavior

**Table 1.** Relative frequencies of behaviors in hamadryas and olive males[a]

| Behavior | Relative frequency | |
|---|---|---|
| | Olive | Hamadryas |
| Follow estrous females | High | Intermediate |
| Follow nonestrous females | Low | Low |
| Herd estrous females | Intermediate | High |
| Herd nonestrous females | Low | High |
| Lead estrous females | Low | Intermediate |
| Lead nonestrous females | Low | High |

[a]Frequencies are relative to each other and represent a summary of information from Kummer (1968a), Ransom (1981), Smuts (1985) and Nystrom (1992).

by taking residuals from the regression analysis of *infant-directed behavior* and the number of consorts for each male that resulted in pregnancy. Values for these residuals were not correlated with *herding* ($r = 0.15$, $p = 0.59$) nor did they differ between OMU and non-OMU males ($F = 0.22$, $p = 0.64$). Values did, however, differ between older (above median age) and younger (below median age) males (ages were based on dental condition (Phillips-Conroy et al., 2000)), with older males exhibiting more infant-directed behavior than younger males ($F = 07.62$, $p = 0.02$).

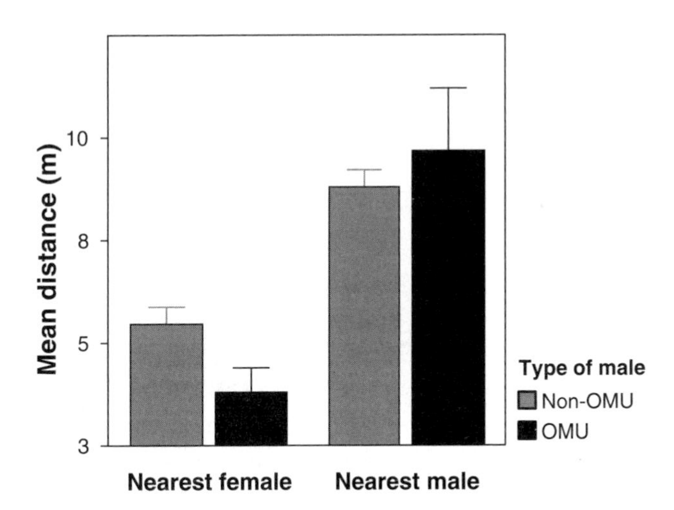


**Figure 3.** Mean distances (+SEM, in meters) to nearest male and female for OMU and non-OMU males when not in consort. OMU males did not differ from non-OMU males in mean distance to an adult male ($F = 0.02$, $p = 0.88$), but OMU males did have a lower mean distance to an adult female ($F = 6.26$, $p = 0.03$).

### 3.3. Subgroups

For all males except one, membership in a single-male subgroup (i.e., a subgroup consisting of adult females and juveniles but no other males) was extremely rare despite the fact that subgroup formation was very common. Males did not isolate themselves from the group even when in consort and therefore were at risk of "losing" their consort partner to other males. However, one male (an OMU male) frequently split from the rest of the group, taking his OMU females with him. Once separated from the rest of the group, this OMU/single-male subgroup often did not join up with the group again for several days. This same male, on the far right of Figure 1, had the largest, most cohesive OMU and was the only male that was followed by females more than he followed them. He also had the highest rate of herding females. Primarily as the result of this one male, there was a significant correlation between *herding* and the proportion of days that males spent in a single-male subgroup ($r = 0.85$, $p < 0.001$, Figure 4).

### 3.4. Consortships

Across 17 estrous cycles, non-OMU females were in consort for 15.3 (± 3.39) days per cycle (range: 10–20 days), representing an average of 85 percent (range: 71–100 percent) of the days they showed any signs of sexual swelling (18.3 days per cycle). These females were engaged in consort activity over a greater portion of the estrous cycle than observed in other baboons (chacma: mean 3.5 ± 2.1 consort days per cycle, 34 percent of swollen days, Seyfarth, 1978; anubis: mean 5.6 ± 2.5 consort days per cycle, 30 percent of swollen days, Bercovitch, 1991).

## 4. DISCUSSION

This analysis generates results consistent with the predictions of hypothesis 3, that hamadryas male behavior evolved as an extension of the attraction to females seen in nonhamadryas consortships. First, even outside of OMUs, group H females spent a greater percentage of their sexually swollen period in consort than either olive or chacma females (Seyfarth, 1978; Bercovitch, 1991). Thus, interest in small sexual swellings may represent a "first step" from the nonhamadryas male attraction to estrous females to the hamadryas male interest in females regardless of their reproductive stage. Second, all males frequently followed nonestrous females—even males that did not herd

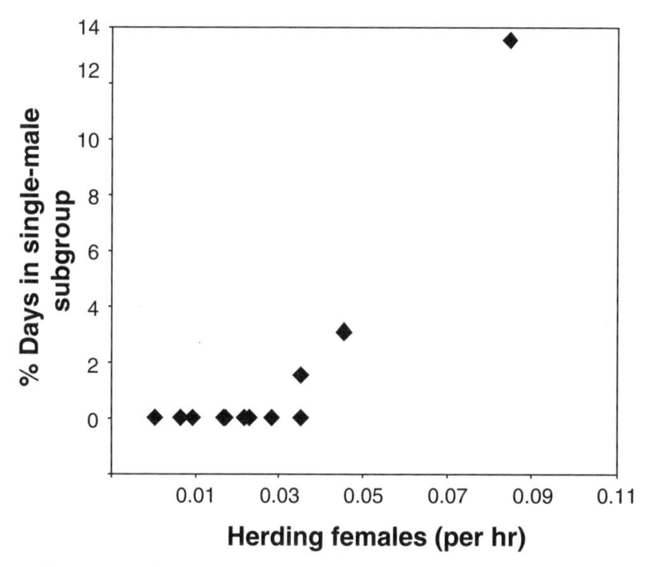


**Figure 4.** Relationship between male rate of herding nonestrous females (per hour) and the percentage of days a male was in a subgroup with no other males ($r = 0.85$, $p < 0.001$). Each point represents an individual.

females. Following nonestrous females appears to be the first sign that a male's attraction to females extends beyond the peak of sexual swelling. Third, the behavior of OMU males resembled that of consorting males. Consequently, just as the mating behavior of hybrid males was extended over a much greater proportion of a female's estrus period, the overall attraction to females was extended beyond the fertile stage—even to nonestrous females.

The data presented here do not support the hypothesis that hamadryas male behavior represents an extension of herding behavior. First, males rarely herded females: Herding was observed only in the most hamadryas-behaving males and was not observed during intergroup encounters. Males followed nonestrous females much more often than they herded them. Second, there was no evidence that herding females corresponded to a desire to avoid other males. Herding females appeared to be unrelated to proximity to other males, and proximity to other males did not differ between OMU and non-OMU males. Third, males rarely split into their own subgroup to sequester estrous females away from other males. As with herding behavior, only males expressing the most hamadryas-like behavior (i.e., males with cohesive OMUs) split from the rest of the group. Thus, separating from the group appeared to be a consequence rather than a cause of hamadryas-like behavior.

Results also fail to support the hypothesis that male–female friendships were the precursors to hamadryas male–female bonds. First, females did not follow males (in the absence of herding) as they do in nonhamadryas friendships. Rather, females only followed the males that herded them, suggesting that herding is necessary before females will follow a male. Second, there was no relationship between infant-directed behavior and herding females or between infant-directed behavior and OMU behavior. In other words, more hamadryas-behaving males did not express an elevated interest in paternal care. While there was some variation in infant-directed behavior, it was associated with male age and not reproductive behavior. In fact, males in group H demonstrated remarkably little interest in infants overall. Thus, there was no evidence that hamadryas-like behavior is associated with a heightened interest in or care of infants. In contrast, there is some evidence that male associations with nonestrous females represent mating investment rather than paternal investment in group H. Associating with nonestrous females is correlated with higher mating success (Bergman and Beehner, 2003), and the majority of consorts (74 percent) are between a male and a female with a prior history of association (Bergman and Beehner, 2004). Associating with nonestrous females in this group may thus confer a reproductive advantage by ensuring future mating.

While analyses of this hybrid group support the extended consortships' hypothesis, it should be noted that the three hypotheses are not mutually exclusive and multiple behavioral precursors may have led to hamadryas OMU behavior. For example, while these results suggest that paternal care is an unlikely source of hamadryas OMU behavior, this does not preclude an evolutionary link between the paternal care that hamadryas males provide and paternal care behavior in nonhamadryas males. Furthermore, some authors have described the herding by chacma males during intergroup encounters as mate guarding (Henzi et al., 1998), indicating that herding behavior may share underlying attributes with male consort behavior. While the results reported here suggest that herding behavior was not the source of the hamadryas reproductive strategy, herding behavior in hamadryas may still be derived from the herding that occurs in consortships *and* intergroup encounters.

### 4.1. Broader Implications

These results suggest that the hamadryas mating strategy is based on extended consort behavior, raising the question of what conditions might have favored the evolution of this behavior. There is some consensus that the first step for the hamadryas behavioral split must have involved an ecological transition to drier habitat that reduced effective group size (e.g., Kummer 1968a), but there remains disagreement about how the single-male unit evolved in this context. Did groups grow subsequently smaller, due to foraging pressure, until they eventually had only one male ("the single-male hypothesis")? Or, did the hamadryas OMU evolve in the context of a larger group that frequently split up into subgroups ("the subgroup hypothesis")?

Some authors have proposed that the small, single-male chacma groups may represent an evolutionary intermediate similar to a stage that may have occurred in hamadryas evolutionary history (Byrne et al., 1987, 1989). However, the results of the present analysis are inconsistent with the single-male hypothesis. With the exception of intergroup encounters, there is little potential for mate competition or infanticide in single-male groups. Therefore, males gain no reproductive advantage by extending their interest to nonestrous females. Rather, for males living in a single-male chacma group, it would seem that they actually could afford to be "choosy" and invest in females only when they are most likely to conceive. A comparison among chacma baboon groups should show that females in single-male groups are not in consort for a greater proportion of their cycle than females in multi-male groups.

While it has received little attention, analyzing the proportion of estrus days a female is in consort is likely a good estimate of how males "value" an estrous female. In many cases, this measure may be more informative than more common estimates, such as the length of individual consortships between a male and a female. Individual consortships are longer in chacma than in olive baboons, lasting for days rather than hours (Ransom, 1981; Bulger, 1993; Weingrill et al., 2000), a characteristic described as a similarity between chacma and hamadryas baboons (Henzi and Barrett, 2003). However, long consortships in chacma baboons derive from high reproductive skew (i.e., dominant males do most of the consorting), a strict male dominance hierarchy, and a lack of coalitions or affiliative behavior among males (Bulger, 1993; Henzi, 1996; Weingrill et al., 2000). Furthermore, in chacma baboons there is no evidence that male–female friendships contribute to later mating opportunities. Rather, estrous females move up the hierarchy of males as they become increasingly "attractive" (Bulger, 1993). These are all ways in which chacma baboons are *less* similar to hamadryas than are olive baboons, probably reflecting the greater phylogenetic distance between chacma and hamadryas baboons than between hamadryas and olive baboons (Newman et al., 2004).

The social system of the current study group suggests that the subgroup hypothesis might be a better explanation for the evolution of a broadened attraction to females. Olive baboons living in arid regions frequently form temporary subgroups when foraging (Aldrich-Blake et al., 1971), and Anderson (1983, 1989) has previously suggested that the formation of temporary subgroups within a larger group provides the impetus for both males and females to form male–female bonds. Based on studies of multimale chacma baboon groups that form seasonal subgroups, Anderson (1983) hypothesized that such a situation favors long-term bonds between males and females because there are fewer potential female mates and because long-term bonds increase paternity certainty (Anderson, 1989). Consequently, a greater investment in each potential mate and subsequent offspring is favored. Furthermore, Anderson (1989) argued that once male–female bonds form and mating becomes semiexclusive, selection serves to increase mating exclusivity.

Results from this study support Anderson's findings. In a regularly subdividing group, there may be an adaptive advantage to an extended period of male interest in estrous females. With subgroup formation, group members may go for days without seeing other group members, rendering males unable to make daily assessments of females' reproductive condition. Thus, males may benefit from entering into consortships at an earlier stage. Otherwise, a male might miss a female's ovulatory period or have to compete for a female he could have "claimed" earlier.

Under the subgroup hypothesis, the evolutionary scenario of hamadryas OMU behavior may have looked something like this: Ecological conditions necessitated smaller foraging parties and groups began to form subgroups. As the frequency of subgroups increased, males initiated an interest in estrous females for a greater proportion of the estrus period and, subsequently, they began to show interest in nonestrous females. As male interest in nonestrous females increased, behaviors such as following and herding appeared—behaviors previously shown only with estrous females.

Male–female bonds strengthened and unattached nonestrous females dwindled, selectively favoring males with the motivation (and ability) to simultaneously form strong bonds with several females. The ability to coerce females into following became advantageous and, as in this study, leading may have been found only among males that herded at high rates. Thus, competition for unattached females probably intensified a male's interest in and herding of nonestrous females, culminating in the rigid OMU behavior of hamadryas males observed today.

The subgroup hypothesis has other corollaries. Many authors have suggested that male baboons provide protection for females (from predators, other males, or other females; Smuts, 1985; Palombit, 2001; Swedell, 2002a, 2006). Henzi and Barrett (2003) argue that subgroup formation was the probable starting point for the differentiation of hamadryas baboons, emphasizing that this would increase the potential for infanticide and the need for paternal care. While this study does not suggest a primary role for paternal behavior in hamadryas males, the protection that males can provide may still be important, particularly from the female's perspective. In group H, females were very rarely in a subgroup without an adult male. The few times this was observed, the females appeared frightened, moved quickly, gave repeated contact calls (cf. Rendall et al., 2000), and climbed tall trees to scan the area. The risk of ending up in a subgroup without a male clearly creates an incentive for females to bond to males rather than females (for further discussion of the benefits of male–female bonding from the female perspective, see Bergman and Beehner, 2003 and Beehner and Bergman, this volume).

### 4.2. Conclusions and prospects for future research

This study provides support for the hypothesis that the hamadryas male reproductive strategy derives from consort behavior. At the same time, this analysis offers little or no support for the hypotheses that intergroup herding or paternal care were precursors of hamadryas male behavior. Rather, herding behavior and paternal care, two behavioral traits that obviously are important for hamadryas males, are more likely consequences of an attraction to nonestrous females and long-term intersexual bonds, respectively. Based on the results of this study, hamadryas males may be viewed as being in a state of "permanent consortship." This perspective may provide a framework for future comparative research into hamadryas biology and behavior. While definitive answers about the evolution of hamadryas baboon behavior remain elusive, studies of behavior in hybrid baboons provide an important baseline for further explorations of behavioral differentiation in baboons.

## ACKNOWLEDGMENTS

I would first like to thank Larissa Swedell and Steve Leigh for the invitation to contribute to this volume. I also thank my counterparts in Ethiopia for facilitating my work in the field: the Ethiopian Wildlife Conservation Department (EWCD) and the biology department of Addis Ababa University (AAU). In particular, at EWCD I want to thank the manager, Ato Tesfaye Hundessa, and the wardens and staff of the Awash National Park. At AAU I would especially like to thank the chairs of the biology department and my faculty associate, Dr. Solomon Yirga. I thank the codirectors of the Awash Baboon Project, Dr. Clifford Jolly and Dr. Jane Phillips-Conroy, for their help both in the field and in the preparation of this manuscript. I also thank two anonymous reviewers for their helpful comments. This work was supported by Boise Fund, the National Geographic Society, the National Science Foundation, Sigma Xi, Washington University, and the Wenner-Gren Foundation.

## REFERENCES

Abegglen, J. J., 1984, *On socialization in hamadryas baboons: a field study*, Associated University Presses, London.

Alberts, S., Watts, H., and Altmann, J., 2003, Queuing and queue-jumping: Long-term patterns of reproductive skew in male savannah baboons, *Papio cynocephalus*, *Anim. Behav.* **65**:821–840.

Aldrich-Blake, F. P. G., Dunn, T. K., Dunbar, R. I. M., and Headley, P. M., 1971, Observations on baboons, *Papio anubis*, in an arid region in Ethiopia, *Folia Primatol.* **15**:1–35.

Altmann, J., 1974, Observational study of behavior: Sampling methods, *Behaviour* **49**:229–267.

Anderson, C. M., 1983, Levels of social organization and male–female bonding in the genus *Papio*, *Am. J. Phys. Anthropol.* **60**:15–22.

Anderson, C. M., 1989, The spread of exclusive mating in a chacma baboon population, *Am. J. Phys. Anthropol.* **78**:355–360.

Bercovitch, F. B., 1986, Male rank and reproductive activity in savanna baboons, *Int. J. Primatol.* 7:533–550.

Bercovitch, F. B., 1988, Female choice, male reproductive success, and the origin of one male groups in baboons, in: *Baboons: Behaviour and Ecology, Use and Care*, M. Thiago de Mello, A. Whiten, and R. W. Byrne, eds., Brasilia, Brasil, pp. 61–76.

Bercovitch, F. B., 1991, Mate selection, consortship formation, and reproductive tactics in adult female savanna baboons, *Primates* **32**:437–452.

Bergman, T. J., 2000, Mating behavior and reproductive success of hybrid male baboons (*Papio hamadryas hamadryas* × *Papio hamadryas anubis*). Ph.D. Dissertation, Washington University, St. Louis, MO.

Bergman, T. J. and Beehner, J. C., 2003, Hybrid zones and sexual selection: Insights from the Awash baboon hybrid zone (*Papio hamadryas anubis* × *P. h. hamadryas*), in: *Sexual Selection and Primates: New Insights and Directions*, C. Jones, ed., American Society of Primatologists, Norman, OK, pp. 500–537.

Bergman, T. J. and Beehner, J. C., 2004, Social system of a hybrid baboon group (*Papio anubis* × *P. hamadryas*), *Int. J. Primatol.* **25**:1313–1330.

Biquand, S., Biquand-Guyot, V., Boug, A., and Gautier, J.-P., 1992, Group composition in wild and commensal hamadryas baboons: A comparative study in Saudi Arabia, *Int. J. Primatol.* **13**:533–543.

Buchan, J. C., Alberts, S. C., Silk, J. B., and Altmann, J., 2003, True paternal care in a multi-male primate society. *Nature* **425**:179–180.

Bulger, J. B., 1993, Dominance rank and access to estrous females in male savanna baboons. *Behaviour* **127**:67–103.

Byrne, R. W., Whiten, A., and Henzi, S. P., 1987, One-male groups and inter-group interactions of mountain baboons, *Int. J. Primatol.* **8**:615–633.

Byrne, R. W., Whiten, A., and Henzi, S. P., 1989, Social relationships of mountain baboons: Leadership and affiliation in a non-female bonded monkey, *Am. J. Primatol.* **18**:191–207.

Caljan, V. G., Meishvili, N. V., and Vancatova, M. A., 1987, Sexual behaviour of hamadryas baboons, *Anthropologie* **25**:183–187.

Cheney, D. L. and Seyfarth, R. M., 1977, Behaviour of adult and immature male baboons during inter-group encounters, *Nature* **269**:404–406.

Colmenares, F., 1992, Clans and harems in a colony of hamadryas and hybrid baboons: Male kinship, familiarity and the formation of brother-teams, *Behaviour* **121**:61–94.

Hamilton, W. J., III, Buskirk, R. E., and Buskirk, W. H., 1975, Chacma baboon tactics during intertroop encounters, *J. Mammal.* **56**:857–870.

Henzi, S. P., 1996, Copulation calls and paternity in chacma baboons, *Anim. Behav.* **51**:233–234.

Henzi, P. and Barrett, L., 2003, Evolutionary ecology, sexual conflict, and behavioral differentiation among baboon populations, *Evol. Anthropol.* **12**:217–230.

Henzi, S. P., Lycett, J. E., and Weingrill, T., 1998, Mate guarding and risk assessment by male mountain baboons during inter-troop encounters, *Anim. Behav.* **55**:1421–1428.

Kummer, H., 1968a, *Social Organization of Hamadryas Baboons: A Field Study*, University of Chicago Press, Chicago.

Kummer, H., 1968b, Two variations in the social organization of baboons, in: *Primates: Studies in Adaptation and Variability*, P. C. Jay, ed., Holt, Rinehart, and Winston, New York, pp. 293–312.

Kummer, H., and Kurt, F., 1965, A comparison of social behavior in captive and wild hamadryas baboons, in: *The Baboon in Medical Research*, H. Vagtborg, ed., University of Texas Press, Austin, pp. 65–80.

Kummer, H., Banaja, A. A., Abo-Khatwa, A. N., and Ghandour, A. M., 1985, Differences in the social behavior between Ethiopian and Arabian hamadryas baboons, *Folia Primatol.* **45**:1–8.

Nagel, U., 1973, A comparison of anubis baboons, hamadryas baboons and their hybrids at a species border in Ethiopia, *Folia Primatol.* **19**:104–165.

Newman, T. K., Jolly, C. J., and Rogers, J., 2004, Mitochondrial phylogeny and systematics of baboons (*Papio*), *Am. J. Phys. Anthropol.* **124**:17–27.

Nystrom, P. D. A., 1992, Mating success of hamadryas, anubis and hybrid male baboons in a "mixed" social group in the Awash National Park, Ethiopia. Ph.D. Dissertation, Washington University, St. Louis, MO.

Packer, C., 1979a, Male dominance and reproductive activity in *Papio anubis*, *Anim. Behav.* **27**:37–45.

Packer, C., 1979b, Inter-troop transfer and inbreeding avoidance in *Papio anubis*, *Anim. Behav.* **27**:1–36.

Palombit, R. A., 2001, Male infanticide in wild savanna baboons: Adaptive significance and intraspecific variation, in: C. Jones, ed., *Sexual Selection and Reproductive Competition in Primates: New Perspectives and Directions*, American Society of Primatologists, New York.

Palombit, R. A., Seyfarth, R. M., and Cheney, D. L., 1997, The adaptive value of friendships to female baboons: Experimental and observational evidence, *Anim. Behav.* **54**:599–614.

Palombit, R. A., Cheney, D. L., Fischer, J., Johnson, S., Rendall, D., Seyfarth, R. M., and Silk, J. B., 2000, Male infanticide and defense of infants in wild chacma baboons, in: *Infanticide by Males and Its Implications*, C. P. van Schaik, and C. H. Janson, eds., Cambridge University Press, Cambridge, pp. 123–152.

Phillips-Conroy, J. E. and Jolly, C. J., 1986, Changes in the structure of the baboon hybrid zone in the Awash National Park, Ethiopia, *Am. J. Phys. Anthropol.* **71**:337–350.

Phillips-Conroy, J. E., Bergman, T., and Jolly, C. J., 2000, Quantitative assessment of occlusal wear and age estimation in Ethiopian and Tanzanian baboons, in: *Old World Monkeys*, P. F. Whitehead, and C. J. Jolly, eds., Cambridge, Cambridge University Press, pp. 321–340.

Ransom, T. W., 1981, *Beach Troop of the Gombe*, Bucknell University Press, London.
Rendall, D., Cheney, D. L., and Seyfarth, R. M., 2000, Proximate factors mediating "contact" calls in adult female baboons (*Papio cynocephalus ursinus*) and their infants, *J. Comp. Psychol.* **114**:36–46.
Seyfarth, R. M., 1978, Social relationships among adult male and female baboons. I. Behaviour during sexual consortship, *Behaviour* **64**:204–226.
Sigg, H., 1980, Differentiation of female positions in hamadryas one-male units, *Z. Tierpsychol.* **53**:265–302.
Smuts, B. B., 1983, Special relationships between adult male and female olive baboons: Advantages, in: *Primate Social Relationships: An Integrated Approach*, R. A. Hinde, ed., Blackwell, London, pp. 262–266.
Smuts, B. B., 1985, *Sex and Friendship in Baboons*, Aldine, New York.
Strum, S. C., 1982, Agonistic dominance in male baboons: An alternative view, *Int. J. Primatol.* **3**:175–202.
Swedell, L., 2002a, Affiliation among females in wild hamadryas baboons (*Papio hamadryas hamadryas*), *Int. J. Primatol.* **23**:1205–1226.
Swedell, L., 2002b, Ranging behavior, group size and behavioral flexibility in Ethiopian hamadryas baboons (*Papio hamadryas hamadryas*), *Folia Primatologica* **73**:95–103.
Swedell, L., 2006, *Strategies of Sex and Survival in Hamadryas Baboons: Through a Female Lens*, Pearson Prentice Hall, Upper Saddle River, NJ.
Weingrill, T., Lycett, J. E., and Henzi, S. P., 2000, Consortship and mating success in chacma baboons (*Papio cynocephalus ursinus*), *Ethology* **106**:1033–1044.
Whiten, A., Byrne, R. W., and Henzi, S. P., 1987, The behavioral ecology of mountain baboons, *Int. J. Primatol.* **8**:367–388.

CHAPTER FIVE

# The Social and Ecological Flexibility of Guinea Baboons: Implications for Guinea Baboon Social Organization and Male Strategies

*Anh Galat-Luong, Gérard Galat, and Suzanne Hagell*


## CHAPTER SUMMARY

The social organization and behavioral ecology of Guinea baboons is poorly understood compared to other baboon taxa. Most data contributing to our current knowledge of their behavior come from either very short field studies or captive populations. In this chapter, we attempt to augment the knowledge base of Guinea baboon behavior with data from a wild population of Guinea baboons inhabiting the Niokolo Koba National Park in Senegal.


---


**Anh Galat-Luong** • **Gerard Galat** • UR 136 Aires protégées, Institut de Recherche pour le Développement (IRD), France **Suzanne Hagell** • City University of New York and the New York Consortium in Evolutionary Primatology, New York, NY, USA

Our results indicate that Guinea baboons have adapted to a wide range of habitats with many different climates and that they vary in their social structure over time depending on habitat and season. Apparently, Guinea baboons have a multilevel social structure that is superficially similar to that seen in hamadryas baboons. The basic social group is the one-male unit, but when necessary these small groups aggregate into successively larger groups. This may occur through a combination of female flexibility and male–male tolerance and cooperation. Fission and fusion of groups during the day are components of foraging and antipredation strategies. Seasonal changes are also possible, as the number of individuals in each of the intermediate group structures is flexible as well. In this way Guinea baboons optimize their group size given their highly variable habitat without placing undue demands on individual social time budgets and risking permanent fragmentation of the one-male unit. It seems likely that Guinea baboon social organization has evolved independently into a multilevel structure that is different from both hamadryas baboons and other savanna baboons. Moreover, Guinea baboons are unique in their response to the demands of the diversity of West African habitats.

## 1. INTRODUCTION

The Guinea baboon of West Africa, *Papio hamadryas papio,* is both the least studied and poorly understood of the five major subspecies of *P. hamadryas* (Henzi and Barrett, 2003). Although their total distribution area is small, ca. 250,000 km$^2$, Guinea baboons have a wide north–south spread (Figure 1), and, as a consequence, they inhabit a wide range of habitats with many different climates. These include sahelian steppe in Mauritania, soudanian shrubby savannas in Senegal and Mali, subguinean mosaic woodlands in Senegal, and secondary high forest in Guinea. They live at sea level in the mangrove forests of Senegal as well as at altitudes of more than 1,000 m in the foothills of the Fouta Djalon mountains of Guinea. The annual rainfall of Guinea baboon habitats varies from less than 200 mm in Mauritania to more than 1,400 mm in the south of their range in Guinea, and the mean daily maximum temperature ranges from ca. 20 to 50°C. Unfortunately, despite being the third most abundant large mammal in the Mafou protected area of the Haut Niger National Park in Guinea (Brugière et al., 2002), several recent surveys have indicated that Guinea baboon distribution overall has declined in recent years (Galat et al., 2002; Galat-Luong and Galat, 2003a).

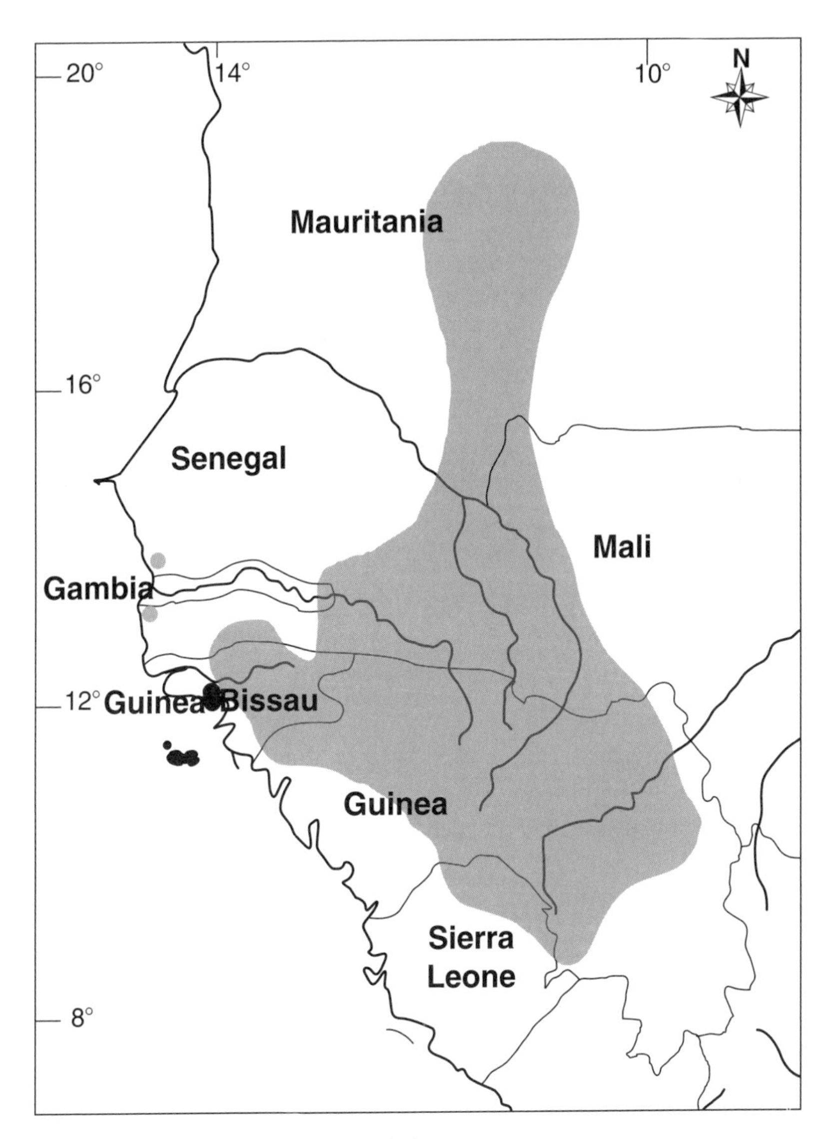


**Figure 1.** Distribution of the Guinea baboon.

How the social organization and mating system of Guinea baboons compares to that of other baboon subspecies is still somewhat unclear. Olive baboons (*Papio h. anubis*), the immediately neighboring baboon population to Guinea baboons, exemplify the social organization common to most other *Papio* subspecies: multimale and multifemale groups in which females are philopatric. In these societies, social and mating behavior is relatively unstructured and indiscriminate among group members. Hamadryas baboons (*Papio h. hamadryas*), on the other hand, are unique among *Papio* baboons in having a highly structured, multilevel social system based around individual one-male units (OMUs) in which females are forcibly herded into permanent consortships with a leader male. The mating strategy of hamadryas males would seem to be to maintain continuous reproductive access to females by controlling social interactions throughout their reproductive cycle. The strategies of females in hamadryas baboon groups have been less discussed, but it seems that females may also benefit from associating with a single protective male (see Beehner and Bergman, this volume; Swedell and Saunders, this volume; Swedell, 2006). Previous studies suggest that Guinea baboons lie somewhere between these two extremes, with a social system that is somewhat intermediate between that of hamadryas and olive baboons (Boese, 1973, 1975).

Most data concerning Guinea baboon behavior and ecology come from one location, the Niokolo Koba National Park in Senegal (e.g., Dekeyser, 1956; Fady, 1972; Dunbar and Nathan, 1972; Boese, 1973; Sharman, 1981; Anderson and McGrew, 1984). Interestingly, although all these authors collected data at the same location, they disagree about the social structure and social organization of these animals. Both Anderson and McGrew (1984) and Dunbar and Nathan (1972) observed sleeping aggregations of Guinea baboons and reported OMU-like subgroupings, but they also witnessed a degree of female flexibility in social relationships not seen in hamadryas. Both of these sets of authors concluded that the Guinea social system more closely resembles olive than hamadryas baboons.

Boese (1973, 1975), who conducted by far the longest study of Guinea baboons, also saw females interacting more freely than in hamadryas, but he observed permanent one-male subgroups, strong male–female bonds, and herding behavior as well. Boese concluded from his observations that Guinea baboons have an OMU system that is intermediate between olive and hamadryas baboon social organization. Boese suggested that male Guinea baboons maintain sexual exclusivity with particular females but are more tolerant of extrasubgroup interactions than hamadryas males. Close social bonds between females and males, as in hamadryas, were attributed to a habitat in which female and immature animals periodically require male protection

(Boese, 1975). Boese's conclusions, however, were drawn mainly from his observations of Guinea baboons in captivity at the Brookfield Zoo in Chicago. Captivity has the potential to enhance aggressive behavior, strengthen dominance hierarchies, and allow the spread of idiosyncratic behaviors within groups, hence observations of zoo populations should be corroborated by observations of wild populations as well.

In this chapter, we report original data collected at the Niokolo Koba National Park in Senegal and compare our results to other studies of Guinea baboons as well as other baboon subspecies and sympatric cercopithecines. We attempt to describe the current state of knowledge concerning Guinea baboon ecology and social organization, and draw some general conclusions about the origins of Guinea baboon social organization and its relationship to male social and mating strategies.

## 2. METHODS

### 2.1. Ecology

Data were collected by Galat-Luong and Galat at the Niokolo Koba National Park during surveys of large mammals conducted in Senegal and Guinea between 1975 and 2001. Using the line-transect method (and Distance software from Laake et al., 1996), we estimated the density of Guinea baboons and compared the results of 1990–1993 censuses with those from 1994 to 1998 to estimate changes in Guinea baboon abundance over time. We also recorded the habitat in which the animals were found: shrubby savanna, arboreal savanna, forest, or open grassland, and additionally described the area as "unburned," "recently burned," or "burned with secondary grass growth." "Recently burned" areas were covered with ashes, with no visible green grasses or leaves. "Burned with secondary grass growth," which replaces "recently burned" areas in 1–10 days, refers to areas in which ash was accompanied by fresh, recently grown grasses. Visibility conditions in these two types of areas are similar and better than in unburned areas. Use of a particular type of area was determined by measuring the percentage of encounters in each area. As the absolute area of each habitat has not been determined and visibility varies for each of these habitats, our determination of "preference" for particular habitats is valid only in comparison with data collected on other species during this study.

### 2.2. Social Structure, Organization, and Behavior

The number of instantaneously visible individuals was counted during each group encounter and used as a comparative index of group sizes. This index is sensitive to variations in visibility and it is thus mainly used for the same site and for the same period of the year (mid-February). Age/sex classes were based on Boese (1973) as well as the authors' own experience.

Time budgets of one population of baboons in the park were estimated from hourly scan samples of individuals (ca. 205 observation hours). These observations were made when the baboons were highly visible at the transition period between the dry and rainy seasons in May and June 1997, when the groups frequented the same water pool.

The authors also opportunistically recorded social interactions among adult baboons (37 observation hours) focusing on interactions related to social organization, i.e., affiliative and agonistic behavior as well as submissive/dominant interactions (Table 1). These behaviors were defined as in Boese (1973, 1975). For example, a male "prance–rump–push" was recorded when a male pushed a female using his rump, a behavior that occurred as part of the "prance" stereotypical display described by Boese (1975).

## 3. RESULTS

### 3.1. Ecology

The results of our analysis of habitat use at Niokolo Koba are very similar to those of Sharman (1981). Guinea baboons at Niokolo Koba spent about 50 percent of their observed time in shrubby savanna, one-third in tree-savanna, and the balance in forest or open grassland (Figure 2). In the savanna, baboons were found most often in recently burned areas (56 percent of encounters, $N$=333) and were encountered less often in unburned areas (25 percent) or in burned areas with secondary grass growth (19 percent). As other large mammals did not show the same preferences, it is unlikely that this result reflects a visibility bias. Grimm's bush duiker (*Sylvicapra grimmia*), for example, was observed to prefer areas with secondary grass growth, whereas the red-flanked duiker (*Cephalophus rufilatus*) did not. Among the other sympatric primates, the green monkey (*Cercopithecus (aethiops) sabaeus*) showed the same preferences as the Guinea baboons for recently burned land, whereas the patas monkeys (*Erythrocebus patas*) did not. Patas were found

**Table 1.** List of behavioral interactions recorded

| Friendly/submissive | Agonistic/dominant | Herding/corralling |
|---|---|---|
| Approach (go and sit nearby another animal) | Fight | Adult male leads OMU |
| Present (lateral to rear presenting) | Chase | Adult male leads group |
| Standing Present (abdominal to inguinal presenting, with or without hands on shoulder or head of receiver; usually, the receiver sniffs the abdominal or inguinal area of the initiator) | Kick | Group of adult males leads group |
| | Run Away | Corral (change direction, accelerate using shaking, jumping, prancing, or running) |
| | Social Mount | Adult male recruit |
| | Solicit social mount | Follow adult male |
| | Prance and supplant | Shrieking recruit |
| Groom | Supplant | |
| Solicit groom | | |
| Head lunge (with or without look away) | Sexual | Male Maternal Behavior |
| Prance–rump–push | | |
| Wait | Sexual mount | Adult male manipulates infant |
| Embrace | Sexual mount and sit | Adult male kidnaps infant with ventral carry |
| Hold | Solicit sexual mount | |
| Muzzle | Adult female dart (withdraw after sexual mount) | |
| Sniff | Adult female sniffs male genital area | |
| | Masturbation | |

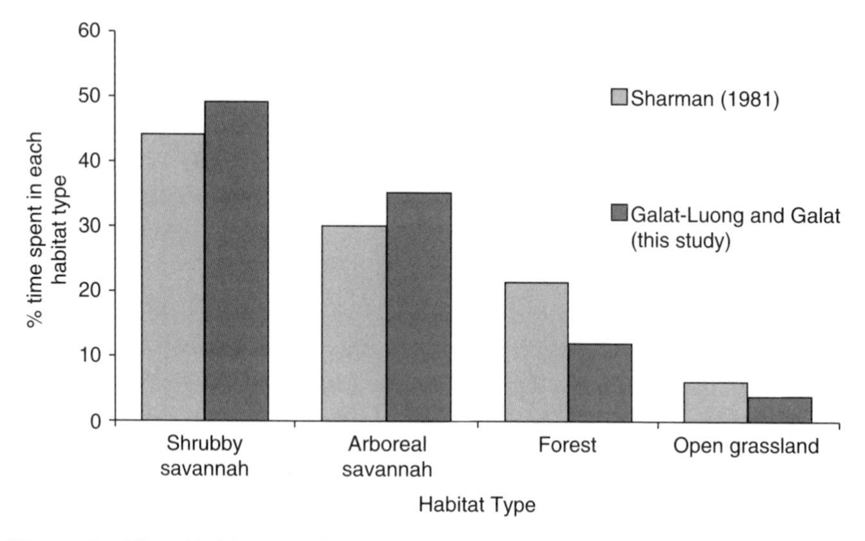


**Figure 2.** Use of habitat types by Guinea baboons in the Niokolo Koba National Park.

most commonly in areas of secondary grass growth (58 percent, $N = 107$). Green monkeys were encountered on 56 percent of occasions in recently burned savanna, 26 percent in unburned areas, and 18 percent in burned areas with second grass growth vegetation ($N = 237$).

Predation risk at Niokolo Koba is high, and the hunting of Guinea baboons by lions (*Panthera leo*) in particular has increased since 1994 (Galat-Luong and Galat, unpublished data). During this study, an attack by a spotted hyena (*Crocuta crocuta*) was observed as well as the barking and shrieking of baboons when lions were close by. On one occasion, a troop of green monkeys was observed to direct alarm calls at a hyena, which then ran away. The baboons, however, were never observed to mob predators in this way.

### 3.2. Social Structure, Organization, and Behavior

The variation in size of the Guinea baboon population in the Niokolo Koba National Park is shown in Figure 3. Observations made during this study point to a multilevel social structure in Guinea baboons similar to that described by Kummer (1968) for hamadryas baboons, in which four hierarchical levels can be distinguished. The smallest subunits, and basic social groups, were composed of 8–10 individuals and resembled hamadryas OMUs. These subgroups were most obvious during feeding, foraging, and

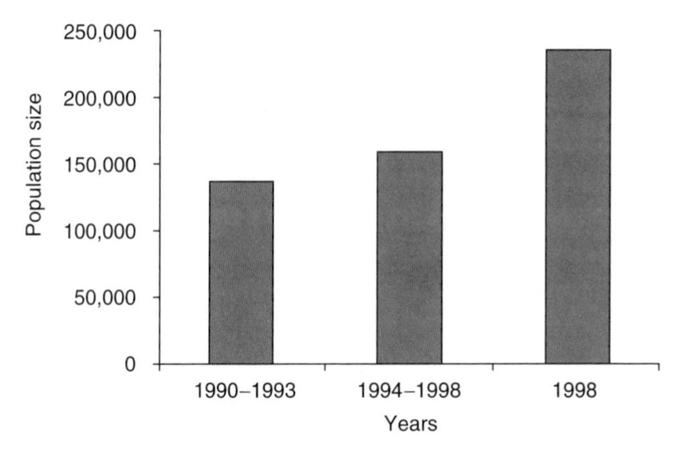


**Figure 3.** Population size changes of Guinea baboons in the Niokolo Koba National Park, Senegal.

resting during the day. When resting, an adult male was frequently in the center of the group. When moving, each OMU was led by an adult male. These OMU-like groups were visibly (spatially) distinct from temporary females or immature parties.

OMUs joined into larger, second-level subgroups when beginning to move or when sleeping at night. The mean size of these second-level subgroups was 19 individuals (5–65; $N = 45$). At night, the second-level subgroups slept either spatially separated from or together with other second-level subgroups. During longer periods of movement, second-level groups were still spatially distinguishable as they walked in long columns with other second-level groups. Several of these second-level groups comprised larger, third-level groups. At Niokolo Koba, the mean size of the third-level group was 62 (22–249; $N = 111$). During a survey outside the Park in 1988, we observed a mean third-level group size of 72 individuals (24–200; $N = 14$). Several third-level groups were observed to share the same sleeping site, forming a fourth-level group. Occasionally, subgroups of females and immatures as well as individual juveniles, adult males and females were seen moving through these larger groups. The number of individuals within groups varied from year to year as well as by time of day.

Time budgets do not have absolute values here and comparisons are limited to the studied population at the waterhole. The Guinea baboons spent more time feeding at the beginning of the rainy season than at the end of the

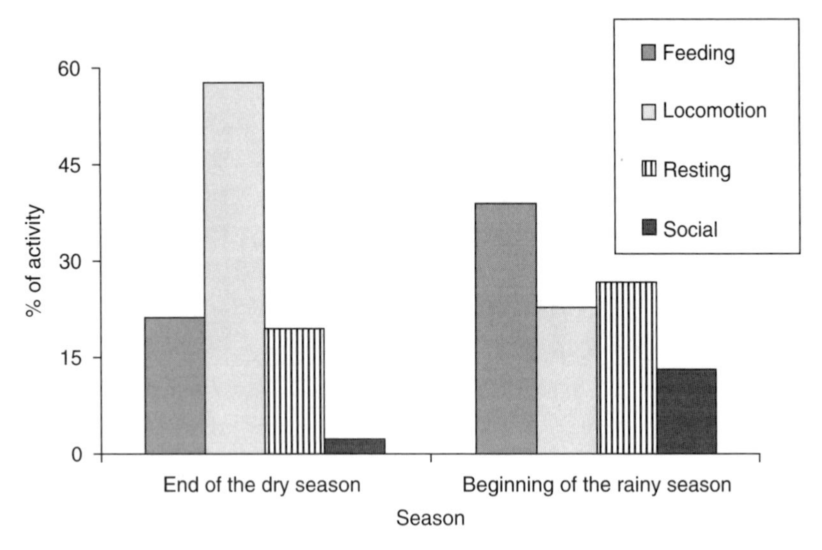


**Figure 4.** Time budgets of Guinea baboons at a waterhole in the Niokolo Koba National Park in May and June 1997.

dry season (Figure 4). The transition between seasons is marked by a change in the distribution and density of grasses, which are widely available, high and dense during the rainy season and reduced and restricted to water pools during the dry season. As grasses became widely available, Guinea baboons were observed to decrease their time spent in locomotion and to increase their social time. Time spent resting did not change during the observation period.

The social behaviors observed in this study were classified as friendly/submissive (e.g., approach, groom), agonistic/dominant, herding/corralling, male maternal behavior, or sexual interactions (Table 1). The frequency of interactions among adults (Table 2) shows a high level of male tolerance. Only three instances of aggressive chasing or fighting were observed during the study, and most agonistic behavior consisted of dominance interactions (e.g., mounts, supplants). Although grooming appeared to be evenly distributed among adults of both sexes, males gave and received most other friendly interactions in the group. Males initiated friendly contact with both males and females equally, and initiated most of the friendly contacts with other males as well. Males initiated copulation only slightly more often than females did. Males were more often affiliative than agonistic with other males.

Herding with violent neck-biting behavior, typical of hamadryas baboons (e.g., Kummer, 1968) and described in Guinea baboons by Boese (1975), was not observed in this study. All kinds of individuals were seen to move

**Table 2.** Adult social interactions. Number of interactions recorded in 37 hr of observation

| Receiver/initiator | Adult female | Adult male | Total | Adult female | Adult male | Total |
|---|---|---|---|---|---|---|
| Friendly/submissive | | | | Presenting interactions | | |
| Adult Females | 3 | 21 | 24 | 1 | 4 | 5 |
| Adult males | 17 | 18 | 35 | 2 | 5 | 7 |
| Total | 20 | 39 | 59 | 3 | 9 | 12 |
| Agonistic/dominant | | | | Approach followed by "contact" | | |
| Adult females | 1 | | 1 | | 4 | 4 |
| Adult males | 6 | 19 | 25 | 7 | 16 | 23 |
| Total | 7 | 19 | 26 | 7 | 20 | 27 |
| Sexual interactions | | | | Solicitation interactions | | |
| Adult females | | 9 | 9 | 2 | 4 | 6 |
| Adult males | 13 | | 13 | 5 | 6 | 11 |
| Total | 13 | 9 | 22 | 7 | 10 | 17 |
| Grooming interactions | | | | | | |
| Adult females | 1 | 9 | 10 | | | |
| Adult males | 6 | 2 | 8 | | | |
| Total | 7 | 11 | 18 | | | |

through subgroups while foraging. Males were, however, observed to control group movements with a behavior that we describe as "corralling," that is, by running to speed the progression of the group and by shaking, jumping, and prancing to change the direction of group movement. Adult males were observed leading OMUs ($N = 3$ occurrences) and larger groups ($N = 3$) as well as corralling a larger group ($N = 2$). Three males led a larger group together once and several males corralled a large group together once. Adult males were also observed to wait for other individuals while assuming vigilance when at the edge of an exposed location. Subgroups were observed to wait for alternate access to a restricted water source.

Specific behaviors observed included adult males presenting to a juvenile male as well as juveniles and subadults presenting to adult males. Adult males were also seen in contact with nonadults, both giving and receiving grooming. Adult males also handled infants ($N = 3$), and on one occasion a male kidnapped an infant, carrying it ventrally. An adult female was observed to carry a deceased infant for 3 days (cause of death unknown).

Postcopulatory behaviors included males and females sitting together ($N = 2$), females running away ($N = 2$), or a resumption of their previous activity, e.g., foraging or moving ($N = 11$). On two occasions a male was unsuccessful at soliciting copulation when the female was carrying an infant ventrally. Adult female solicitations for copulation were successful twice and unsuccessful twice.

If we analyze our data with reference to the "obedience" test used by Nagel (1971) in his comparative study of olive and hamadryas baboons, we rarely observed males and females to look behind at an individual while walking away from them following a period of contact (i.e., in order to check if this individual is actually following). This did occur, though, if a supplant seemed likely to occur, such as when moving to a water pool to drink.

## 4. DISCUSSION

The data presented here demonstrate the opportunistic way in which Guinea baboons use their habitats, changing their diet and time budgets in response to seasonally changing resources. We believe this to be a factor behind the fact that Guinea baboon populations in the Niokolo Koba Park have not declined despite the decrease in population size of most other large mammals in the park (Galat-Luong and Galat, 2003a,b). Another factor behind their success may be that the baboons are not hunted, while ungulates are. Guinea baboons, however, as shown by the 1998 outburst of reproductive success (Figure 3), also appear to show a fast reactivity to fluctuating environmental conditions, which may indicate an adaptation for ecological flexibility. This coincides with the flexibility seen in their social organization.

While we cannot say anything yet about the stability of Guinea baboon social units through time or the pattern of sex-biased dispersal, we can draw some parallels between Guinea baboon social structure and that of hamadryas baboons. Both subspecies are characterized by multiple levels of social structure that fragment and coalesce depending on ecological conditions. Additionally, in both subspecies the smallest social unit consists of one adult male and several females. Boese (1973) suggested that the Guinea baboon social structure represents an evolutionary precursor to the more rigid multilevel social structure seen in hamadryas. Sharman (1981) suggested, however, that Guinea baboon OMUs likely represent matrilineal kin groups rather than the male-policed harems found in hamadryas baboons and that the two subspecies are thus only superficially comparable. We would suggest that the two systems are indeed *not* homologous and that the use of Kummer's terminology for hamadryas baboons—i.e., harem, clan, band, and troop—may thus not be appropriate.

If Guinea baboons do represent a social organization intermediate between the relaxed, multimale societies of olive baboons and the multileveled structure of hamadryas, then this would accord nicely with the provisional findings of Jolly and Phillips-Conroy (this volume) in which testis size in captive Guinea baboons was found to be intermediate between that of hamadryas and olive baboons. Jolly and Phillips-Conroy's findings would suggest that Guinea baboons have a mating system that is less polyandrous than olive baboons but more so than hamadryas, which would make sense given a social system that, while organized around OMUs, also includes a greater degree of female flexibility than in hamadryas society.

Though limited, our data seem to confirm the suggestion by Boese (1975) that the multileveled system seen in Guinea and hamadryas baboons is rooted in male–male tolerance. Such tolerance results in the fusion of OMUs into large troops when predation pressure is highest, particularly when traveling through more risky areas in the terrain. This suggests that leader males may coordinate their units so that each OMU in the troop receives protective benefits. At Niokolo Koba, Guinea baboons are exposed to lions concealed in the high grasses of the savanna, as well as to leopards when in gallery forests along rivers, and it is when traveling in these areas that Guinea baboons are observed to form large columns. Additionally, Boese (1975) has suggested that adult males permit the presence of subadult males in OMUs so that they may assist in vigilance. In this way, the OMU structure that Guinea baboons form during the day allows for a reduction in food competition as well as moderate protection from predation. Tolerance of nonadult males within an OMU has also been described by Kummer (1968) and Swedell (2006) for hamadryas baboons.

Dunbar (1992, 1994) proposed that the mean size of social groups is influenced by the social budget allowed under local ecological conditions. According to this model, as group size increases, so does the burden on individual group members to balance vigilance and time spent feeding with time for socializing. At a certain group size, these individuals are no longer able to maintain cohesiveness and the group will fragment. Guinea baboons represent another strategy, in which group structure can be adjusted with a given season and time of day to optimize the number of individuals. Leader males may retain semiexclusive access to mates, while cooperating with subadult males within the group for assistance with vigilance. Females and immatures (including the subadult males) benefit from forming strong associations with a male who will protect them from predation and extragroup harassment. Both females and subadult males are also provided a certain degree of freedom to interact outside the group.

Seasonal flexibility can be shown by comparing group counts at different times of the year. Sharman (1981) recorded that the size of the troop (third-level structure) changed from 50-90 individuals during the dry season to 135–250 during the rainy season. This may reflect an adjustment of group size to food availability, which is highest in the rainy season. Heavy rainfalls occurred in 1997, 2.4 times more than during the preceding years, and in 1998 the Guinea baboons reacted with an outburst of reproductive success (Figure 3) that has subsequently caused crop depredations (Galat-Luong and Galat, 2003a).

Galat and Galat-Luong surveyed group sizes of green monkeys along the same north–south gradient in Senegal, and these monkeys show interesting parallels to Guinea baboons. Green monkeys were observed to live in small territorial groups of 8–33 individuals when in relatively species-rich areas of high and constant vegetal diversity (Galat and Galat-Luong, 1976). In the more arid areas of northern Senegal, the green monkeys lived in larger groups of up to 174 individuals with little territorial behavior (Galat and Galat-Luong, 1977). In this habitat, vegetal diversity is severely impoverished, deciduous trees have disappeared, and *Acacia nilotica* is the dominant species. When the rainy season approached, OMU-like subgroups formed and males showed a tendency to herd females; half of these females were pregnant or carrying newborn infants. As with the Guinea baboons, the OMUs were most visible when foraging or resting during midday in trees. At twilight these OMUs aggregated into large clumps for sleeping. Coincidentally, Guinea baboons are no longer found in the northerly area of Senegal where green monkeys also form these larger aggregates.

The Niokolo Koba National Park is located at the center of the Guinea baboons' north–south range. Because the park encompasses several habitat types, the environment permits adaptations to both desert and forest habitats. We would predict that in more northerly populations, where resources are more scattered and unpredictable, Guinea baboons would show more conspicuous OMUs and larger third-level aggregations. In the south, on the other hand, where the forest is denser, the populations should exhibit less conspicuous OMUs as well as larger second-level groups and age-graded units as seen in green monkeys in these areas. In green monkeys, OMUs and second-level groupings may be more efficient for foraging in separate trees, whereas larger groups would not be able to maintain cohesiveness due to restricted visibility. In fact, Brugière et al. (2002) reported the mean number of instantaneously visible individuals of Guinea baboons (which we think to be mainly tied to the size of the second-level groups) in the Haut Niger National Park, approximately 300 km south of Niokolo Koba, to be 27 ($N$=3), which is much greater than at Niokolo Koba (6–15), though the figure should be biased toward a lower value as visibility in more dense vegetation is likely reduced.

In conclusion, the social organization of Guinea baboons can be viewed as a highly adaptable social network capable of responding quickly to ecological fluctuations. Although this multilevel social organization is superficially similar to that seen in hamadryas baboons, most of the behavioral elements we have observed suggest that it may not be homologous. The fusion and coordination of separate subgroups is only possible through the tolerant and affiliative behavior of males. Some males receive submissive behaviors from individuals of different OMUs. Some males are able to lead small OMUs as well as larger groups. Males also cooperate in the leading and corralling of these groups. In this study, males were not observed to herd females in their OMU in the manner seen in hamadryas baboons, and female movements through groups appeared to be less rigid. Based on the limited observations reported so far, Guinea baboons do not appear to possess the same restrictive, harem structure seen in hamadryas baboons. Rather, they are characterized by a multilevel system that, while it includes one-male harem-like groupings, also includes a greater degree of female flexibility than seen in hamadryas. It seems more likely that Guinea social organization has evolved independently toward a fission–fusion structure that is different from both hamadryas baboons and other savanna baboons and that they are unique in their response to the demands of the diversity of the West African habitats they occupy.

## ACKNOWLEDGMENTS

The census of the large terrestrial fauna in Senegal was made in collaboration with the Senegalese National Parks, whom we thank for their kind permission to perform field research in the Niokolo Koba National Park. We also thank

B. Lavocat of the University of Corte, who collected time budget data. We are grateful to Larissa Swedell and two anonymous referees for their valuable help and comments. The study was supported by Orstom-IRD (the French Research Institute for Development), FFEM (the French Foundation for Global Environment), AFD (the French Development Agency), and DDR-Sodefitex (the Senegalese Rural Development Management).

## REFERENCES

Anderson, J. R. and McGrew, W. C., 1984, Guinea baboons at a sleeping site, *Am. J. Primatol.* **6**:1–14.

Boese, G. K., 1973, Behavior and social organization of the Guinea baboon *(Papio papio)*. Ph.D. Dissertation, Johns Hopkins University.

Boese, G. K., 1975, Social behavior and ecological considerations of West African baboons, in: R. H. Tuttle, ed., *Socioecology and Psychology of Primates*, Mouton, The Hague, pp. 205–230.

Brugière, D., Dia, M., Magassouba, B., Diakité, S., Gbansara, M., Mamy, M., and Saliou, B., 2002, *Dénombrement des Moyens et Grands Mammifères de la Zone Intégralement Protégée Mafou du Parc National du Haut-Niger, Guinée*, Unpublished report to AGIR, Conakry.

Dekeyser, P. L., 1956, Mammifères, *Mémoires de l'Institut Français d'Afrique Noire.* **48**:35–79.

Dunbar, R. I. M. D., 1992, Time: A hidden constraint on the behavioral ecology of baboons, *Behav. Ecol. Sociobiol.* **31**:35–49.

Dunbar, R. I. M., 1994, Ecological constraints on group size in baboons, in: *Animal Societies: Individuals, Interactions, and Social Organization*, P. Jarman and A. Rossiter, eds., Holt, New York, pp. 253–292.

Dunbar, R. I. M. and Nathan, M. F., 1972, Social organization of the Guinea baboon, *Papio papio*, *Folia Primatol.* **17**:321–334.

Fady, J. C., 1972, Absence de coopération de type instrumental en milieu naturel chez *Papio papio*, *Behaviour* **43**:157–164.

Galat, G. and Galat-Luong, A., 1976, La colonisation de la mangrove par *Cercopithecus aethiops sabaeus* au Senegal, *Rev. Ecol. Terre Vie* **30**:3–30.

Galat, G. and Galat-Luong, A., 1977, Démographie et régime alimentaire d'une troupe de *Cercopithecus aethiops sabaeus* en habitat marginal au nord Senegal, *Rev. Ecol. Terre Vie* **31**:557–577.

Galat, G., Galat-Luong, A., and Keita, Y., 2002, Régression de la Distribution et Statut Actuel du Babouin, *Papio papio*, en Limite d'Aire de Répartition au Sénégal, *Afr. Primates* **4**:69–70.

Galat-Luong, A. and Galat, G., 2003a, La ressource grande faune terrestre du Sénégal oriental: ses potentialités, ses contraintes, in: *Potentialités, contraintes et systèmes d'exploitation au Sénégal Oriental et en Haute Casamance*, UCAD–IRD–DDR, eds., CD-ROM, Dakar.

Galat-Luong, A. and Galat, G., 2003b, Social and ecological flexibility in Guinea baboons as an adaptation to unpredictable habitats, *Am. J. Phys. Anthropol. Suppl.* **36**:98.

Henzi, P. and Barrett, L., 2003, Evolutionary ecology, sexual conflict, and behavioral differentiation among baboon populations, *Evol. Anthropol.* **12**:217–230.

Kummer, H., 1968, *Social Organization of Hamadryas Baboons: A Field Study*, University of Chicago Press, Chicago.

Laake, J. L., Buckland, S. T., Anderson, D. R., and Burnham, K. P., 1996, *DISTANCE 2.2*, Colorado Cooperative Fish and Wildlife Research Unit, Colorado State University, Fort Collins, CO.

Nagel, U., 1971, Social organization in a baboon hybrid zone, *Proc. 3rd Int. Cong. Primatol. Zurich* **3**:48–57.

Sharman, M. J., 1981, Feeding, ranging, and social behavior of the Guinea baboon. Ph.D. Dissertation, University of St. Andrews.

Swedell, L., 2006, *Strategies of Sex and Survival in Hamadryas Baboons: Through a Female Lens*, Prentice Hall, Upper Saddle River, NJ.

CHAPTER SIX

# Social Organization, Reproductive Systems, and Mediation of Baboon Endogenous Virus (BaEV) Copy Number in Gelada, Hamadryas, and Other *Papio* Baboons

*Monica Uddin, Jane E. Phillips-Conroy, and Clifford J. Jolly*

## CHAPTER SUMMARY

The social system of geladas (*Theropithecus gelada*) and hamadryas baboons (*Papio hamadryas hamadryas*), characterized by a multilevel structure based upon one-male reproductive units, is unique among primates. Population

---

**Monica Uddin** • Center for Molecular Medicine and Genetics, Wayne State University School of Medicine, Detroit, MI, USA **Jane E. Phillips-Conroy** • Department of Anatomy and Neurobiology, Washington University School of Medicine and Department of Anthropology, Washington University, St. Louis, MO, USA **Clifford J. Jolly** • Department of Anthropology, New York University and the New York Consortium in Evolutionary Primatology, NY, USA

genetic theory predicts that the semiclosed, substructured social groupings exhibited by these taxa should result in increased relatedness among group members, possibly facilitating relatively rapid rates of evolution when compared to taxa without similarly subdivided populations. In this chapter, we test these predictions by quantifying a mobile genetic element known as baboon endogenous virus (BaEV). Quantitative real-time PCR methods were used to assess relative BaEV copy numbers in captive and wild-caught baboons, including samples from geladas, Guinea (*P. h. papio*), olive (*P. h. anubis*), and hamadryas baboons, as well as in individuals with varying degrees of olive and hamadryas ancestry. Results show that geladas possess significantly higher BaEV copy numbers when compared to *Papio* baboons, and that African hamadryas show a trend toward higher BaEV copy numbers within *Papio* baboons. These findings illustrate the impact of social organization and reproductive systems on genome evolution, suggesting that hierarchically organized gelada and hamadryas societies (i) promote relatively greater degrees of inbreeding than in their close papionin relatives and (ii) provide conditions under which maintaining an increased BaEV element number confers a selective advantage.

## 1. INTRODUCTION

Geladas (*Theropithecus gelada*) and hamadryas (*Papio hamadryas hamadryas*) baboons exhibit a complex arrangement of hierarchically organized social groupings that is not observed in other primates (Dunbar, 1984; Kummer, 1968). Although their social organization differs in many important respects, the two taxa share a multilevel society based upon one-male reproductive units (OMUs) in which relatedness among group members may be high compared to other cercopithecines (Dunbar, 1979, 1983a; Woolley-Barker, 1998, 1999; Swedell and Woolley-Barker, 2001; Swedell, 2006). A relatively high degree of relatedness among group members increases the potential for inbreeding (Hartl and Clark, 1997), the genetic effects of which have long been studied from both theoretical (Wright, 1932, 1938; Charlesworth and Wright, 2001) and empirical (Wright, 1977; Morgan, 2001; Wright et al., 2002) perspectives. Here, we address these effects in papionins, testing the relationships among social organization, degree of inbreeding, and copy number of a mobile genetic element known as baboon endogenous virus (BaEV). Analyses of mobile elements such as BaEV provide novel opportunities to examine the impact of social organization and reproductive systems on population genetic structure and genome evolution because these elements have the potential to evolve rapidly. Analysis of this diversity also presents opportunities to investigate fundamental dimensions of endogenous virus evolution.

### 1.1 Social Organization of Geladas and Hamadryas Baboons

Geladas live in one-male units containing one breeding male seeking to monopolize access to an average of three to five females and their offspring (Dunbar and Dunbar, 1975), for a total membership that ranges from 10 to 15 individuals per unit (Dunbar, 1980). A cluster of OMUs that shares a common home range, together with one or more all-male groups (see below), constitutes a *band* (Dunbar and Dunbar, 1975), which often travels together as a coordinated unit of 30–250 animals (Dunbar, 1980, 1984). Females rarely migrate, with the one reported case occurring between OMUs within the same band (Dunbar, 1980). OMUs are marked by an almost exclusive interaction among females with their immediate matrilineal kin (Dunbar, 1979, 1983a), with grooming dyads in these units consisting mainly of mothers and their mature daughters (Dunbar, 1984). Strong female bonds facilitate OMU stability over extended periods of time, with group composition normally maintained through events often disruptive in other species (e.g., births, deaths, and takeovers) (Dunbar, 1979).

Resident OMU males are socially peripheral to gelada OMUs and rarely herd the females (Dunbar, 1983b). In addition, subadult and young adult males who are not OMU members form all-male groups that constitute discrete and long-lasting social groups (Dunbar, 1980). These all-male groups tend to associate with specific bands, which are assumed to be the natal band of the majority of its members (Dunbar, 1980). In addition, most males (~70 percent) appear to acquire reproductive units within their natal bands (Dunbar, 1980). Units of a band may associate more closely with other units in the same band, and such "teams" (akin to the hamadryas "clan"; see below) represent a social grouping intermediate between the reproductive unit (the OMU) and the band (Kawai et al., 1983; Dunbar, 1984). Bands may also travel and forage together in larger (600–700 animals) *herds* (Dunbar, 1980), which consist of one or more units of one band in temporary association with one or more units of a neighboring band(s) (Dunbar, 1984). Most migration out of bands occurs through band fission when part of a band moves into a new home range (Dunbar, 1984). Population density is therefore regulated largely by this aperiodic emigration of entire (sub) sections of its members rather than by a "steady trickle" of individuals (Dunbar, 1980).

Like geladas, hamadryas baboons exhibit a multilevel social organization founded on the one-male unit. Hamadryas OMUs consist of a leader male, one or more adult females, their juvenile offspring (until approximately 2–3 years) and, frequently, one (or more) "follower" male(s) who may be related to the leader and who often assumes the leader position once his (presumed) relative senesces (Kummer, 1968; Abegglen, 1984). OMU cohesion is maintained principally by aggressive herding behavior by a leader male toward his females (Kummer, 1968). OMUs group together into progressively more inclusive units: the *clan*, a foraging unit consisting of two to three OMUs (Abegglen, 1984); the *band*, a cooperative defense unit consisting of either several OMUs or several clans; and the *troop*, an impermanent collection of two to four bands sharing a sleeping site (Kummer, 1968; Abegglen, 1984; Swedell, 2006). Although reports of band sizes do vary, four out of the five reported hamadryas bands at the Filoha site of Awash National Park, Ethiopia between 1996 and 1998 had group sizes of at least 100 (and 2 of the 5 had a minimum of 200) (Swedell, 2002b). Smaller band sizes have been reported for Saudi Arabian and Yemeni populations, with numbers reaching as low as nine for the former (Biquand et al., 1992; Boug et al., 1994) and 22 for the latter (Al-Safadi, 1994); and bands with sizes intermediate between those found at Filoha and on the Arabian peninsula have been reported for other sites in Ethiopia, including Erer Gota and Awash Station (Kummer, 1968; Nagel, 1971; Abegglen, 1984; Sigg and Stolba, 1981).

Members of a hamadryas clan are thought to be close genetic relatives (Abegglen, 1984). In particular, males are believed to remain in their natal clan for life, while females exhibit limited movement between clans or bands. This type of social organization stands in contrast to that of olive (*P. h. anubis*) baboons, who live in large, undifferentiated multimale, multifemale groups characterized by male dispersal and female philopatry (Pusey and Packer, 1987). Genetic data confirm behavioral observations on hamadryas: A nine-locus microsatellite study of Awash baboons shows high levels of mean pairwise interindividual relatedness ($r$) among hamadryas males believed to be from the same social unit (i.e., band) ($r = 0.23$ and $0.31$; compared with mean levels in olives, $r = 0.1$) (Woolley-Barker, 1999). Additional behavioral data on female–female social interactions within OMUs suggest that relatedness among hamadryas females may be higher than previously assumed (Swedell, 2002a, 2006). Finally, hamadryas baboons show significant excess homozygosity and, correspondingly, a level of inbreeding that is more than twice that observed in olive-like groups found in Awash (Woolley-Barker, 1998, 1999).

## 1.2 Inbreeding and the Structure of Population-Level Genetic Variation

The structure of genetic variation is determined by several factors that are related to the type of reproductive system maintained by populations (Wright, 1978; Hartl and Clark, 1997). Both the frequency of heterozygous versus homozygous genotypes (Wright, 1978; Nei 1987; Hartl and Clark, 1997) and the extent of genetic recombination (Hartl and Clark, 1997; Nordborg, 2000) influence both genetic diversity within populations and divergences between populations. Highly inbreeding populations will have increased frequencies of homozygous genotypes; this increased homozygosity, in turn, means that only highly similar genomes recombine (Charlesworth and Wright, 2001). The effective recombination rate, moreover, depends on both the rate of crossing-over within genomes and the rate of outcrossing among individuals. Although even highly inbred organisms do occasionally outcross, the effective recombination rate among polymorphic sites is, in general, relatively less frequent in inbreeding populations (Charlesworth and Wright, 2001). If such populations are isolated from each other, the resulting genetic divergence between them becomes even more pronounced. The potential for a relatively rapid rate of evolution in subdivided, inbreeding populations has been well reviewed (Wright, 1932, 1965, 1977). For example, Sewall Wright's shifting balance theory outlines the process by which populations divided into a number of smaller, partially isolated subpopulations can effectively produce rapid evolutionary change by fixing alternate alleles in each subpopulation through drift followed by selection (Wright, 1932, 1977). This process suggests that different social organizations may have different effects on population genetic diversity. Substructuring of populations through social and reproductive behavior potentially leads to rapid evolution, as a result of both reduced effective deme size and relatively greater selection coefficients operating within demes.

Population genetic theory makes several predictions about the consequences of social organization that can be assessed through analyses of genetic diversity in species that maintain substructured social organizations, such as those exhibited by geladas and hamadryas baboons. Generally, this kind of organization should result in an accumulation of inbreeding due to the increased likelihood of mating pairs having remote relatives in common, even if mates are chosen at random (Hartl and Clark, 1997). Comparisons of genetic data for hamadryas with those for other *Papio* taxa (e.g., olive baboons without similarly substructured social organizations) show higher levels of inbreeding in hamadryas (Woolley-Barker, 1998, 1999). Based on these results, we expect even higher probabilities of common ancestry in a species such as the gelada, characterized by male band philopatry (Dunbar, 1984) and OMUs composed largely of female matrilineal kin (Dunbar, 1979, 1983a, 1984). Hamadryas, in turn, should have a higher probability of common ancestry between mates than olive baboons. Olive baboons exhibit female philopatry (Pusey and Packer, 1987) but lack the OMU structure that produces, through nonrandom mating, individuals sharing many alleles. Given the increased relatedness predicted by the unique reproductive systems, migration patterns, and multilevel social organizations maintained by geladas and hamadryas baboons, and the accompanying potential for rapid evolutionary change outlined above, these taxa should exhibit evidence of more rapid rates of evolution than their close, relatively outbreeding, relatively non-subdivided papionin relatives.

We predict that the hierarchical, semiclosed social organization maintained by geladas and hamadryas baboons will result in relatively greater BaEV copy numbers in these two taxa when compared to other *Papio* baboons. In particular, we compare Guinea (*P. h. papio*), olive, and hybrid baboons with varying degrees of olive and hamadryas ancestry. In addition to providing insight about baboon genetic diversity, this analysis presents novel opportunities to evaluate basic features of endogenous virus evolution in primates.

## 2. MATERIALS AND METHODS

### 2.1 Materials

Genomic DNA samples were derived from numerous sources and geographic locations (Table 1). The samples were obtained from olive baboons (24 from Masai Mara, Kenya, and 1 olive individual from the Awash National Park,

**Table 1.** Group- and taxon-specific data distribution of relative (and estimated absolute) BaEV copies

| Statistic | Olive | African hamadryas | Arabian hamadryas | Awash group G[1] | Awash group H[1] | Guinea | All *Papio* | Gelada |
|---|---|---|---|---|---|---|---|---|
| *N* | 25 | 21 | 4 | 13 | 66 | 10 | 139 | 9 |
| Min. | 0.52 (14.56) | 0.63 (17.64) | 0.85 (23.80) | 0.10 (2.8) | 0.14 (3.92) | 0.12 (3.36) | 0.10 (2.8) | 1.79 (50.12) |
| 25th percentile | 0.90 (25.2) | 1.03 (28.84) | N/A | 0.59 (16.52) | 1.04 (29.12) | 0.65 (18.2) | 0.98 (27.44) | 2.00 (56.00) |
| Median | 1.42 (39.76) | 1.27 (35.56) | 0.97 (27.16) | 1.13 (31.64) | 1.24 (34.72) | 1.24 (34.72) | 1.23 (34.44) | 3.18 (89.04) |
| 75th percentile | 1.83 (51.24) | 1.69 (47.32) | N/A | 1.24 (34.72) | 1.53 (42.84) | 1.75 (49.0) | 1.55 (43.40) | 4.98 (139.44) |
| Max. | 2.87 (80.36) | 3.23 (90.44) | 1.41 (39.48) | 1.40 (39.20) | 2.88 (80.64) | 2.13 (59.64) | 3.23 (90.44) | 5.68 (159.04) |
| Mean | 1.47 (41.16) | 1.46 (40.88) | 1.05 (29.40) | 0.93 (26.04) | 1.27 (35.56) | 1.21 (33.88) | 1.29 (36.12) | 3.44 (96.32) |
| Std. dev. | 0.60 (16.80) | 0.69 (19.32) | 0.25 (7.0) | 0.41 (11.48) | 0.45 (12.60) | 0.62 (17.36) | 0.54 (15.12) | 1.54 (43.12) |
| Std. error | 0.12 (3.36) | 0.15 (4.20) | 0.13 (3.64) | 0.11 (3.08) | 0.06 (1.68) | 0.20 (5.60) | 0.05 (1.40) | 0.51 (14.28) |
| Lower 95 percent CI | 1.22 (34.16) | 1.15 (32.2) | 0.65 (18.20) | 0.68 (19.04) | 1.16 (32.48) | 0.76 (21.28) | 1.20 (33.60) | 2.26 (63.28) |
| Upper 95 percent CI | 1.72 (48.16) | 1.77 (49.56) | 1.45 (40.60) | 1.18 (33.04) | 1.38 (38.64) | 1.65 (46.20) | 1.38 (38.64) | 4.62 (129.36) |

[1] Groups G and H are hybrid groups.

Ethiopia), African hamadryas baboons (from the Awash National Park) and Arabian hamadryas baboons (Yemen and Saudi Arabia), and captive Guinea baboons (10 from the Los Angeles Zoo). Nine gelada samples were also included, originally obtained from captive individuals of unknown provenience, but probably from the Ethiopian highlands north of Addis Ababa. Finally, we analyzed samples from *P. h. hamadryas* × *P. h. anubis* hybrids in the Awash hybrid zone. These samples differed in degree of olive and hamadryas ancestry, presenting opportunities to assess diversity at the population level. We refer to hybrid and hamadryas individuals from the Awash National Park collectively as Awash baboons. Ancestry in these baboons was measured by a phenotypic hybrid index (PHI) (Nagel, 1971; Bergman and Beehner, 2003; Beehner and Bergman, this volume). We use linear regression to explore the relation, if any, between BaEV copy number and PHI, noting that individuals with greater degrees of hamadryas ancestry should show higher BaEV copy number.

## 2.2 Methods

We studied diversity in BaEV, an ideal genetic marker with which to test for evidence of rapid evolutionary rates. BaEV is an endogenous retrovirus, a retrovirus (Figure 1) that has become incorporated into the germ line of its host and is thus vertically transmitted between generations. Germ-line colonization events such as that represented by BaEV are not uncommon: Endogenous viruses and retroviral elements have been found in virtually all vertebrates investigated (Herniou et al., 1998) and are estimated to comprise approximately 7–8 percent of the human genome (Consortium, 2001; Hughes and Coffin, 2001). Furthermore, retroviruses in general (including their endogenous form) belong

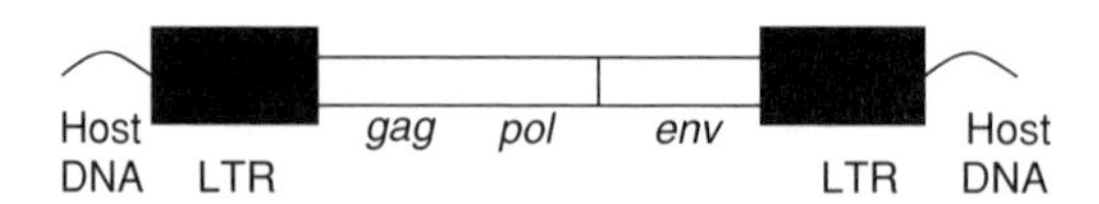


**Figure 1.** Genetic organization of a generalized simple provirus—a retrovirus that has successfully integrated itself into its host's DNA. Major coding regions include *gag*, *pol*, and *env* producing inner core viral particles, a number of viral enzymes, and a viral envelope protein, respectively. Long terminal repeats (LTRs) abut cellular DNA and are synthesized during reverse transcription; they do not exist in viral RNA.

to a larger class of genetic elements known as mobile or transposable elements (TEs), small (100–10 kb) genetic units that have the ability to move about, or transpose, within genomes (Li, 1997). TEs comprise about 45 percent of the human genome (Consortium, 2001) and 38 percent of the mouse (*Mus musculus*) genome (Consortium, 2002). Of the three main classes of TEs, retroviruses belong to class I or "retroelement" category. These are genetic entities that move about by retroposition (transcription of DNA into RNA, followed by reverse transcription into DNA). Class I TEs contain a gene coding for reverse transcriptase, which gives them an intrinsic ability to transpose themselves. This can result in copy number amplification when TEs transcribed from existing elements are reinserted into the host genome.

BaEV is one of the most intensively studied nonhuman primate endogenous retroviruses. Approximately 5–15 copies of the virus are present in the haploid baboon genome (Benveniste and Todaro, 1974). Unlike most endogenous retroviruses, which are believed to be "silent" genomic elements (Boeke and Stoye, 1997), BaEV is transcribed, so that its activity can be detected in baboon placental, testis, uterine, lung, and spleen tissues and, to a lesser extent, in liver tissue as well (Benveniste et al., 1974a,b). Expression is strongest in the placenta (Benveniste et al., 1974a), and new copies are hypothesized to be acquired very early in development (Uddin, 2003), as is the case with other endogenous retroviruses (Jaenisch, 1976; Soriano et al., 1987). Fragments of the virus have been successfully amplified from DNA obtained from all papionin genera except macaques (*Macaca* spp.). Other cercopithecins, except for *Chlorocebus aethiops*, seem to lack virus fragments (van der Kuyl et al., 1995).

Despite a growing literature on endogenous viral detection and dynamics in primates, little is known about their variation in natural primate populations. While the identification of common insertion sites in one or two representative individuals per species can establish a common, vertical mode of inheritance among all members of that species, copy number estimations of particular endogenous retroviruses based on similarly limited sample sizes may miss intrapopulation variation. Such variation has been documented, for example, in wild house mice (*M. musculus* ssp.), one of the few mammalian species in which population-level endogenous retroviral variation has been reported (Kozak and O'Neill, 1987; Inaguma et al., 1991). Consequently, our results have important implications for understanding the evolution of such viruses in primates and other taxa.

Quantitative real-time PCR (QPCR) methods were used to assess BaEV copy numbers among individuals of the same baboon population, different baboon populations, and closely related papionin lineages. QPCR methods used in this study followed protocols designed for the ABI Prism 7700 Sequence Detector. Previously collected BaEV sequence data (Uddin, 2003) were used to determine an invariant portion (p10) of the BaEV genome around which to design primers and probes, and BaEV copy numbers were tested in a total of 148 individuals by calibrating the p10 results to that of a single copy gene, *CCR5*. Relative BaEV copy numbers among individuals were assessed using a standard reference individual for all assays, and absolute copy numbers were estimated using calculations based upon the average difference between individuals' p10 and *CCR5* results. In order to elucidate the parameters affecting BaEV copy number among individuals and across populations, standard statistical methods were used to test QPCR data collected by this study against demographic, phenotypic, and genetic data previously collected from Awash baboons (Bergman, 2000; Jolly and Phillips-Conroy, unpublished data). A detailed experimental strategy (and accompanying results) is described elsewhere (Uddin, 2003).

## 3. RESULTS

Significant differences in BaEV copy number were detected among the samples as grouped in Table 1 (one-way ANOVA, $p < 0.0001$, with African and Arabian hamadryas samples combined to increase sample size). Tukey–Kramer post hoc tests revealed that this result was entirely due to comparisons between geladas and each of the five samples of *Papio* baboons. This difference is observed despite considerable copy number heterogeneity among individuals within groups. We also note that there seem to be no significant differences in BaEV copy number between sexes, among age classes, or among age/sex cohorts within Awash baboons (Uddin, 2003).

Predictions for BaEV copy number in *Papio* baboons are generally confirmed through a trend toward higher BaEV copy numbers in African hamadryas. Minimum, maximum, and upper 95 percent confidence intervals were highest in this taxon (excluding Arabian hamadryas due to small sample size; Table 1). At the population level, Awash baboon males carrying the hamadryas-associated Y chromosome marker (Bergman, 2000) ($N = 14$) showed a slightly but significantly higher average relative BaEV copy number than did individuals carrying the olive-associated Y chromosome marker (1.39 versus 1.09, representing estimated average absolute copy numbers of 38.92 versus 30.52, respectively; $p = 0.049$). Finally, linear regression analysis of 24 Awash adult males, including olive, hamadryas, and hybrid baboons, detected a significant relationship between BaEV copy numbers and phenotypic hybrid index (PHI) scores (Nagel, 1971; Bergman and Beehner, 2003; Beehner and Bergman, this volume). Individuals of more hamadryas-like phenotype had higher relative BaEV copy numbers than their more olive-like counterparts ($p = 0.03$, $R^2 = 0.2$) (Uddin, 2003). A similar analysis of 31 females failed to find a significant relationship between phenotype and BaEV copy number, but this sample did not include any individuals from unmixed hamadryas populations (data not shown).

## 4. DISCUSSION

### 4.1 BaEV Copy Number Variation Among Baboons

These analyses confirm several basic predictions of this study and provide insights into BaEV evolution at a wider level. First, geladas exhibit significantly higher copy numbers of BaEV when compared to other samples included in this study. Second, African hamadryas show a trend toward higher BaEV copy numbers within *Papio* baboons. Third, African hamadryas males show significantly higher copy numbers when compared to their olive counterparts from the Awash National Park.

Overall, the finding of relatively higher BaEV copy numbers in geladas and the trend toward higher copy numbers in African hamadryas is consistent with reports documenting the effects of many generations of inbreeding on the acquisition of novel proviral copies of endogenous retroviruses in laboratory mice (*M. musculus*) (Rowe and Kozak, 1980; Buckler et al., 1982; Herr and Gilbert, 1982; Steffen et al., 1982) and domestic pigs (*Sus scrofa*) (Mang et al., 2001). These findings are also consistent with existing population genetic theory. As noted previously, genetic analyses of hamadryas and olive populations in the Awash National Park show the contrasting effects of a subdivided, semi-closed social organization with that of a relatively undifferentiated multimale, multifemale social organization (Woolley-Barker, 1999). Although comparable genetic data from geladas are currently unavailable, their social organization should promote even higher levels of mean pairwise relatedness and inbreeding than observed in hamadryas because (1) reproductive units consist of related females, which remain in their natal unit for life (Dunbar, 1979, 1983a) and offspring thus share more alleles than in hamadryas OMUs, which consist of (presumably) unrelated females and (2) the majority of males acquire reproductive units in their natal band (Dunbar, 1980), thereby increasing the likelihood of a higher genetic relatedness between themselves and their mating partners. While behavioral data suggest that substructuring into one-male units does occur among Guinea baboons (Dunbar and Nathan, 1972; Galat-Luong et al., this volume), and their testicular growth trajectories are somewhat similar to hamadryas (Jolly and Phillips-Conroy, this volume), behavioral reports also emphasize the flexibility of social structure in this taxon. The contingent nature of these groupings, combined with a lack of herding behavior of males toward females (Dunbar and Nathan, 1972; Galat-Luong et al., this volume), contrasts with the OMU-based social organization and behavioral repertoire of hamadryas baboons, which appear to be fixed even under changing ecological and social conditions (Kummer and Kurt, 1965; Kummer, 1968; Kummer et al., 1981; Nystrom, 1992; Bergman, 2000; Beehner and Bergman, this volume).

Interestingly, mean BaEV copy numbers appear to be comparable in hamadryas and olives from Kenya (Masai Mara). Recent work by Wildman et al. (2004) documents the close relationship between hamadryas and olive baboons from Awash, finding that the latter are more closely related to the former than to Kenyan olive baboons. These authors suggest that this may result from sexually differentiated introgression of olive baboon nuclear genes into a marginal hamadryas population. The contrasting, relatively higher BaEV copy numbers detected in Awash hamadryas versus olives may thus be the result not only of inbreeding in hamadryas but also the historical, extreme "outbreeding" (i.e., hybridization) of hamadryas females with olive males argued by Wildman et al. (2004) to have produced the present-day Awash olive baboons and reduce copy numbers among these individuals.

### 4.2 Models of BaEV Acquisition and Diffusion

That there are more copies of BaEV among some papionin taxa may, at first, appear difficult to explain from a fitness point of view. TE insertions, including those derived from endogenous viruses like BaEV, are generally thought to be harmful (Charlesworth et al., 1994). Nevertheless, copy numbers are known to vary among populations (Kozak and O'Neill, 1987; Inaguma et al.,

1991) and species (Morgan, 2001). Two models have been proposed to account for how TEs might spread throughout populations. The first is a deleterious model, in which element insertions disrupt gene function and therefore decrease organismal fitness. Population-level TE spread under this model has been investigated assuming fitness to be (i) a decreasing function of total element abundance and (ii) a decreasing function of only homozygous (i.e., recessive) elements (Wright and Schoen, 1999). The second model involves ectopic exchange (Figure 2), in which selection is assumed to act against only heterozygous TEs, as they promote unequal crossing-over between nonhomologous insertion sites that can result in harmful chromosomal rearrangements (Morgan, 2001). Empirical data are more consistent with a deleterious model than with an ectopic exchange model. In plants, from which most of the lineage-specific TE data have been collected, selfing (i.e., extremely inbreeding) species have a relatively lower frequency of TEs (although exceptions do exist, such as the 1,000,000 *Ty1*-copia-like elements found in the inbreeding fava bean plant, *Vicia faba*; Morgan, 2001).

We would argue that our results, on the other hand, are best explained by a model of ectopic exchange, or unequal crossing-over between elements at nonhomologous insertion sites (Figure 2). In light of the shifting balance

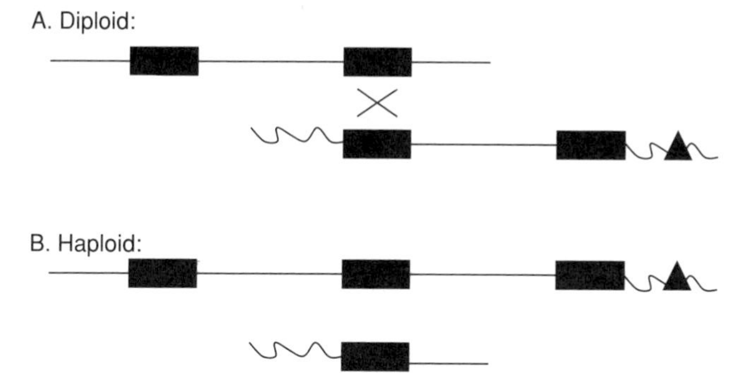


**Figure 2.** Hypothetical ectopic exchange event between proviral insertions at nonhomologous chromosomal locations during meiosis. (A) Unequal crossing-over between LTRs (black rectangles) of nonhomologous proviruses in a diploid cell. Triangles indicate the location of a hypothetical gene. (B) Haploid cells showing the potentially harmful chromosomal rearrangement produced by the ectopic exchange event. The top gamete carries a tandem duplication of the virus, with the gene appearing in a new location, while the bottom gamete carries a solo LTR (with no intervening viral sequence) and has lost the gene.

theory, the finding of higher BaEV copy numbers in geladas and hamadryas baboons is likely to be a consequence of their hierarchically divided social organization and mating system, which would have promoted the fixation of more proviruses in ancestral hamadryas and gelada deme(s). While an increased element number would not immediately appear to have a selective advantage, ectopic exchange may be the selective pressure most likely to produce an increased proviral copy number. The thousands of solo long terminal repeats (LTRs) residing in the mouse and human genomes indicate that proviral loss through recombination may be a mechanism for relieving proviral loads over long periods of time (Boeke and Stoye, 1997). However, no mechanism of precise proviral excision is currently known (Stoye, 2001). Accordingly, genetic exchange between nonhomologous, heterozygous repeat elements can create potentially harmful chromosomal rearrangements, as has been observed in humans (Reiter et al., 1996). Homozygosity, however, may decrease the likelihood of ectopic exchange, because elements will have homologues to pair with during recombination (Morgan, 2001). Like other endogenous retroviruses, novel BaEV copies are likely acquired early in development. As with other polymorphic loci, these novel insertions would be more likely to encounter an identical homologue in a population that tended toward inbreeding. Thus, under the scenario envisioned here, the selective advantage of homozygous versus heterozygous insertions would have produced greater BaEV copy numbers in one or more ancestral hamadryas and gelada demes. This superior equilibrium fitness peak would then have been exported to neighboring demes, resulting in a gradual shift toward the new adaptive pattern in both taxa.

### 4.3 Transposable Elements and Genome Evolution

These results do not easily account for the maintenance of BaEV in the other, relatively outbreeding populations included in this study (i.e., Masai Mara olives, Awash Group G, Guinea baboons, and Awash Group H—the latter relatively inbreeding today but derived from a recent fusion event detectable in samples obtained in 1973 (Newman, 1997; Woolley-Barker, 1999). Although both the deleterious and ectopic exchange models make some allowance for the presence of TEs in populations with relatively low levels of inbreeding, the latter more so than the former, one shortcoming of both models is that they operate from the assumption that TEs are self-propagating elements capable of amplifying in host genomes and producing a range of effects. At the organismal level, these effects are, at best, neutral and, at worst, deleterious. This overly simplistic view has recently been questioned in favor of a model of TE–host relationship that encompasses a continuum from extreme parasitism to mutualism (Kidwell and Lisch, 2000, 2001) in light of the growing number of examples in which TEs evolve neutrally early on at the host level but in subsequent periods of evolution become coopted for new host functions (Britten, 1996; Kidwell and Lisch, 1997, 2001). Genomes appear to be capable of being partitioned into a variety of "ecological niches," which can be exploited in different ways by different types of transposons, most likely as a result of the long-term interaction between host and parasite (Kidwell and Lisch, 2001). For instance, TE involvement is indicated in the human immune system. The human major histocompatibility complex (MHC) region is surrounded by densely clustered areas of retroelements, including endogenous retroviruses, that facilitate genomic plasticity; in addition, the MHC's central portion contains two duplicated segments that are the result of a gene duplication event that occurred prior to the divergence of platyrrhines and catarrhines (Kulski et al., 1997). Examples such as this highlight the profound impact that host recruitment of TEs can have on organismal function even long after their arrival in a particular genome.

An additional consideration is that host genomes themselves differ widely with respect to genome size and TE content. Between the mouse and human genomes, for example, important differences in TE dynamics have been identified. These include a decrease in transposition during the last 40 million years of anthropoid evolution (measured in the human lineage). This contrasts with the fairly constant rate observed in the mouse lineage. Moreover, there are very different distributions of different types of TEs between mouse and human genomes. For instance, the mouse genome possesses many more types of short interspersed nuclear elements (SINEs), while the human genome possesses many more types of DNA transposons (Consortium, 2002). Given that most of the modeling of TE dynamics to date has been undertaken with nonmammalian data, and that even within mammals there appears to be considerable variation in the abundance of particular TE categories, caution should be taken when applying the assumptions of these models directly to primates. Indeed, for animals with relatively large genomes such as primates, a key difference in transposon dynamics may be that some of the restraints on copy number have been lifted for many, but likely not all, TE families

(Kidwell, 2002). Thus, patterns of TE distribution and abundance in organisms such as primates may prove to be much different from those investigated to date, as exemplified by the finding in this study of higher BaEV copy numbers in the relatively inbreeding geladas and hamadryas baboons.

Finally, the possibility that TEs such as BaEV may in fact be experiencing a mix of selective regimes in host genomes, with some heterozygous proviral insertions selected against due to their promotion of deleterious ectopic exchange, and others undergoing a selective sweep toward fixation in particular populations (Hartl and Clark, 1997), which would be particularly pronounced if they were inbreeding (Charlesworth and Wright, 2001), should also be considered. This is especially true for endogenous retroviruses, of which thousands of copies persist in the human genome (Consortium, 2001) and which have yet to be examined under these models.

## 4.4 Summary and Conclusions

In summary, the variation in BaEV copy numbers among papionin taxa appears to be the result of a complex interplay across many levels of biological organization, encompassing the virus itself, the individual in whom it is found, the population of the BaEV host, and even the lineage to which the population belongs. Nevertheless, results presented here demonstrate that social organization also plays a major role in mediating BaEV copy numbers. The finding of significantly higher BaEV copy numbers in geladas and a trend toward higher copy numbers in hamadryas suggest that their hierarchical, semiclosed social systems promote relatively greater degrees of inbreeding than in their close papionin relatives, providing conditions under which maintaining an increased (homozygous) element number would confer a selective advantage. Although it is now well established that TEs have made profound contributions to eukaryotic host genome evolution, investigation into the relationship between transposon dynamics and reproductive systems is still in its infancy. Indeed, of those few studies conducted, one limitation that has been identified is the lack of comparison between closely related species differing in mating system (Morgan, 2001). Results from this study help to close this gap by illustrating, both within a taxon (*Papio*) and between sister taxa (*Papio* and *Theropithecus*), that different types of social organization are correlated with significantly different BaEV copy numbers; that increased average proviral copy numbers appear to be a function of the progressively greater degrees of inbreeding found in nonhamadryas baboons, hamadryas baboons, and geladas; and that these differences are more easily accounted for under a model of ectopic exchange rather than the currently favored model of deleterious selection.

## ACKNOWLEDGMENTS

We would like to thank the many individuals and agencies that contributed samples to this study: Irven DeVore (Masai Mara samples), Rob Hammond and Derek Wildman (Arabian hamadryas samples), and the Los Angeles Zoo (Guinea and gelada samples). We also gratefully acknowledge Thore Bergman, Jacinta Beehner, and Tamsin Woolley-Barker for sharing data that informed this work. Special thanks are due to Todd Disotell for providing resources to do background sequencing necessary to the project, Mechthild Prinz of the Office of the Chief Medical Examiner of New York City for allowing access to their real-time PCR machine, Derek Wildman and Larissa Swedell for helpful comments and discussion, and the many ANPBRP participants who helped to trap animals used in this study. This work was supported by funding from the National Science Foundation (grants #BCS-0087329 and #BCS-9615150), the Wenner-Gren Foundation, the Sigma-Xi Foundation, the Harry Frank Guggenheim Foundation, NYCEP, New York University, and Washington University.

## REFERENCES

Abegglen, J.J., 1984, *On Socialization in Hamadryas Baboons*, Associated University Presses, London.

Al-Safadi, M. M., 1994, The hamadryas baboon, *Papio hamadryas* (Linnaeus, 1758) in Yemen (Mammalia: Primates: Cercopithecidae), *Zool. Middle East* **10**:5–16.

Benveniste, R. E. and Todaro, G. J., 1974, Multiple divergent copies of endogenous C-type virogenes in mammalian cells, *Nature* **252**:170–173.

Benveniste, R. E., Heinemann, R., Wilson, G. L., Callahan, R., and Todaro, G. J., 1974a, Detection of baboon type C viral sequences in various primate tissues by molecular hybridization, *J. Virol.* **14**:56–67.

Benveniste, R. E., Lieber, M. M., Livingston, D. M., Sherr, C. J., and Todaro, G. J., 1974b, Infectious type C virus isolated from a baboon placenta, *Nature* **248**:17–20.

Bergman, T. J., 2000, Mating behavior and reproductive success of hybrid male baboons (*Papio hamadryas hamadryas* × *Papio hamadryas anubis*). Ph.D. Dissertation, Washington University, St. Louis, MO.

Bergman, T. J. and Beehner, J. C., 2003, Hybrid zones and sexual selection: Insights from the Awash baboon hybrid zone (*Papio hamadryas anubis* × *P. h. hamadryas*), in: *Sexual Selection and Primates: New Insights and Directions*, C. Jones, ed., American Society of Primatologists, Norman, OK, pp. 500–537.

Biquand, S., Biquand-Guyot, V., Boug, A., and Gautier, J.-P., 1992, Group composition in wild and commensal hamadryas baboons: A comparative study in Saudi Arabia, *Int. J. Primatol.* **13**:223–243.

Boeke, J. D. and Stoye, J. P., 1997, Retrotransposons, endogenous retroviruses, and the evolution of retroelements, in: *Retroviruses*, J. M. Coffin, S. H. Hughes, and H. E. Varmus, eds., Cold Spring Harbor Laboratory Press, Cold Spring Harbor, pp. 343–435.

Boug, A., Biquand, S., Biquand-Guyot, V., and Kamal, K., 1994, Home range and daily march of commensal *Papio hamadryas* in the Alhada Mountain of Saudi Arabia, *Congr. Intl. Primatol. Soc.* **15**:148.

Britten, R. J., 1996, Cases of ancient mobile element DNA insertions that now affect gene regulation, *Mol. Phyl. Evol.* **5**:13–17.

Buckler, C. E., Staal, S. P., Rowe, W. P., and Martin, M. A., 1982, Variation in the number of copies and in the genomic organization of ecotropic murine leukemia virus proviral sequences in sublines of AKR mice, *J. Virol.* **43**:629–640.

Charlesworth, D. and Wright, S. I., 2001, Breeding systems and genome evolution, *Curr. Opin. Genet. Dev.* **11**:685–690.

Charlesworth, B., Sniegowski, P., and Stephan, W., 1994, The evolutionary dynamics of repetitive DNA in eukaryotes, *Nature* **371**:215–220.

Consortium, I. H. G. S., 2001, Initial sequencing and analysis of the human genome, *Nature* **409**:860–921.

Consortium, M. G. S., 2002, Initial sequencing and comparative analysis of the mouse genome, *Nature* **420**:520–562.

Dunbar, R. M., 1979, Structure of gelada baboon reproductive units. I. Stability of social relationships, *Behaviour* **69**:72–87.

Dunbar, R. M., 1980, Demographic and life history variables of a population of gelada baboons (*Theropithecus gelada*), *J. Anim. Ecol.* **49**:485–506.

Dunbar, R. M., 1983a, Structure of gelada baboon reproductive units. II. Social relationships between reproductive females, *Anim. Behav.* **31**:556–564.

Dunbar, R. M., 1983b, Structure of gelada baboon reproductive units. III. The male's relationship with his females, *Anim. Behav.* **31**:565–575.

Dunbar, R. I. M., 1984, *Reproductive Decisions: An Economic Analysis of Gelada Baboon Social Strategies*, Princeton University Press, Princeton.

Dunbar, R. M. and Dunbar, E. P., 1975, *Social Dynamics of Gelada Baboons*, Karger, Basel.

Dunbar, R. I. M., and Nathan, M. F., 1972, Social organization in the Guinea baboon, *Papio papio*, *Folia Primatol.* **17**:321–334.

Hartl, D. L. and Clark, A. G., 1997, *Principle of Population Genetics.* Sinauer Associates, Inc., Sunderland, MA.

Herniou, E., Martin, J., Miller, K., Cook, J., Wilkinson, M., and Tristem, M., 1998, Retroviral diversity and distribution in vertebrates, *J. Virol.* **72**:5955–5966.

Herr, W. and Gilbert, W., 1982, Germ-line MuLV reintegrations in AKR/J mice, *Nature* **296**:865–868.

Hughes, J. F. and Coffin, J. M., 2001, Evidence for genomic rearrangements mediated by human endogenous retroviruses during primate evolution, *Nat. Genet.* **29**:487–489.

Inaguma, Y., Miyashita, N., Moriwaki, K., Huai, W. C., Mei-Lei, J., Xinqiao, H., and Ikeda, H., 1991, Acquisition of two endogenous ecotropic murine leukemia viruses in distinct Asian wild mouse populations, *J. Virol.* **65**:1796–1802.

Jaenisch, R., 1976, Germ line integration and Mendelian transmission of the exogenous Moloney leukemia virus, *Proc. Natl. Acad. Sci. USA* **73**:1260–1264.

Kawai, M., Dunbar, R. I. M., Ohsawa, H., and Mori, U., 1983, Social organisation of gelada baboons: Social units and definitions, *Primates* **24**:1–13.

Kidwell, M. G., 2002, Transposable elements and the evolution of genome size in eukaryotes, *Genetica* **115**:49–63.

Kidwell, M. G., and Lisch, D., 1997, Transposable elements as sources of variation in animals and plants, *Proc. Natl. Acad. Sci. USA* **94**:7704–7711.

Kidwell, M. G. and Lisch, D. R., 2000, Transposable elements and host genome evolution, *Trends Ecol. Evol.* **15**:95–99.

Kidwell, M. and Lisch, D., 2001, Perspective: Transposable elements, parasitic DNA, and genome evolution, *Evolution* **55**:1–24.

Kozak, C. A. and O'Neill, R. R., 1987, Diverse wild mouse origins of xenotropic, mink cell focus-forming, and two types of ecotropic proviral genes, *J. Virol.* **61**:3082–3088.

Kulski, J. K., Gaudieri, S., Bellgard, M., Balmer, L., Giles, K., Inoko, H., and Dawkins, R. L., 1997, The evolution of MHC diversity by segmental duplication and transposition of retroelements, *J. Mol. Evol.* **45**:599–609.

Kummer, H., 1968, *Social Organization of Hamadryas Baboons: A Field Study*, University of Chicago Press, Chicago.

Kummer, H. and Kurt, F., 1965, A comparison of social behavior in captive and wild hamadryas baboons, in: *The Baboon in Medical Research*, H. Vagtborg, ed., University of Texas Press, Austin, pp. 65–80.

Kummer, H., Banaja, A. A., Abo-Khatwa, A. N., and Ghandour, A. M., 1981, A survey of hamadryas baboons in Saudi Arabia, *Fauna Saudi Arabia* **3**:441–471.

Li, W., 1997, *Molecular Evolution.* Sinauer Associates Inc., Sunderland.

Mang, R., Maas, J., Chen, X., Goudsmit, J., van Der Kuyl, A. C., 2001, Identification of a novel type C porcine endogenous retrovirus: Evidence that copy number of endogenous retroviruses increases during host inbreeding, *J. Gen. Virol.* **82**:1829–1834.

Morgan, M. T., 2001, Transposable element number in mixed mating populations, *Genet. Res.* 77:261–275.

Nagel, U., 1971, Social organization in a baboon hybrid zone, *Proc. 3rd Int. Congr. Primat. Zurich* **3**:48–57.

Nei, M., 1987, *Molecular Evolutionary Genetics*, Columbia University Press, New York.

Newman, T. K., 1997, Mitochondrial DNA analysis of intraspecific hybridization in *Papio hamadryas anubis, P. h. hamadryas* and their hybrids in the Awash National Park, Ethiopia. Ph.D. Dissertation, New York University.

Nordborg, M., 2000, Linkage disequilibrium, gene trees and selfing: An ancestral recombination graph with partial self-fertilization, *Genetics* **54**:923–929.

Nystrom, P., 1992, Mating success of hamadryas, anubis and hybrid male baboons in a "mixed" social group in the Awash National Park, Ethiopia. Ph.D. Dissertation, Washington University.

Pusey, A. E. and Packer, C., 1987, Dispersal and philopatry, in: *Primate Societies*, B. B. Smuts, D. L. Cheney, R. M. Seyfarth, R. W. Wrangham, and T. T. Strushaker, (eds.), University of Chicago Press, Chicago, pp. 250–266.

Reiter, L. T., Murakami, T., Koeuth, T., Pentao, L., Muzny, D., Gibbs, R. A., and Lupski, J. R., 1996, A recombination hotspot responsible for two inherited peripheral neuropathies is located near a *mariner* transposon-like element, *Nat. Genet.* **12**:288–297.

Rowe, W. P. and Kozak, C. A., 1980, Germ-line reinsertions of AKR murine leukemia virus genomes in Akv-1 congenic mice, *Proc. Natl. Acad. Sci. USA* 77:4871–4874.

Sigg, H. and Stolba, A., 1981, Home range and daily march in a Hamadryas baboon troop, *Folia Primatol.* **36**:40–75.

Soriano, P., Gridley, T., and Jaenisch, R., 1987, Retroviruses and insertional mutagenesis in mice: Proviral integration at the Mov 34 locus leads to early embryonic death, *Genes Dev.* **1**:366–375.

Steffen, D. L., Taylor, B. A., and Weinberg, R. A., 1982, Continuing germ line integration of AKV proviruses during the breeding of AKR mice and derivative recombinant inbred strains, *J. Virol.* **42**:165–175.

Stoye, J. P., 2001, Endogenous retroviruses: Still active after all these years? *Curr. Biol.* **11**:R914–R916.

Swedell, L., 2002a, Affiliation among females in wild hamadryas baboons (*Papio hamadryas hamadryas*), *Int. J. Primatol.* **23**:1205–1226.

Swedell, L., 2002b, Ranging behavior, group size and behavioral flexibility in Ethiopian hamadryas baboons (*Papio hamadryas hamadryas*), *Folia Primatol.* **73**:95–103.

Swedell, L., 2006, *Strategies of Sex and Survival in Hamadryas Baboons: Through a Female Lens*, Pearson Prentice Hall, Upper Saddle River, NJ.

Swedell, L. and Woolley-Barker, T., 2001, Dispersal and philopatry in hamadryas baboons: A re-evaluation based on behavioral and genetic evidence, *Am. J. Phys. Anthropol. Suppl.* **32**:146.

Uddin, M. 2003, Baboon endogenous virus (BaEV) variation in natural populations of cercopithecine primates. Ph.D. Dissertation, New York University.

van der Kuyl, A. C., Dekker, J., and Goudsmit, J., 1995, Distribution of baboon endogenous virus among species of African monkeys suggests multiple ancient cross-species transmissions in shared habitats, *J. Virol.* **69**:7877–7887.

Wildman, D. E., Bergman, T. J., al-Aghbari, A., Sterner, K. N., Newman, T. K., Phillips-Conroy, J. E., Jolly, C. J., Disotell, T. R., 2004, Mitochondrial evidence for the origin of hamadryas baboons, *Mol. Phylogenet. Evol.* **31**:287–296.

Woolley-Barker, T., 1998, Genetic structure reflects social organization in hybrid hamadryas and anubis baboons, *Am. J. Phys. Anthropol. Suppl.* **26**:235.

Woolley-Barker, T., 1999, Social organization and genetic structure in a baboon hybrid zone, Ph.D. Dissertation, New York University.

Wright, S., 1932, The roles of mutation, inbreeding, cross-breeding, and selection in evolution, *Proc. VI Int. Congr. Genet.* **1**:356–366.

Wright, S., 1938, Size of population and breeding structure in relation to evolution, *Science* **87**:430–443.

Wright, S., 1965, The interpretation of population structure by *F*-statistics with special regard to systems of mating, *Evolution* **19**:395–420.

Wright, S., 1977, *Evolution and the Genetics of Populations. Vol. 3. Experimental Results and Evolutionary Deductions*, University of Chicago Press, Chicago.

Wright, S., 1978, *Evolution and the Genetics of Populations. Vol. 4. Variability Within and Among Populations*, University of Chicago Press, Chicago.

Wright, S. I. and Schoen, D. J., 1999, Transposon dynamics and the breeding system, *Genetica* **107**:139–148.

Wright, S. I., Lauga, B., and Charlesworth, B., 2002, Rates and patterns of molecular evolution in inbred and outbred *Arabidopsis*, *Mol. Biol. Evol.* **19**:1407–1420.

PART II

# Life History, Development, and Parenting Strategies

CHAPTER SEVEN

# Reproduction, Mortality, and Female Reproductive Success in Chacma Baboons of the Okavango Delta, Botswana

*Dorothy L. Cheney, Robert M. Seyfarth, Julia Fischer, Jacinta C. Beehner, Thore J. Bergman, Sara E. Johnson, Dawn M. Kitchen, Ryne A. Palombit, Drew Rendall, and Joan B. Silk*

## CHAPTER SUMMARY

Predation, food competition, and infanticide all negatively impact female reproductive success. Female dominance rank may mitigate these effects, if

---

**Dorthy L. Cheney** • Department of Biology, University of Pennsylvania, Philadelphia, PA, USA **Robert M. Seyfarth** • Department of Psychology, University of Pennsylvania, Philadelphia, PA, USA **Julia Fischer** • German Primate Center, Goettingen, Germany **Jacinta C. Beehner** • Departments of Psychology and Anthropology, University of Michigan, Ann Arbor, MI, USA **Thore J. Bergman** • Department of Psychology, University of Michigan, Ann Arbor, MI, USA **Sara E. Johnson** • Department of Anthropology, California State University, Fullerton, CA, USA **Dawn M. Kitchen** • The Ohio State University, Columbus, OH, USA **Ryne A. Palombit** • Department of Anthropology, Rutgers University, New Brunswick, NJ, USA **Drew Rendall** • Department of Psychology and Neuroscience, University of Lethbridge, Lethbridge, Canada **Joan B. Silk** • Department of Anthropology, University of California, Los Angeles, CA, USA

competitive exclusion allows high-ranking females to gain priority of access to critical food resources. It may also exacerbate them, if low-ranking females are forced to feed or rest in marginal habitats where they are at increased risk. In this chapter, we present data on reproduction, mortality, and female reproductive success from a 10-year study of free-ranging chacma baboons (*Papio hamadryas ursinus*) in the Okavango Delta of Botswana and examine the influence of predation, infanticide, and dominance rank on female reproductive success. Predation appeared to be the cause of most deaths among adult females and juveniles, whereas infanticide was the most likely cause of deaths among infants. Seasonality strongly affected both births and mortality: The majority of conceptions occurred during the period of highest rainfall. Mortality due to predation and infanticide was highest during the 3-month period when flooding was at its peak, most likely because the group was more constrained to move along predictable routes during this time. Those reproductive parameters most likely to be associated with superior competitive ability—interbirth interval and infant growth rates—conferred a slight fitness advantage on high-ranking females. This fitness advantage was counterbalanced, however, by the effects of infanticide and predation. Infanticide affected high- and low-ranking females more than middle-ranking females, while predation affected females of all ranks relatively equally. As a result, there were few rank-related differences in estimated female lifetime reproductive success.

## 1. INTRODUCTION

The relative importance of predation and food competition for the evolution of sociality in baboons (*Papio hamadryas* spp.) and other nonhuman primates is a subject of considerable debate. According to some hypotheses, predation has exerted the primary selective pressure on sociality, because the costs of intragroup feeding competition would otherwise prevent females from living in groups (e.g., van Schaik, 1983, 1989; Janson, 1988, 1998; Sterck et al., 1997; Hill and Dunbar, 1998). Other hypotheses, however, place more weight on the benefits derived from intergroup resource competition (e.g., Wrangham, 1980; Isbell, 1991; Cheney, 1992) or protection against infanticide (Janson and van Schaik, 2000; see also Sterck et al., 1997).

Whatever the benefits of sociality, group life also imposes costs in the form of increased competition for resources and breeding opportunities.

When predation pressure, resource availability, or limited dispersal options make group life essential, breeding may become skewed, with one or a few females monopolizing reproduction (reviewed by, e.g., Vehrencamp, 1983; Keller and Reeve, 1994; Johnstone, 2000). High reproductive skew is especially likely to evolve under severe ecological constraints, when a single female may not be able to produce offspring without the help of nonbreeding adults (e.g., dwarf mongooses, *Helogale parvula*: Creel and Waser, 1997; suricates, *Suricata suricatta*: O'Riain et al., 2000; wild dogs, *Lycaon pictus*: Creel and Creel, 2002; naked mole rats, *Heterocephalus glaber*: Jarvis, 1991). In contrast, low reproductive skew, with relatively egalitarian breeding among all females, is expected to occur when ecological constraints are relaxed and dominant females do not benefit from suppressing the reproduction of subordinates (e.g., lions, *Panthera leo*: Packer et al., 2001; banded mongoose, *Mungos mungo*: Cant, 2000; DeLuca and Ginsberg, 2001; elephants, *Loxodonta africana*: Moss, in press).

The extent of reproductive skew varies widely in the primate order. In the callitrichids, dominant females monopolize reproductive activity and suppress reproduction in subordinates (Goldizen, 1987; Goldizen et al., 1996; French, 1997). In other primates, reproductive skew is considerably reduced, and nearly all adult females produce offspring at more or less regular intervals. Even in species manifesting low reproductive skew reproductive performance may be mediated by dominance rank and access to food resources (e.g., Janson, 1985; Barton, 1993).

Female savannah baboons can be characterized as displaying low, but not insignificant, reproductive skew. Like females in such Old World monkey species as macaques (*Macaca* spp.) and vervet monkeys (*Cercopithecus aethiops*), female baboons form linear, nepotistic dominance hierarchies in which high-ranking females and their close relatives enjoy priority of access to resources (reviewed by Melnick and Pearl, 1987; Silk, 1987, 1993). Although there is no evidence of complete reproductive suppression, low-ranking females may experience reduced fecundity and offspring survival. High-ranking females often give birth at earlier ages (baboons: Altmann et al., 1988; Altmann and Alberts, 2003), have shorter interbirth intervals (baboons: Bulger and Hamilton, 1987; Smuts and Nicholson, 1989; Packer et al., 1995; Barton and Whiten, 1993; Wasser et al., 1998; Altmann and Alberts, 2003; Japanese macaques, *Macaca fuscata*: Sugiyama and Ohsawa, 1982; long-tailed macaques, *M. fascicularis*: van Noordwijk and van Schaik, 1999), or experience higher offspring survival than low-ranking females (baboons: Bulger and Hamilton, 1987; Rhine et al., 1988; Packer et al., 1995; long-tailed macaques: van Noordwijk and van Schaik, 1999; vervet monkeys: Whitten, 1983; Japanese macaques: Sugiyama and Ohsawa, 1982; see also review by Silk, 1993). Each of these reproductive variables is influenced by nutritional condition and access to food, which is often mediated by dominance rank (Wrangham, 1981; Whitten, 1983; Barton and Whiten, 1993; Barton et al., 1996).

Not all studies, however, have documented significant differences in lifetime reproductive success between high- and low-ranking females (baboons: Altmann et al., 1988; Packer et al. 1995; vervets: Cheney et al., 1988a,b; reviewed by Silk, 1993). The lack of a consistently strong relationship between female rank and reproductive success is probably a consequence of at least two factors. First, the effects of food competition on reproduction are likely to be most evident under extreme ecological conditions, such as during severe drought (Alberts et al., 2005). At other times these effects may be smaller and more difficult to measure, especially over the short term (van Noordwijk and van Schaik, 1999). Second, causes of mortality unrelated to food competition, such as infanticide and predation, may affect high- and low-ranking females relatively equally, thereby masking the impact of food competition.

The interaction of feeding competition and predation pressure may especially be costly to low-ranking animals. Because low-ranking animals may be excluded from safe feeding sites toward the center of the group and forced to feed in more peripheral areas, they may suffer increased predation. Again, however, data in support of this hypothesis are inconsistent. Although some studies have documented a relationship between low rank and increased vulnerability (baboons: Ron et al., 1996; long-tailed macaques: van Noordwijk and van Schaik, 1987), others have not (baboons: Bulger and Hamilton, 1987; vervet monkeys: Cheney et al., 1988a,b).

We provide here a descriptive account of female reproduction and mortality over a 10.5-year period (July 1992–December 2002) in one group of chacma baboons inhabiting the Moremi Reserve in the Okavango Delta of Botswana. Because baboon females can live for over 20 years in the wild (Packer et al., 1995; Altmann and Alberts, 2003), our data do not provide a complete analysis of the factors that influence lifetime reproductive success. They do, however, permit evaluations of the relative influences of predation, infanticide, interbirth intervals, and infant survival on the reproductive success of females of different ages and dominance ranks.

## 2. STUDY GROUP AND HABITAT

The focus of this study was a group of free-ranging chacma baboons (*Papio hamadryas ursinus*) inhabiting the Moremi Game Reserve in the Okavango Delta of Botswana. Grasslands in the delta flood annually (usually between June and October), leaving elevated "islands" edged with woodland. Baboons feed extensively on a number of tree species in these edged woodlands, including wild or strangler figs (*Ficus thonningii*), sycamore figs (*F. sycamorus*), sausage trees (*Kigelia africana*), African mangosteens (*Garcinia livingstonei*), jackalberries or African ebonies (*Diospyros mespiliformis*), marula trees (*Sclerocarya birrea*), camelthorn acacias (*Acacia erioloba*), candle-pod acacias (*Acacia hebeclada*), knobthorn acacias (*Acacia nigrescens*), and real fan palms (*Hyphaene ventricosa*) (Hamilton et al., 1976; Bulger and Hamilton, 1987; Ross, 1987; Ellery et al., 1993; Roodt, 1998). Islands can be less than one to hundreds of hectares in size. During floods, baboons ford the submerged plains and move between islands throughout an approximately 5 $km^2$ range. The population density of baboons in this area is considerably higher than in other areas of Africa (approximately 24/$km^2$; Hamilton et al., 1976, unpublished data).

Predation is the most important cause of mortality for juveniles and adults in Moremi (described in detail in Cheney et al., 2004; see also Busse, 1982; Cowlishaw, 1994). Most predation is due to leopards (*Panthera pardus*) and lions (*Panthera leo*). Although we have also observed several attacks by crocodiles (*Crocodilus niloticus*), we have not been able to confirm any deaths due to crocodile predation. Other potential, but unconfirmed, predators include hyenas (*Crocuta crocuta*), wild dogs (*Lycaon pictus*), and pythons (*Python sebae*).

The study group, C, has been observed since 1978, with almost daily observations since mid-1992. The ages and matrilineal relatedness of all natal animals are known, as are the origins and destinations of many immigrant males. During the course of this study, the group averaged around 75 individuals. The number of adult females in the group ranged from 19 to 26, while the number of adult males ranged from 3 to 12. As in most other baboon populations, females remain in their natal groups throughout their lives, but males typically emigrate to neighboring groups after attaining sexual maturity.

Dominance ranks among adult females were determined by the direction of approach–retreat interactions (Silk et al., 1999). Female dominance ranks remained stable over the 10-year period of this study, with daughters assuming ranks similar to those of their mothers (Silk et al., 1999, unpublished data). Younger sisters have typically risen in rank over older sisters, while ranks between mothers and daughters have not been as predictable. Some mothers have continued to rank higher than their adult daughters, while others have dropped below their daughters (see Combes and Altmann, 2001). One female, which was orphaned at 9 months, achieved a rank different from (in this case, higher than) that of her closest female relatives.

Females were assigned ranks according to the proportion of females dominated, which largely controls for variation in the number of females present in the group across time. When assessing the effects of female rank on particular demographic events (e.g., infant death), we used each female's rank at the time of the event. When considering lifetime reproductive success, we calculated the female's mean rank across years. For the purposes of the analyses described here, we divided females into high-, middle-, and low-rank categories. No females changed from one rank category to another across time.

Infants were defined as animals under the age of 1 year. Juveniles were animals aged 1–5 years in the case of females and 1–6 years in the case of males. Females were considered to be adult at 6 years of age, and males at 7 years of age. No male was known to emigrate from his natal group before 8.5 years of age.

Most of the data presented in this chapter were gathered between 1992 and 2002. For one analysis that estimates lifetime reproductive success, however, we also used demographic data on births and deaths gathered prior to 1992. These data were collected by W. J. Hamilton, J. B. Bulger, and colleagues during the 1980s and early 1990s. Because these records did not always include the exact dates of all demographic events or precise information on causes of mortality, we have not used them for the bulk of our analysis. Disappearances were classified into several categories (Table 1), with additional categories for cases of infanticide (Table 2).

**Table 1.** Classification of disappearances and causes of mortality

| | |
|---|---|
| Ill | An animal disappeared after appearing to be ill or listless within the previous 24 hr. Individuals that disappeared after appearing ill were not included as victims of predation or infanticide |
| Confirmed predation | The predation event was witnessed by observers, or a predator was observed with the carcass of a known individual who had been observed, apparently healthy, within the previous 24 hr, or the carcass of an individual known to be healthy was found in conjunction with predator tracks and feces containing baboon remains |
| Suspected predation | A baboon who had been seen, apparently healthy, within the previous 24 hr disappeared in contexts in which predation was strongly suspected. These contexts included alarm calls and the sighting of predators or predator tracks in close proximity to the baboons at the time of the individual's disappearance |
| Disappear apparently healthy | The disappearance of an apparently healthy animal within 24 hr of being seen. Animals were classified as "apparently healthy" if they were not obviously diseased or listless at the time of their disappearance. Clearly, however, it remains possible that such individuals might have suffered from illnesses that were not detectable to observers. Infants, juveniles, and adult females that disappeared were presumed to have died. We have observed no cases of female emigration in this study group. Baboons are able to swim even at a very young age, so it is unlikely that any disappearances during times of flooding were due to drowning |

Annual mortality rates were calculated for 10 years (1 August 1992–31 July 2002), based on the number of individuals in each age/sex class on 1 August of each year. All individuals aged less than 1 year as of 1 August plus all individuals born between 1 August and 31 July were included in the "infant" class for that year. Thus the same individual might be counted as an infant in two successive years.

**Table 2.** Classification of cases of infanticide

| | |
|---|---|
| Confirmed infanticide | The infanticidal attack was witnessed by observers |
| Strongly suspected infanticide | An infant disappeared after a fight involving a male and female, and after sustaining wounds suggestive of a baboon bite, or at the same time that its mother sustained wounds suggestive of a baboon bite. In one suspected case, a male, which had been observed killing other infants, was seen eating a carcass that appeared to be that of an infant baboon |
| Suspected infanticide | An apparently healthy infant disappeared at around the same time that a male was confirmed to have killed other infants |

## 3. FEMALE REPRODUCTION

### 3.1. Seasonality

Although baboons produce offspring throughout the year, a number of studies have documented birth peaks that are correlated with seasonal fluctuation in rainfall. Most conceptions occur during the rainy season, and births tend to peak during the dry season or winter months (Mikumi, Tanzania: Rhine et al., 1988, 1989; Amboseli, Kenya: Alberts et al., 2005; de Hoop and Drakensberg Mountains, South Africa: Barrett et al., this volume). In Moremi, births showed seasonal variation, with 76 percent of 122 births occurring in the 6 months from July through December (Figure 1; two-tailed chi square one-sample test, df = 1, $\chi^2 = 33.58$, $P < 0.001$). This seasonal effect held for females of all ranks, although high-ranking females' births appeared to be slightly less seasonal than those of middle- and low-ranking females. While 83 percent of middle- and low-ranking females' births occurred from July through December, only 65 percent high-ranking females' infants were born during this period.

As in other baboon populations, birth seasonality in Moremi appeared to be influenced by seasonal fluctuations in rainfall (Figure 1). Rainfall in the Okavango Delta is highly variable, but most rain falls during the months of November–March, usually peaking in January and February. Baboon gestation periods are approximately 6 months in length. Thus, if females were most likely to become pregnant after periods of high rainfall when food was more plentiful, births would be expected to peak in the months immediately following July and August.

### 3.2. First Birth

Females first gave birth at an average age of 6 years, 9 months (range 5 years, 9 months to 7 years, 11 months; median age: 6 years, 6 months; $N = 28$ females). In contrast to some other baboon populations (see "Introduction"), age at first birth appeared to be unrelated to female rank. Females in the top third of the female dominance hierarchy first gave birth at a mean age of 6 years, 7 months (median: 6 years, 7 months; $N = 11$). Those in the mid-third gave birth at a mean age of 7 years, 0 months (median: 6 years, 11 months; $N = 9$), and those in the bottom third at a mean age of 6 years, 8 months (median: 6 years, 6 months; $N = 8$).

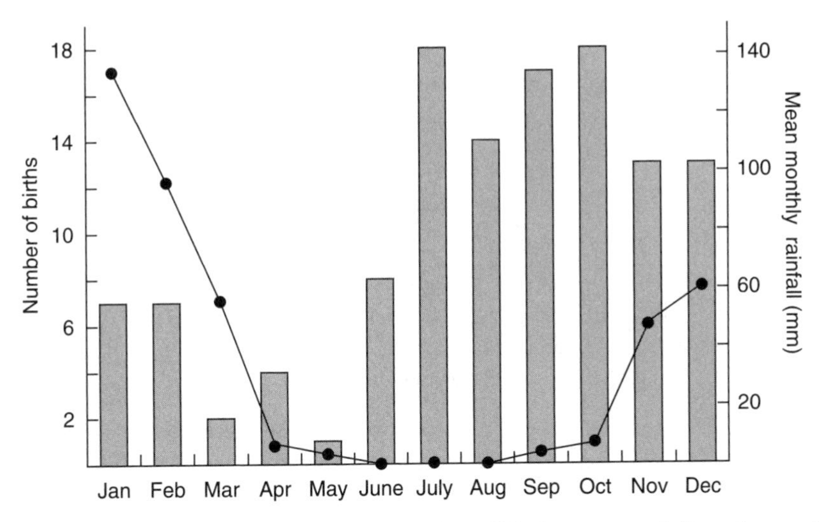


**Figure 1.** Relationship between the number of births per month from August 1992 through July 2002 (bars) and the mean monthly rainfall for each month (in mm; solid line) (two-tailed chi square one-sample test, df = 1, $\chi^2 = 33.58$, $P < 0.001$). Data on rainfall are based on records from 1992 to 1999 (http://tao.atmos.washington.edu/data_sets/wilmott). Reprinted from Cheney et al., "Factors affecting reproduction and mortality among baboons in the Okavango Delta, Botswana," *International Journal of Primatology* 25:401–428, with the kind permission of Springer.

The presence of adult maternal kin (i.e., mothers and/or maternal sisters) appeared to influence age at first birth, but this effect varied according to female rank. Whereas high- or middle-ranking females with a living mother or adult sister gave birth for the first time at significantly younger ages than females without close kin, the reverse was true for low-ranking females: Low-ranking females with living kin gave birth at older ages than those without kin (Figure 2; general linear model (GLM) with rank and presence of female kin; significant interaction between presence of female kin and rank, $F = 10.1$, df = 2, 22; $P < 0.01$).

### 3.3. InterBirth Intervals

We used a GLMM (general linear mixed effect model; maximum likelihood estimation) to assess whether sex of the previous infant, infant survival to 1 year, presence of female kin, rank, and female age had an effect on interbirth interval. With this analysis, repeated measures from the same individual can be analyzed. We used data from 20 females (using female identity as a

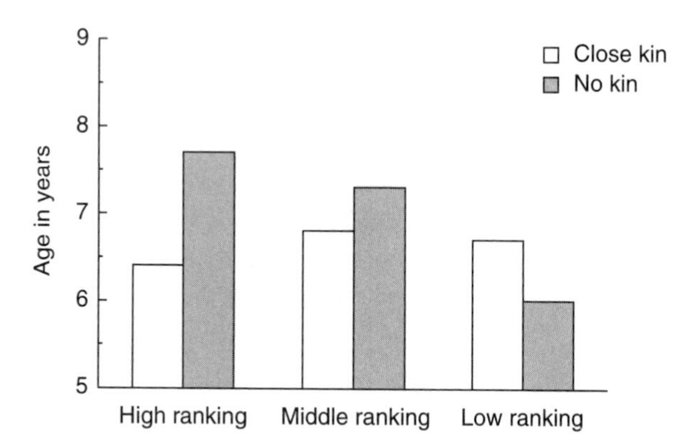


**Figure 2.** Mean age at first birth for females with and without living adult kin. High-ranking females: $N=11$; middle-ranking female: $N=9$; low-ranking females: $N=8$. A general linear model (GLM) with rank and presence of female kin yielded a significant interaction between presence of female kin and rank ($F=10.1$, df=2, 22; $P < 0.01$). Reprinted from Cheney et al., "Factors affecting reproduction and mortality among baboons in the Okavango Delta, Botswana," *International Journal of Primatology* 25:401–428, with the kind permission of Springer.

random factor in the analysis) and randomly selected one data point per female in the "infant survived" and one in the "infant died before 1 year of age" category. We first calculated the initial model with all factors and then examined different models using Akaike's information criterion (Pinheiro and Bates, 2000) to identify the best one.

Although there was no significant effect of infant sex on interbirth interval, there was a significant interaction between infant survival and infant sex: If the infant survived and was male, the interbirth interval was longer (759 ± 84 days as opposed to 718 ± 85 days for female infants). However, if the infant died and was male, the interbirth interval was significantly shorter (429 ± 91 as opposed to 506 ± 78 days for female infants; $F=4.6$, df=1, 34.4, $P < 0.05$). Interbirth intervals for females aged 15 years or older were somewhat longer (812 ± 106 days, $N=3$ females) than for younger females (721 ± 76 days, $N=17$), but only if the infant survived to 1 year of age (if the infant died the average interbirth interval was 428 ± 85 versus 470 ± 94 days).

As in several previous studies of baboons and macaques (see "Introduction"), interbirth intervals for females whose previous infant survived tended to be shorter for high-ranking females (mean = 682 ± 78 days, $N = 6$) than for middle-ranking (mean = 759 ± 66 days, $N=8$) and low-ranking females (mean

= 754±101 days, $N=6$). If a female's previous infant had died, rank-related differences in interbirth intervals were less obvious. The mean interbirth interval for high-ranking females whose previous infant had died was 475 days, compared with 447 and 464 for middle- and low-ranking females, respectively. Interbirth intervals for females that had living adult female kin were shorter (724±80, $N=14$) than for females without such kin (760±96, $N=6$). However, none of these effects reached statistical significance.

### 3.4. Infant Sex Ratios

A long-term study of a population of baboons in Amboseli, Kenya, has documented a female-biased sex ratio among high-ranking females, possibly reflecting local competition for resources (Altmann et al., 1988). A similar bias has been reported for captive bonnet macaques (*M. radiata*; Silk, 1983). Rank-related biases have not consistently been documented, however, in other baboon populations (Rhine et al., 1992) or in other populations of monkeys (e.g., vervet monkeys: Cheney et al., 1988a,b). In fact, rank-related sex ratio biases may result largely from the errors inherent in small samples, as the effects of maternal rank on birth sex ratios disappear as sample size increases (Brown and Silk, 2002).

In Moremi, overall birth sex ratios were almost even ($N=70$ female births, 62 male births, and 1 birth of unknown sex from June 1992 through December 2002). There was no evidence for a rank-related difference in birth sex ratios. The mean female/male infant sex ratio for high-ranking females ($N=14$) was 0.52, compared with 0.51 for middle-ranking females ($N=15$) and 0.55 for low-ranking females ($N=13$).

## 4. MORTALITY

### 4.1. Age-Specific Mortality

Figure 3 presents data on age-specific mortality rates for males and females from birth to 10 years of age. The highest mortality rates occurred among infants, with juveniles and young adults experiencing relatively low mortality.

### 4.2. Adult Female Mortality

Confirming the observations made by Bulger and Hamilton (1987) in an earlier study of the same group, almost all adult female deaths appeared to be

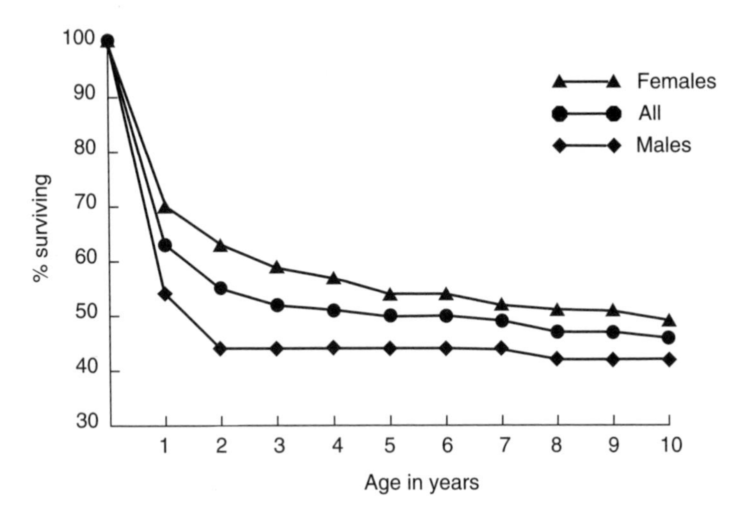


**Figure 3.** Age-specific mortality estimates for males and females from birth through 10 years of age. Reprinted from Cheney et al., "Factors affecting reproduction and mortality among baboons in the Okavango Delta, Botswana," *International Journal of Primatology* 25:401–428, with the kind permission of Springer.

due to predation (Table 3). Only one female was ill or listless at the time of her disappearance. Leopard predation on four females was confirmed, and in six other cases predation by leopards ($N=2$) and lions ($N=4$) was strongly suspected. An additional thirteen females disappeared after having been observed, apparently healthy, on the previous day.

Female mortality peaked between the months of July and September, during maximal flooding. Overall, 58 percent of the 19 known or suspected cases of predation on females occurred in these months, significantly more than

**Table 3.** Causes of mortality among adult females, juveniles, and infants

| | *N* | Ill | Confirmed predation | Suspected predation | Disappear healthy | Confirmed infanticide | Suspected infanticide | M's death |
|---|---|---|---|---|---|---|---|---|
| Adult females | 24 | 1 | 4 | 6 | 13 | | | |
| Juv. females | 12 | 0 | 1 | 2 | 9 | | | |
| Juv. males | 7 | 0 | 0 | 1 | 6 | | | |
| Infants | 46 | 7 | 0 | 1 | 6 | 11 | 14 | 7 |

See Tables 1 and 2 for definitions. "M's Death" stands for infant died following mother's death. "Suspected Infanticide" includes both strongly suspected and suspected cases of infanticide. An additional three infants disappeared when observers were absent for more than 24 hr. Reprinted from Cheney et al., "Factors affecting reproduction and mortality among baboons in the Okavango Delta, Botswana," *International Journal of Primatology* 25:401–428, with the kind permission of Springer.

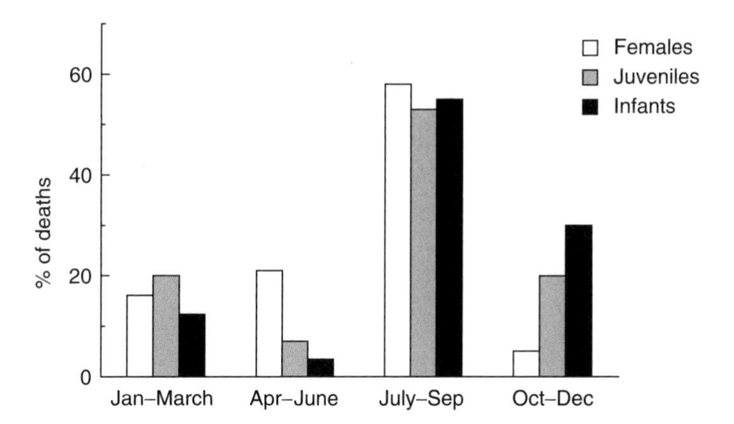


**Figure 4.** Proportion of all deaths among adult females, juveniles, and infants between August 1992 and July 2002 that occurred in each 3-month period. Of the 19 known or suspected cases of predation on adult females, 58 percent occurred in July–September, significantly more than would be expected by chance (chi square one-sample test, $\chi^2=8.99$, df=1, $P < 0.01$); of 15 juvenile deaths between August 1992 and July 2002, 53 percent occurred in July–September ($\chi^2=6.43$, df=1, $P < 0.05$); and of the 33 infants that disappeared healthy, 55 percent disappeared in July–September ($\chi^2=15.37$, df=1, $P < 0.01$). Analysis excludes individuals that disappeared ill and, in the case of infants, after the deaths of their mothers. Reprinted from Cheney et al., "Factors affecting reproduction and mortality among baboons in the Okavango Delta, Botswana," *International Journal of Primatology* 25:401–428, with the kind permission of Springer.

would have been expected by chance (Figure 4; chi square one-sample test, $\chi^2=8.99$, df=1, $P < 0.01$).

Adult females experienced an annual mortality rate of 0.09 (s.d.=0.04). Mortality rates fluctuated between 0.04 and 0.16, indicating that females encountered substantially higher predation rates in some years than in others. For example, during 2002 seven of 29 females in the group disappeared (24 percent), all apparently victims of predation. A drought and widespread fires during this year caused the group to scatter over large distances while foraging, possibly increasing females' vulnerability to predation. Bulger and Hamilton (1987) noted a similar surge in female mortality in 1980–1981, when 40 percent of the group's females died, all apparently due to predation.

Little rank-related variation in mortality rates was apparent. Of 24 female deaths, 29 percent occurred among high-ranking females, 38 percent among middle-ranking females, and 33 percent among low-ranking females. Females ranked in the top third of the dominance hierarchy experienced a mean annual mortality rate of 0.08 (sd=0.11). Females ranked in the mid-third experienced a mortality rate of 0.11 (sd=0.13), and those in the bottom third a rate of 0.11 (s.d.=0.13) (Figure 5; Kruskal–Wallis one-way analysis of variance, $H$=1.46, df=2, $P > 0.10$).

The average age of death for all females was 13 years, 11 months (median age: 13 years, 7 months). Five females are known to have been over 20 years of age at the time of death. High-ranking females tended to die at slightly younger ages than middle- and low-ranking females, although these differences were not significant. The average age of death for high-ranking females was 13 years, 5 months (median: 10 years, 4 months; $N$=7), compared with 13 years, 7 months (median: 13 years, 7 months; $N$=9) for middle-ranking females and 14 years, 8 months (median: 14 years, 8 months; $N$=8) for low-ranking females.

Pregnant females may have been particularly vulnerable to predation. At least 13 of the 23 females (56 percent) that died of apparent predation were pregnant when they died. By comparison, only 4 (19 percent) females with infants under the age of 6 months, the approximate length of gestation in baboons, died.

### 4.3. Juvenile Mortality

All juvenile deaths appeared to be due to predation (Table 3). Of the 19 juveniles that disappeared, 1 was killed by a lion, 2 were suspected to have been killed by lions, and 1 disappeared after sustaining injuries that appeared to be crocodile-inflicted. An additional 15 juveniles disappeared when healthy. As in the case of adult females, juvenile mortality increased during the period of greatest flooding; 53 percent of 15 juvenile deaths between August 1992 and July 2002 occurred during the months of July–September (Figure 4; $\chi^2 = 6.43$, df = 1, $P < 0.05$).

Mortality rates among juveniles were generally lower than those for adult females and infants (Figure 5). Juveniles of both sexes experienced an average annual mortality rate of 0.04 (females = 0.04, sd=0.04; males = 0.04, sd = 0.04).

Previous research on this study group has suggested that the juvenile offspring of low-ranking mothers may be more vulnerable to predation than other juveniles, because they are more likely to feed in peripheral areas (Johnson, 2003). Indeed, the juvenile offspring of low-ranking females suffered higher levels of mortality. Ten (53 percent) of the juveniles that died of apparent predation had mothers who ranked in the bottom third of the female dominance hierarchy, compared with five (26 percent) and four (21 percent) for offspring of high- and middle-ranking females, respectively.

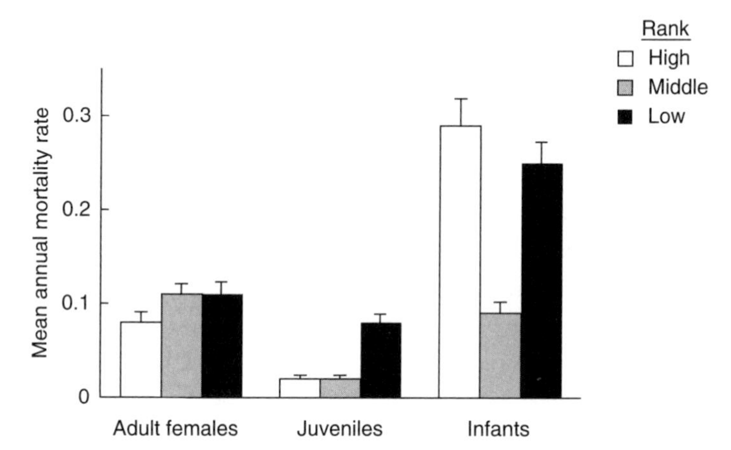


**Figure 5.** Mean annual mortality rates for high-, middle-, and low-ranking females and their offspring (Kruskal–Wallis one-way analysis of variance; adult females: $H=1.46$, df=2, $P > 0.10$; juveniles: $H= 6.78$, df=2, $P < 0.05$; infants: $H=7.25$, df=2, $P < 0.05$). Reprinted from Cheney et al., "Factors affecting reproduction and mortality among baboons in the Okavango Delta, Botswana," *International Journal of Primatology* 25:401–428, with the kind permission of Springer.

The average annual mortality rate among the juvenile offspring of low-ranking females was 0.08, compared with a rate of 0.02 for the offspring of both high- and middle-ranking females (Figure 5; $H=6.78$, df=2, $P < 0.05$).

Though sample sizes are small, there was some indication that juvenile male mortality decreased with increasing age, whereas juvenile female mortality did not. Only one (14 percent) of the seven deaths among juvenile males occurred after 2 years of age. In contrast, eight (67 percent) of the twelve juvenile females who died were over 2 years of age (two-tailed Fisher exact probability test, $P < 0.05$). After about the age of 2 years, juvenile males begin to dominate juvenile females in competitive interactions, regardless of their mothers' relative ranks (Pereira, 1988; Lee and Johnson, 1992; unpublished Moremi data). The juvenile daughters of low-ranking females may have been forced to forage in more peripheral areas than juvenile males, thereby increasing their vulnerability to predation.

## 4.4. Infant Mortality

### *4.4.1. Causes and Patterns of Infant Mortality*

Of 120 infants born in the group, 46 (38 percent) died before reaching the age of 1 year (Figure 3). Seven infants either died on the day of birth or appeared to be ill at the time of death (Table 3). Seven others died when their mothers were killed by predators. Three infants disappeared while observers were absent for more than 24 hr and three infants disappeared within 24 hr of having been observed apparently healthy. Infanticide accounted for the majority of infant deaths, ranging from at least 23 percent to as much as 70 percent annually (see also Palombit et al., 2000). Observers witnessed 11 cases of infanticide by resident adult males and another nine cases of infanticide were strongly suspected (Table 3). Five other apparently healthy infants disappeared when other infants were known to have been killed by adult males. The majority (64 percent) of the 25 infants known or suspected to have been killed by males were under 3 months of age and still being carried by their mothers.

As with predation, infanticide was most frequent during the period of peak flooding (July–September). Six (55%) of the eleven cases of observed infanticide occurred from July through September, while 44 percent of the nine infants strongly suspected to have been killed by infanticide also disappeared during this time. Overall, a significant proportion (55 percent) of the 33 infants that disappeared healthy between August 1992 and July 2002 died during these 3 months (Figure 4; $\chi^2 = 15.37$, df = 1, $P < 0.01$).

Seasonal peaks in the frequency of infanticide were not obviously related to seasonal fluctuations in male immigration patterns. Of the 40 males that immigrated into the group from July 1992 through December 2002, 30 percent immigrated in January–March, 32 percent in April–June, 25 percent in July–September, and 13 percent in October–December.

Mortality rates among infants fluctuated more widely than mortality in other age classes, ranging from 0.03 to 0.57 annually. The average annual mortality rate for infants was 0.21 (sd=0.15).

The infants of high-ranking females were not more likely to survive than the infants of lower-ranking females. Of the 42 infants born to high-ranking females, 45 percent died before the age of 1, compared with 31 percent of infants born to middle-ranking females ($N=43$), and 37 percent of infants born to low-ranking females ($N=35$). The average annual mortality rate for infants of high-ranking females was 0.29, higher than the mean mortality rate for low-ranking (0.25) and especially middle-ranking females (0.09) (Figure 5; $H=7.25$, df=2, $P < 0.05$).

The infants of middle-ranking females appeared to be less vulnerable to infanticide than the infants of high- and low-ranking females. Only 12 percent of the 25 infants suspected of being killed by infanticide were the offspring of middle-ranking mothers. In contrast, 53 percent were the offspring of high-ranking females and 36 percent the offspring of low-ranking females.

Primiparous females experienced higher infant mortality (44 percent; $N = 25$ births) than multiparous females (35 percent; $N = 97$). This was particularly true for low-ranking primiparous females. A majority (71 percent) of the infants born to primiparous low-ranking females died before reaching 1 year of age ($N = 7$), compared with 30 percent ($N = 10$) and 38 percent ($N = 8$) of the infants born to high-ranking and middle-ranking primiparous females, respectively. This difference, however, did not reach statistical significance.

### 4.4.2. *Factors affecting infant survival*

We used logistic regression to identify factors that predict infant survival to 1 year of age, using the –2 log likelihood (Tabachnik and Fidell, 2001) to identify the best model. We examined the following factors: infant sex, age category of the mother (5–7.9, 8–14.9, 15, and older), season when infant was born (quarter of the year), parity of the mother (primiparous or multiparous), rank of the mother (high, middle, low), and the presence of close adult female kin. The analysis included the 119 infants for whom there were no missing values.

Mother's age had a significant effect on infant survival (likelihood ratio test, $\chi^2 = 8.65$, df=2, $P = 0.013$), with middle-aged females experiencing lower infant mortality rates (27 percent; $N=64$) than young (41 percent; $N = 29$) or old (62 percent; $N=26$) females. The mortality rate was significantly different between young and middle-aged females ($\chi^2=3.83$, df=1, $P = 0.05$, exp $(B) = 0.303$), and between middle-aged and old females ($\chi^2 = 8.143$, df=1, $P < 0.01$, exp(B) = 0.299).

Season of birth tended to have some effect on infant survival (likelihood ratio, $\chi^2=6.67$, df= 3, $P = 0.083$). Mortality was highest among infants born in the second quarter of the year, which was also the time of the fewest births (see above). There were significant differences in the mortality rates between the second and the fourth quarter $\chi^2= 6.11$, df= 1, $P < 0.05$, $\exp(B) = 0.174$).

As in other populations of baboons (Altmann and Alberts, 2003), males tended to experience higher mortality than females (likelihood ratio, $\chi^2$) = 3.03, df=1, $P = 0.082$): 46 percent of males ($N=56$) and 30 percent of females ($N=63$; one infant could not be sexed before death) died before the age of one year $\chi^2=2.97$, df= 1, $P < 0.10$, $\exp(B) = 2.08$).

Neither maternal rank, parity, nor the presence of living kin contributed significantly to infant survival in this analysis (Cheney et al., 2004).

## 5. FEMALE LIFETIME REPRODUCTIVE SUCCESS

Because we have not yet followed a large sample of females from birth to death, we cannot specify precisely the extent to which lifespan, fecundity, and offspring survival influence female lifetime reproductive success. As a first pass at such an analysis, we calculated the number of breeding years, fecundity, and offspring survival for all females present in the group between 1992 and 2002, including all offspring produced by these females prior to 1992. Ideally, this analysis would have been restricted to females whose entire reproductive history was known. However, given small sample sizes and the relatively short duration of the study, our first analysis included all females present in the group between 1992 and 2002 for whom complete birth records were available ($N = 42$ females, average age of all females = 13 years, 7 months). In a separate analysis, we removed females that were still alive and restricted our sample to females whose reproductive lives were completed ($N = 24$ females, average age = 13 years, 11 months).

Similarly, our calculations of offspring survival would ideally have included only those offspring that survived to adulthood. Again due to small sample size, however, we calculated offspring survival in terms of infant survival to one year. Table 4 presents mean values of $L$ (number of breeding years), $F$ (fecundity, or number of offspring/$L$), and two measures of $S$ (the proportion of offspring surviving to 1 year, $S_i$, and to reproductive age, $S_a$) for all 42 breeding females, divided into high-, middle-, and low-rank categories. Table 5 presents the same data for the 24 deceased females. Both tables also show values for mean lifetime reproductive success, calculated as the product of $LFS$.

We detected few consistent rank-related differences in reproductive success, particularly when the analysis was restricted to deceased females whose reproductive lives were complete (Tables 4 and 5). Fecundity was higher among high-ranking females, while middle-ranking females on average experienced higher offspring survival. Lifetime reproductive success, however, was similar for high- and middle-ranking females. Indeed, what seems most notable about the calculations is the high degree of variance within every rank category for each measure of reproductive success. This variance was particularly striking for high- and low-ranking females.

**Table 4.** Mean (+variance) values for life history traits for all breeding females present in group between 1992 and 2002

| | All ($N$ = 42) | High ranking ($N$ = 14) | Middle ranking ($N$ = 15) | Low ranking ($N$ = 13) | Kruskal–Wallis $H$ (df =2) |
|---|---|---|---|---|---|
| $L$ | 7.35 (23.38) | 7.66 (29.24) | 6.61 (16.18) | 7.87 (28.24) | 7.07 |
| $F$ | 0.63 (0.04) | 0.66 (0.05) | 0.64 (0.02) | 0.58 (0.05) | 9.91* |
| $S_i$ | 0.62 (0.09) | 0.53 (0.08) | 0.75 (0.05) | 0.55 (0.11) | 3.41 |
| $S_a$ | 0.39 (0.07) | 0.34 (0.11) | 0.50 (0.06) | 0.34 (0.04) | 1.80 |
| $LFS_i$ | 2.79 (3.58) | 3.01 (4.65) | 2.73 (2.15) | 2.62 (4.61) | 7.43 |
| $LFS_a$ | 2.17 (2.99) | 2.32 (5.25) | 2.22 (1.34) | 1.95 (2.68) | 0.63 |

L is the number of breeding years; F is the fecundity; Si is the percentage of offspring surviving to 1; Sa is the percentage of offspring surviving to adult. Data exclude five females over the age of 6 that had either not produced offspring or whose offspring had not yet reached 1 year of age by 31 December 2002. Kruskal–Wallis one-way analysis of variance test compares high-, middle-, and low-ranking females. * = P < 0.05, after Bonferroni adjustment. See text for description of each life history trait. Reprinted from Cheney et al., "Factors affecting reproduction and mortality among baboons in the Okavango Delta, Botswana," *International Journal of Primatology* 25:401–428, with the kind permission of Springer.

**Table 5.** Mean (+variance) values for four life history traits for all now-deceased adult females present in group between 1992 and 2002

| | All ($N$ = 24) | High ranking ($N$ = 7) | Middle ranking ($N$ = 9) | Low ranking ($N$ = 8) | Kruskal–Wallis $H$ (df =2) |
|---|---|---|---|---|---|
| $L$ | 7.85 (30.52) | 6.96 (48.13) | 7.40 (18.39) | 9.14 (35.06) | 0.51 |
| $F$ | 0.58 (0.05) | 0.63 (0.10) | 0.60 (0.01) | 0.51 (0.05) | 2.28 |
| $S_i$ | 0.53 (0.09) | 0.37 (0.09) | 0.69 (0.05) | 0.46 (0.08) | 11.57* |
| $S_a$ | 0.43 (0.07) | 0.30 (0.09) | 0.55 (0.07) | 0.39 (0.04) | 2.46 |
| $LFS_i$ | 2.58 (4.13) | 2.43 (7.63) | 2.69 (1.99) | 2.76 (5.96) | 0.33 |
| $LFS_a$ | 2.33 (3.30) | 2.38 (6.86) | 2.34 (1.84) | 2.27 (3.14) | 0.87 |

Legend as in Table 4. Reprinted from Cheney et al., "Factors affecting reproduction and mortality among baboons in the Okavango Delta, Botswana," *International Journal of Primatology* 25:401–428, with the kind permission of Springer.

We next subjected these calculations to the model developed by Brown (1988) for partitioning variance in lifetime reproductive success into additive contributions due to variation in single components, and to variation and covariation among groups of components (for discussions of this method, see Brown, 1988; Grafen, 1988). Table 6 presents the estimated overall variance in lifetime reproductive success among all females who survived to breeding age (approximately 56 percent). Table 7 presents the same calculations for all deceased females. In both tables, values show the percentage of $V(LFS)$ accounted for by variation in $L$, $F$, $S$, and their products, with the diagonal containing single component contributions. Due to small sample sizes, $S$ was calculated as the proportion of offspring reaching 1 year of age ($S_i$), a value that overestimates actual $S$.

**Table 6.** Percentage contribution of the components of lifetime reproductive success to variation in *LRS* for all females of breeding age between 1992 and 2002

| | *L* | *F* | *S* |
|---|---|---|---|
| *L* | 49.59 | | |
| *F* | −7.82 | 5.52 | |
| $S_i$ | −21.64 | −5.73 | 29.84 |
| *LFS* | −2.673 | | |

N = 42 females. See text for explanation of analysis. Reprinted from Cheney et al., "Factors affecting reproduction and mortality among baboons in the Okavango Delta, Botswana," *International Journal of Primatology* 25:401–428, with the kind permission of Springer.

**Table 7.** Percentage contribution of the components of lifetime reproductive success to variation in *LRS* for all deceased females of breeding age between 1992 and 2002

| | *L* | *F* | *S* |
|---|---|---|---|
| *L* | 38.58 | | |
| *F* | −0.45 | 4.90 | |
| $S_i$ | −8.57 | −11.78 | 30.54 |
| *LFS* | −0.34 | | |

N = 24 females. See text for explanation of analysis. Reprinted from Cheney et al., "Factors affecting reproduction and mortality among baboons in the Okavango Delta, Botswana," *International Journal of Primatology* 25:401–428, with the kind permission of Springer.

Even when only those females that survived to breeding age were considered in the analysis and offspring survival was overestimated, *L*, or the number of breeding years, accounted for the highest proportion of variance in lifetime reproductive success. Offspring (infant) survival accounted for the second highest proportion. Both fecundity and offspring survival negatively covaried with lifespan, with the result that their effect on the product reduced its variance rather than increasing it. The same was true of the covariance between fecundity and survival.

These trends remained the same when calculations were performed separately for high-, middle-, and low-ranking females. Because most adult and juvenile mortality was due to predation, and most infant mortality was due to infanticide, these two factors appeared to be the most important variables influencing female lifetime reproductive success.

## 6. DISCUSSION

Female baboons, like many other species of Old World monkeys, form nepotistic dominance hierarchies. Regardless of their rank, however, all females are

able to breed. As a result, differences in the reproductive success of high- and low-ranking females are not as obvious as they are in species with high reproductive skew, in which some females do not breed at all (see "Introduction").

Because one of the primary advantages of high rank is increased access to food, the reproductive benefits associated with high rank are likely to be most obvious in times of food scarcity or drought (Alberts et al., 2005). In the Moremi population, linear dominance relations among adult females were unambiguous, and there was competition among adult females and juveniles for food (Johnson, 2003). There was little evidence, however, of rank-related illness or malnutrition, and any functional effects of food competition on female reproductive success were subtle. As in many other populations of Old World monkeys (see "Introduction"), high-ranking females experienced somewhat shorter interbirth intervals than mid- or low-ranking females. Like baboons in Amboseli, Kenya (Altmann and Alberts, 2003), their infant and juvenile daughters also grew at faster rates and achieved higher weights for their age (Johnson, 2003, this volume). In contrast to Amboseli, however, these faster growth rates did not result in a younger age at first birth.

In Amboseli, female baboons with extensive social networks have significantly higher offspring survival than females without such networks (Silk *et al.*, 2003). Although this effect was not evident in the Moremi population, we did find that females with close female kin gave birth at younger ages and had shorter interbirth intervals. This pattern, however, was apparent only in high- and middle-ranking females. In captive vervet monkeys, primiparous females with living mothers experience significantly reduced infant mortality (Fairbanks and McGuire, 1986). This effect is particularly striking among high-ranking females, whose mothers take a more active role in care-giving than lower-ranking grandmothers (Fairbanks, 1988). Similarly, in baboons (Cheney, 1992; Silk et al., 1999, 2003, 2004) and rhesus macaques (*M. mulatta*; Berman, 1980), female members of high-ranking matrilines tend to maintain greater proximity, groom each other at higher rates, and support each other at higher rates than the members of low-ranking matrilines. The presence of female kin may accelerate reproduction in high- and middle-ranking individuals, for example, by enhancing access to resources or reducing stress levels. These effects may be less evident in less cohesive low-ranking matrilines, where the presence of kin may even decrease females' competitive access to food.

Unlike females in some other populations of baboons and Old World monkeys (see "Introduction"), high-ranking females in this study did not experience higher infant survival, probably because infant survival was not entirely condition dependent. Infanticide was the primary cause of infant mortality, and it affected females regardless of their competitive abilities. Adult male baboons have nearly twice the body mass of adult females, and even high-ranking females are unable to defend themselves or their infants against infanticidal males. Female alliances appear to be equally ineffective in resisting such attacks (Palombit et al., 2000). Perhaps as a result, high-ranking females and females with adult kin did not experience higher infant survival than other females.

Although females of all ranks were vulnerable to infanticide, middle-ranking females appeared to be less susceptible than high- and low-ranking females. Several factors may have contributed to high- and low-ranking females' increased risk of infanticide. First, males and high-ranking females enjoy priority of access to high-quality food resources, with the result that high-ranking females may be more likely than other females to feed in close proximity to a potentially infanticidal male. Second, high-ranking females are more successful than lower-ranking females at competing for "friendships" with high-ranking males (Palombit et al., 2001). These males, however, are also more likely than lower-ranking males to compete with other males for access to estrous females, which may again expose high-ranking females to potentially infanticidal males. Low-ranking females, in contrast, are more likely to feed or take refuge in more peripheral areas (van Noordwijk and van Schaik, 1987; Ron et al., 1996; Johnson, 2001) and to form friendships with low-ranking males who may be unable to resist attacks by higher-ranking males. Both factors may increase the vulnerability of low-ranking females to infanticide.

As in previous studies of this same population, predation was the primary cause of mortality among juveniles and adults (Busse, 1982; Bulger and Hamilton, 1987) and appeared to be a major determinant of female lifetime reproductive success. The intensity of predation fluctuated strongly, even over short periods of time. In some years, females experienced relatively low mortality rates. In one year, however, 24 percent of the group's adult females died of apparent predation (see also Bulger and Hamilton, 1987).

For female baboons at Moremi, two factors in particular accounted for the highest proportions of variance in lifetime reproductive success: a female's number of breeding years and the survival of her infants. This finding is unsurprising: Baboons are long-lived mammals that give birth to single offspring requiring extensive maternal investment. Indeed, similar analyses on other populations of baboons (Rhine et al., 2000; Altmann and Alberts, 2003), other monkeys (e.g., Cheney et al., 1988a,b), and other mammals (e.g., lions: Packer et al., 1988; elephant seals, *Mirounga angustirostris:* Le Boeuf and Reiter, 1988) have also shown that lifespan (number of breeding years) and offspring survival contribute substantially more to lifetime reproductive success than fecundity. The negative covariations between the products of $L$, $F$, and $S$ shown in Tables 6 and 7 are also expected. It is well known that offspring survival decreases female fecundity, and that investment in offspring may reduce the number of breeding years. More interesting, perhaps, is the lack of any effect of female dominance rank on the components contributing to lifetime reproductive success. High-ranking females did not live significantly longer than low-ranking females, and their infants suffered relatively high mortality rates. Both factors appeared to offset any advantages in increased fecundity that high-ranking females might have gained through priority of access to food.

There were strong seasonal effects on mortality. Most deaths among females, juveniles, and infants occurred during the 3 months of peak seasonal flooding. Throughout this period, the baboons took highly predictable routes with limited numbers of escape options when moving between islands, which may have increased their vulnerability to both predation and infanticide. Infanticide was not obviously correlated with the timing of male immigration, however, probably because not all immigrant males assumed high rank and those that did achieved this status at varying intervals after immigrating. Moreover, male immigration was not itself seasonal.

In a previous study of baboons in Amboseli, lactating females were found to experience higher mortality than females in other reproductive states (Altmann et al., 1988). In this study, by contrast, mortality appeared to be highest among pregnant females. One possible cause for this difference is that the friendships formed between adult males and lactating female baboons in Moremi are particularly strong and enduring (Palombit et al., 1997). In addition to reducing the probability of infanticide, friendships may have also reduced lactating females' own vulnerability to predation (Busse, 1984; Cowlishaw, 1999). In contrast, pregnant females spend more time feeding and less time engaged in social activities than other females (Silk, 1987), which may in turn increase their vulnerability to predation.

The findings reported here may be compared with those of an earlier 8-year study of the same group living at a similar population density (Bulger and Hamilton, 1987). In that study, high-ranking females also had shorter interbirth intervals. In this study, however, high-ranking females also experienced lower infant mortality and higher adult mortality. These differences in rank-related correlates of reproductive success serve to emphasize that the factors influencing reproduction and survival are both facultative and variable, even over short periods of time.

In sum, those reproductive parameters most likely to be associated with superior competitive ability—interbirth interval and infant growth rates—conferred a slight fitness advantage on high-ranking females. This fitness advantage was counteracted, however, by the effects of infanticide and predation. Infanticide affected high- and low-ranking females more than middle-ranking females, while predation affected females of all ranks relatively equally. As a result, there were few rank-related differences in estimated lifetime reproductive success.

## ACKNOWLEDGMENTS

We thank the Office of the President and the Department of Wildlife and National Parks of the Republic of Botswana for permission to conduct research in the Moremi Game Reserve. We are particularly grateful to W. J. Hamilton for providing essential background data and for making Baboon Camp available to us. For help in the field, we thank J. Bock, R. Boyd, K. Hammerschmidt, M. Kgosiekae, T. McNutt, M. Metz, J. Nicholson, K. Rendall, and K. Seyfarth. We especially thank M. Mokopi for his indispensable expertise in locating and identifying individuals. We also thank A. Dunham for writing the computer program for Brown's model. Research was supported by grants from the National Science Foundation, the National Institutes of Health, the DFG and KFN, the NSERC of Canada, the National Geographic Society, the LSB Leakey Foundation, and the University of Pennsylvania. The research presented here was described in Animal Research Protocol No. 190-1, approved annually since April 1992 by the Institutional Animal Care and Use Committee of the University of Pennsylvania.

## REFERENCES

Alberts, S. C., Altmann, J., Hollister-Smith, J., Mututua, R. S., Sayialel, S. N., Muruthi, P. M. and Warutere, J. K., 2005, Seasonality in a constantly changing environment, in: *Primate Seasonality: Implications for Human Evolution*, D. K. Brockman, and C. P. van Schaik, eds., Cambridge University Press, Cambridge.

Altmann, J. and Alberts, S. C., 2003, Intraspecific variability in fertility and offspring survival in a nonhuman primate: Behavioral control of ecological and social sources, in: *Offspring: The Biodemography of Fertility and Family Behavior*, K. W. Wachter, and R. A. Bulatao, eds., National Academy Press, Washington, DC, pp. 140–169.

Altmann, J., Hausfater, G., and Altmann, S. A., 1988, Determinants of reproductive success in savannah baboons *Papio cynocephalus*, in: *Reproductive Success*, T. H. Clutton-Brock, ed., University of Chicago Press, Chicago, pp. 403–418.

Barton, R.A., 1993, Sociospatial mechanisms of feeding competition in female olive baboons, *Papio anubis*, *Anim. Behav.* **46**:791–802.

Barton, R. A. and Whiten, A., 1993, Feeding competition among female olive baboons, *Papio anubis*, *Behav. Ecol. Sociobiol.* **38**:321–329.

Barton R. A., Byrne, R. W., and Whiten, A., 1996, Ecology, feeding competition and social structure in baboons, *Behav. Ecol. Sociobiol.* **38**:321–329.

Berman, C. M., 1980, Early agonistic experience and rank acquisition among free-ranging infant rhesus monkeys, *Int. J. Primatol.* **1**:152–170.

Brown, D., 1988, Components of lifetime reproductive success, in: *Reproductive Success*, T. H. Clutton-Brock, ed., University of Chicago Press, Chicago, pp. 439–453.

Brown, G. R. and Silk, J. B., 2002, Reconsidering the null hypothesis: Is maternal rank associated with birth sex ratios in primate groups? *Proc. Natl. Acad. Sci. USA* **99**:11252–11255.

Bulger, J. and Hamilton, W. J., 1987, Rank and density correlates of inclusive fitness measures in a natural chacma baboon (*Papio ursinus*) population, *Int. J. Primatol.* **8**:635–650.

Busse, C., 1982, Leopard and lion predation upon chacma baboons living in the Moremi Wildlife Reserve, *Botswana Notes Rec.* **12**:15–21.

Busse, C. D., 1984, Spatial structure of chacma baboon groups, *Int. J. Primatol.* **5**:247–261.

Cant, M. A., 2000, Social control of reproduction in banded mongooses, *Anim. Behav.* **59**:147–158.

Cheney, D. L., 1992, Within-group cohesion and inter-group hostility: The relation between grooming distributions and inter-group competition among female primates, *Behav. Ecol.* **3**:334–345.

Cheney, D. L., Andelman, S. J., Seyfarth, R. M., and Lee, P. C., 1988a, The reproductive success of vervet monkeys, in: *Reproductive Success*, T. H. Clutton-Brock, ed., University of Chicago Press, Chicago, pp. 384–402.

Cheney, D. L., Seyfarth, R. M., Andelman, S. J., and Lee, P. C., 1988b, Reproductive success in vervet monkeys, in: *Reproductive Success: Studies of Individual Variation in Contrasting Breeding Systems*, T. H. Clutton-Brock, ed., University of Chicago Press, Chicago, pp 384–402.

Cheney, D. L., Seyfarth, R. M., Fischer, J., Beehner, J., Bergman, T., Johnson, S. E., Kitchen, D. M., Palombit, R. A., Rendall, D., Silk, J. B., 2004, Factors affecting reproduction and mortality among baboons in the Okavango Delta, Botswana, *Int. J. Primatol.* **25**:401–428.

Combes, S. L. and Altmann, J., 2001, Status change during adulthood: Life-history byproduct or kin-selection based on reproductive value? *Proc. R. Soc. Lond. B* **268**:1367–1373.

Cowlishaw, G., 1994, Vulnerability to predation in baboon populations, *Behaviour* **131**:293–304.

Cowlishaw, G., 1999, Ecological and social determinants of spacing behaviour in desert baboon groups, *Behav. Ecol. Sociobiol.* **45**:67–77.

Creel, S. and Creel, N. M., 2002, *The African Wild Dog: Behavior, Ecology, and Conservation*, Princeton University Press, Princeton.

Creel, S. R. and Waser, P. M., 1997, Variation in reproductive suppression among dwarf mongooses: Interplay between mechanisms and evolution, in: *Cooperative breeding in mammals*, N. G. Solomon, and J. A. French, eds., Cambridge University Press, Cambridge, pp. 150–170.

De Luca, D. W. and Ginsberg, J. R., 2001, Dominance, reproduction and survival in banded mongooses: Towards an egalitarian social system? *Anim. Behav.* **61**:17–30.

Ellery W. N., Ellery, K., and McCarthy, T. S., 1993, Plant distribution in island of the Okavango Delta, Botswana: Determinants and feedback interactions, *Afr. J. Ecol.* **31**:118–134.

Fairbanks, L. A., 1988, Vervet monkey grandmothers: Interactions with infant grand-offspring, *Int. J. Primatol.* **9**:425–441.

Fairbanks, L. A. and McGuire, M. T., 1986, Age, reproductive value, and dominance-related behaviour in vervet monkey females: Cross-generational influences on social relationships and reproduction, *Anim. Behav.* **34**:1710–1721.

French, J. A., 1997, Singular breeding in callitrichid primates, in: *Cooperative Breeding in Mammals*, N. G. Soloman, and J. A. French, eds., Cambridge University Press, Cambridge, pp. 34–75.

Goldizen, A. W., 1987, Tamarins and marmosets: Communal care of offspring, in: *Primate Societies*, B. B. Smuts, D. L. Cheney, R. M. Seyfarth, R. W. Wrangham, and T. T. Struhsaker, eds., University of Chicago Press, Chicago, pp. 34–43.

Goldizen, A. W., Mendelson, J., van Vlaardingen, M., and Tergorgh, J., 1996, Saddle-back tamarin (*Saguinus fuscicollis*) reproductive strategies: Evidence from a thirteen-year study of a marked population. *Am. J. Primatol.* **38**:57–83.

Grafen, A., 1988, On the uses of data on lifetime reproductive success, in: *Reproductive Success*, T. H. Clutton-Brock, ed., University of Chicago Press, Chicago, pp. 454–471.

Hamilton, W. J., Buskirk, R. E., and Buskirk, W. H., 1976, Defense of space and resources by chacma (*Papio ursinus*) baboon troops in an African desert and swamp, *Ecology* **57**:1264–1272.

Hill, R. A. and Dunbar R. I. M., 1998, An evaluation of the roles of predation rate and predation risk as selective pressures on primate grouping behaviour, *Behaviour* **135**:411–430.

Isbell, L. A., 1991, Contest and scramble competition: Patterns of female aggression and ranging behavior among primates, *Behav. Ecol.* **2**:143–155.

Janson, C., 1985, Aggressive competition and individual food consumption in wild brown capuchin monkeys (*Cebus apella*), *Behav. Ecol. Sociobiol.* **18**:125–138.

Janson, C. H., 1988, Intra-specific food competition and primate social structure: A synthesis, *Behaviour* **105**:1–17.

Janson, C., 1998, Testing the predation hypothesis for vertebrate sociality: Prospects and pitfalls, *Behaviour* **135**:389–410.

Janson, C. H. and van Schaik, C. P., 2000, The behavioral ecology of infanticide by males, in: *Infanticide by Males and Its Implications*, C. van Schaik and C. Janson, eds., Cambridge University Press, Cambridge, pp. 469–494.

Jarvis, J. U. M., 1991, Reproduction in naked mole-rats, in: *The Biology of the Naked Mole-rat*, P. W. Sherman, J. U. M. Jarvis, and R. D. Alexander, eds., Princeton University Press, Princeton, pp. 384–425.

Johnson, S. E., 2001, Modeling the trade-off between energy acquisition and predation risk: Effects on individual variation in growth and mortality among baboons (*Papio hamadryas ursinus*) in the Okavango Delta, Botswana. Ph.D. Dissertation, University of New Mexico.

Johnson, S. E., 2003, Life history and the competitive environment: Trajectories of growth, maturation, and reproductive output among Chacma baboons, *Am. J. Phys. Anthropol.* **120**:83–98.

Johnstone, R. A., 2000, Models of reproductive skew: A review and synthesis, *Ethology* **106**:5–26.

Keller, L. and Reeve, H. K., 1994, Partitioning of reproduction in animal societies, *Trends Ecol. Evol.* **9**:98–103.

Le Boeuf, B. J. and Reiter, J., 1988, Lifetime reproductive success in northern elephant seals, in: *Reproductive Success*, T. H. Clutton-Brock, ed., University of Chicago Press, Chicago, pp.344–362.

Lee, P. C. and Johnson, J. A., 1992, Sex differences in alliances, and the acquisition and maintenance of dominance status among immature primates, in: *Coalitions and Alliances in Primates and Other Animals*, A. Harcourt, and F. de Waal, eds., Oxford University Press, Oxford, pp. 391–414.

Melnick, D. J. and Pearl, M. C., 1987, Cercopithecines in multimale groups: Genetic diversity and population structure, in: *Primate Societies*, B. B. Smuts, D. L. Cheney,

R. M. Seyfarth, R. W. Wrangham, T. T. Struhsaker, eds., University of Chicago Press, Chicago, pp. 121–134.

Moss, C. J., ed., in press, *The Elephants of Amboseli*, University of Chicago Press, Chicago.

van Noordwijk, M. A. and van Schaik, C. P., 1987, Competition among female long-tailed macaques, *Macaca fascicularis*, *Anim. Behav.* **35**:577–589.

van Noordwijk, M. A. and van Schaik, C. P., 1999, The effects of dominance rank and group size on female lifetime reproductive success in wild long-tailed macaques, *Macaca fascicularis*, *Primates* **40**:105–130.

O'Riain, M. J., Bennett, N. C., Brotherton, P. N., McIlrath, G., and Clutton-Brock, T. H., 2000, Reproductive suppression and inbreeding avoidance in wild populations of cooperatively breeding meerkats (*Suricata suricatta*), *Behav. Ecol. Sociobiol.* **48**:471–477.

Packer, C., Herbst, L., Pusey, A. E., Bygott, J. D., Hanby, J. P., Cairns, S. J., and Borgerhoff Mulder, M., 1988, in: *Reproductive Success*, T. H. Clutton-Brock, ed., University of Chicago Press, Chicago, pp.363–383.

Packer, C., Collins, D. A., Sindimwo, A. and Goodall, J., 1995, Reproductive constraints on aggressive competition in female baboons, *Nature* **373**:60–63.

Packer, C., Pusey, A. E., and Eberly, L. E., 2001, Egalitarianism in female African lions, *Science* **293**:690–693.

Palombit, R. A., Seyfarth, R. M., and Cheney, D. L., 1997, The adaptive value of "friendships" to female baboons: Experimental and observational evidence, *Anim. Behav.* **54**:599–614.

Palombit, R., Cheney, D., Seyfarth, R., Rendall, D., Silk, J., Johnson, S., and Fischer, J., 2000, Male infanticide and defense of infants in chacma baboons, in: *Infanticide by Males and Its Implications*, C. van Schaik, and C. Janson, eds., Cambridge University Press, Cambridge, pp. 123–152.

Palombit, R. A., Cheney, D. L., and Seyfarth, R. M., 2001, Female–female competition for male "friends" in wild chacma baboons (*Papio cynocephalus ursinus*), *Anim. Behav.* **61**:1159–1171.

Pereira, M., 1988, Agonistic interactions of juvenile savanna baboons. I. Fundamental features, *Ethology* **79**:195–217.

Pinheiro, J. C. and Bates, D. M., 2000, *Mixed-effect models in S and S-Plus*, Springer, New York.

Rhine, R. J., Wasser, S. K., and Norton, G. W., 1988, Eight-year study of social and ecological correlates of mortality among immature baboons of Mikumi National Park, Tanzania, *Am. J. Primatol.* **16**:199–212.

Rhine, R. J., Norton, G. W., Wynn, G. M., and Wynn, R. D., 1989, Plant feeding of yellow baboons (*Papio cynocephalus*) in Mikumi National Park, Tanzania, and the relationship between seasonal feeding and immature survival, *Int. J. Primatol.* **10**:319–342.

Rhine, R. J., Norton, G. W., Rogers, J., and Wasser, S. K., 1992, Secondary sex ratio and maternal dominance rank among wild yellow baboons (*Papio cynocephalus*) of Mikumi National Park, Tanzania, *Am. J. Primatol.* **27**:261–273.

Rhine, R. J., Norton, G. W., and Wasser, S. K., 2000, Lifetime reproductive success, longevity, and reproductive life history of female yellow baboons (*Papio cynocephalus*) of Mikumi National Park, Tanzania, *Am. J. Primatol.* **51**:229–241.

Ron, T., Henzi, S. P. and Motro, U., 1996, Do female chacma baboons compete for a safe spatial position in a southern woodland habitat? *Behaviour* **133**:475–490.

Roodt, V., 1998, *Trees and Shrubs of the Okavango Delta: Nutritional Uses and Nutritional Value*, Shell Oil Botswana, Gaborone, Botswana.

Ross, K., 1987, *Okavango: Jewel of the Kalahari*, Macmillan, New York.

van Schaik, C. P., 1983, Why are diurnal primates living in groups? *Behaviour* **87**:120–144.

van Schaik, C. P., 1989, The ecology of social relationships amongst female primates, in: *Comparative Socioecology: The Behavioural Ecology of Humans and Other Animals*, V. Standen and R. A. Foley, eds., Blackwell, Oxford, pp. 195–218.

Silk, J. B., 1983, Local resource competition and facultative adjustment of sex ratios in relation to competitive abilities, *Am. Nat.* **121**:56–66.

Silk, J. B., 1987, Activities and feeding behavior of free ranging pregnant baboons, *Int. J. Primatol.* **8**:593–613.

Silk, J. B., 1993, The evolution of social conflict among female primates, in: *Primate Social Conflict*, W. A. Mason and S. P. Mendoza, eds., State University of New York Press, Albany, pp. 49–83.

Silk, J. B., Seyfarth, R. M., and Cheney, D. L., 1999, The structure of social relationships among female baboons in the Moremi Reserve, Botswana, *Behaviour* **136**:679–703.

Silk, J. B., Alberts, S. C., and Altmann, J., 2003, Social bonds of female baboons enhance infant survival, *Science* **302**:1231–1234.

Silk, J. B., Alberts, S. C., and Altmann, J., 2004, Patterns of coalition formation by adult female baboons in Amboseli, Kenya, *Anim. Behav.* **67**:573–582.

Smuts, B. and Nicolson, N., 1989, Reproduction in wild female olive baboons, *Am. J. Primatol.* **19**:229–246.

Sterck, E. H. M., Watts, D. P., and van Schaik, C. P., 1997, The evolution of female social relationships in nonhuman primates, *Behav. Ecol. Sociobiol.* **41**:291–309.

Sugiyama, Y. and Ohsawa, H., 1982, Population dynamics of Japanese monkeys with special reference to the effects of artificial feeding, *Folia primatol.* **39**:238–263.

Tabachnik, B. G. and Fidell, L. S., 2001, *Using Multivariate Statistics*, Allyn and Bacon, Boston.

Vehrencamp, S. L., 1983, A model for the evolution of despotic versus egalitarian societies, *Anim. Behav.* **31**:667–682.

Wasser, S. K., Norton, G. W., Rhine, R. J., Klein, N., and Kleindorfer, S., 1998, Ageing and social rank effects on the reproductive system of free-ranging yellow baboons (*Papio cynocephalus*) at Mikumi National Park, Tanzania, *Hum. Reprod. Update* **4**:430–438.

Whitten, P. L., 1983, Diet and dominance among female vervet monkeys (*Cercopithecus aethiops*), *Am. J. Primatol.* **5**:139–159.

Wrangham, R. W., 1980, An ecological model of female-bonded primate groups, *Behaviour* **75**:262–300.

Wrangham, R. W., 1981, Drinking competition in vervet monkeys, *Anim. Behav.* **29**:904–910.

CHAPTER EIGHT

# Maternal Characteristics and Offspring Growth in Chacma Baboons: A Life History Perspective

*Sara E. Johnson*


## CHAPTER SUMMARY

The pace of growth and development of offspring is a critical component of female reproductive success. Optimality models assume selection favoring maternal characteristics associated with a growth trajectory of maximum rate and minimum cost. Moreover, these attributes vary with social and ecological context. This chapter examines the interaction between maternal and offspring characteristics that affect immature growth in chacma baboons. The analyses evaluate offspring growth, measuring the probability of a weight measurement below the weight-for-age curve for the population. As maternal age increases beyond 10–14 years, offspring growth rate significantly decreases, controlling for maternal rank. Male offspring with younger mothers grow more slowly than others, but the trend is not significant. Maternal rank significantly affects female growth independent of maternal age variation. The female offspring of


**Sara E. Johnson** • Department of Anthropology, California State University, Fullerton, Fullerton, CA, USA

lower-ranking mothers grow more slowly than their higher-ranking peers. Maternal rank appears to have a weaker and opposing effect on male offspring growth. Male offspring of the lowest-ranking mothers seem to grow slightly faster than their higher-ranking counterparts. However, this effect disappears when maternal age is included in the model, suggesting minimal impacts of maternal rank on male growth. The effects of maternal rank on female growth are consistent and strong, controlling for variation in maternal age and variation in rainfall. These analyses indicate that the benefits of rank to reproductive success in terms of female offspring growth are most pronounced under conditions of environmental stress. These findings contribute to our understanding of the complex dynamics relating maternal characteristics of rank and age to the growth trajectories of male and female offspring. Identifying mediating variables that affect trajectories of growth and development allows us to synthesize the allocation rules that affect maternal decision making and the pattern of investment. The addition of environmental variation permits an evaluation of the varying intensity with which maternal rank affects reproductive success.

## 1. INTRODUCTION

The setting of individual life history trajectories is a complex interweaving of parental and offspring characteristics shaped by the physical and social environment (Lee and Kappeler, 2003; Altmann and Alberts, 2003; Pereira and Leigh, 2003). The costs and benefits of a particular growth trajectory have been discussed in terms of ecological risk (Janson and van Schaik, 1993), social complexity (Leigh, 1995), and maternal strategies (Ross and MacLarnon, 1995; Altmann and Alberts, 2003; Lee, 1999). This chapter focuses on maternal and environmental conditions affecting immature growth in chacma baboons within a life history framework. The study population, consisting of chacma baboons with a multimale social structure inhabiting a resource-rich, highly competitive environment with high levels of predation, provides significant opportunities to examine the features of the physical and social environment that affect growth. The analytical portion of this chapter uses hypotheses grounded in life history theory to examine maternal strategies and their effects on offspring growth trajectories within both a social and an ecological context.

Maternal characteristics that influence resource acquisition and energy transfer, as well as hormonal state, affect the parental investment strategies available to individual adult females. This has been discussed within the framework of life history theory by examining the factors that impact the trade-off between current and future reproduction (Lycett et al., 1998; Lee and Kappeler, 2003). For example, greater access to resources and a large initial investment in lactation shorten the interbirth interval by increasing infant growth rate (Lee et al., 1991). The cost of this reproductive strategy varies both with access to resources and the ability of the mother to maintain a positive energy balance. Offspring and maternal characteristics also interact to affect the investment strategies of individual females, and this has been particularly well studied with regard to offspring sex (Trivers and Willard, 1973; Silk, 1983; Takahata et al., 1995). This chapter evaluates the effects of maternal age and rank as independent characteristics differentially influencing the developmental trajectories of female and male offspring. This analysis also evaluates the mother's role in offspring growth by testing for an effect of maternal presence or absence. Furthermore, the role of ecological variation on growth is addressed by examining season of birth in relation to offspring growth trajectories. The present analysis complements a previous study of the Okavango population, which showed that the female infants of high-ranking mothers were significantly heavier for their age than the female infants of low-ranking mothers. High-ranking mothers also had shorter interbirth intervals than low-ranking mothers (Johnson, 2003). The present study adds to our knowledge of baboon life history by investigating in more detail the interaction between maternal and offspring characteristics, including offspring sex, in the context of environmental variation.

### 1.1. Maternal Condition

The effects of maternal condition on maternal style and patterns of maternal investment have been studied in many primate species, spanning across a range of maternal body sizes and reproductive patterns (Berman, 1988, 1990; Ross and MacLarnon, 1995; Lycett et al., 1998; Tardif et al., 2002). Maternal condition has been operationalized in a number of ways (e.g., body fat (Berman, 1988); maternal age, rank, and weight (Fairbanks and McGuire, 1995). Similarly, the outcome of maternal investment has been measured as infant and juvenile survival, weaning age, and growth rate. This study disaggregates maternal condition into maternal rank and maternal age, then tests for separate effects of these variables based on the differential roles each variable is hypothesized to play with respect to immature growth.

Characteristics of individual offspring, such as sex and mass gain, interact with maternal condition to shape maternal investment strategies. As mentioned above, the pace of infant mass gain affects the interbirth interval (IBI; see Barrett et al., this volume) and ultimately, the reproductive success of mothers (Johnson, 2003). Both the Trivers–Willard hypothesis (1973) as well as the Local Resource Competition model predict that maternal condition will affect the production of sons versus daughters. Schino and colleagues (1999) extend this model to predict variation in postnatal investment. They measured investment through lactation and infant carrying in their study of rhesus macaques, reporting that high-ranking mothers with male infants have significantly more ventroventral contact and carry them longer. On the other hand, low-ranking mothers carried female infants longer, with no differences by sex in time spent in ventroventral contact. In the current study, differential postnatal investment is measured in terms of growth.

### 1.2. Maternal Rank

Maternal rank, or position in the social hierarchy, contributes to differential access to resources, depending on both the distribution and abundance of resources and the level of intragroup competition (Sterck et al., 1997; Barton et al., 1996; Cheney et al., this volume). Lactation and infant carrying comprise two major routes of maternal investment (Altmann and Samuels, 1992) that influence offspring energy balance and growth. The effect of maternal rank on the growth of male versus female offspring could occur through either, or both, avenues of investment. For example, increasing maternal rank may positively affect offspring growth through several means, including higher-quality milk (see Roberts et al., 1985), more efficient energy transfer through less interrupted suckling bouts, greater access to weaning foods, or priority of access to high-quality food patches. Higher resource availability may be associated with shorter periods of investment (cf. Barrett et al., this volume). For example, higher-ranking mothers have been observed to devote less time to infant carrying and may shorten interbirth intervals without increasing the risk of infant mortality by shortening the period of lactation (Altmann and Samuels, 1992).

In a previous study of this population, the infant mass at which high- and low-ranking mothers resumed cycling was relatively constant, but high-ranking mothers were able to bring their infants to that mass sooner (Johnson, 2003). Rather than focusing solely on the constraints facing low-ranking mothers, the role of maternal rank on trajectories of growth and development will be considered as an important determinant of investment strategy that is also affected by offspring characteristics such as sex.

### 1.3. Maternal Age

Reproductive value generally declines with age during the reproductive period. Thus, a prediction consistent with life history theory is that maternal investment in any single bout of reproduction should increase with age (Roff, 1992). In solving the trade-off between current and future offspring, each increase in maternal age should increase the value of the current offspring by diminishing the potential of future offspring. This predicts a positive relationship between maternal age and offspring investment in terms of growth. However, other aspects of aging may have opposing effects on offspring growth. Senescence of either the reproductive system or soma, as it relates to energy balance, may negatively affect offspring outcome in terms of low birth mass or low infant growth rate. At some threshold age, the effects of senescence may overwhelm the effects of age-specific reproductive value, leading to the hypothesis that the relationship between maternal age and offspring achieved mass-for-age is not linear but follows a more complex function. As maternal age increases, the probability of an immature being below a population-specific growth curve decreases. This occurs until some threshold age, when the probability of being below the growth curve begins to increase with maternal age as a result of diminishing maternal access to resources through the progression of senescence.

### 1.4. Environmental Conditions and Seasonality

Maternal investment strategies are shaped, in part, by the demands of the physical environment (see Barrett et al., this volume). In highly seasonal environments, due to variation in plant productivity or predation pressure, certain times of the year may be associated with better offspring outcome (Barrett et al., this volume; DiBitetti and Janson, 2000). Clumping of reproductive events, rather than randomly distributed births, raises the possibility that the timing of reproduction maximizes maternal fecundity, energy transfer to offspring through lactation, or the availability of weaning foods (Lancaster and Lee, 1965; Altmann, 1980; DiBitetti and Janson, 2000).

Long-term data from the Okavango population demonstrate a seasonal pattern to reproduction, with births concentrated from July through September (Cheney et al., 2004, this volume). Because birth season potentially affects maternal condition, patterns of investment, and offspring foraging competence through the availability of appropriate transition foods, it may play a significant role in immature growth. Infants weaned in the dry season may face shortages of appropriate weaning foods because many of the fruiting tree species flower only during the wet season. Maternal condition may be enhanced by dry season pregnancy if the initial period of lactation occurs after the winter floods, a period of high plant productivity (Ross, 1989).

### 1.5. Predictions from Life History Theory

Maternal characteristics and offspring competence interact to produce different trajectories of offspring growth and development (Johnson, 2003; Johnson and Bock, 2004). The optimal solution to the central life history trade-off between current and future offspring varies with maternal characteristics such as rank and age, and thereby shapes patterns of maternal investment. Maternal rank and age may have independent effects on offspring achieved mass-for-age, especially as it pertains to energy transfer because factors mediating this relationship are assumed to be different. For example, a consequence of rank—such as differential access to food—is an immediate constraint on maternal investment, whereas the effect of age on reproductive value patterns the overall level of investment in a particular offspring. While all older females are expected to favor investment in current offspring rather than future reproduction, higher-ranking females will have fewer limits on the amount of investment available. In this study, I examine the following hypotheses and predictions, all derived from life history theory:

*Hypothesis 1.* Maternal age affects offspring growth differently at different points in the mother's life course. *Prediction 1.* There is a nonlinear relationship between maternal age and offspring growth.

*Hypothesis 2.* Environmental conditions at the birth of the individual affect offspring growth. *Prediction 2.* The level of rainfall at the birth of an individual affects the probability of being below the mass-for-age curve.

*Hypothesis 3.* Increases in maternal rank are associated with higher rates of offspring growth. *Prediction 3.* Immature offspring of higher- and medium-ranking females are more likely to be above the mass-for-age curve.

*Hypothesis 4.* The death of a mother negatively affects offspring growth. *Prediction 4.* Mother absence is a significant risk factor for a mass measurement below the mass-for-age curve.

A previous study demonstrated that maternal rank had a significant bearing on female growth but no significant consequence for male growth (Johnson, 2003). The current analysis examines maternal rank over three levels because, although the results of the previous study were not significant with respect to male growth, the regression of average residual on maternal rank suggested that the effect of low rank may differ with medium and high maternal rank. Additionally, the lighter weaning mass of males versus females suggests variation in maternal investment that merits further study. These results, coupled with the substantial literature on sex-biased patterns of investment, suggest that the growth of male and female immatures should be considered separately. In this study, the effect of maternal rank is evaluated while controlling for other important predictive variables contributing to maternal condition and offspring competence.

## 2. METHODS

### 2.1. Data Collection

Weights (as a measure of mass) of 42 immature baboons were obtained without sedation or baiting from a troop of well-habituated chacma baboons in the Moremi National Wildlife Reserve of the Okavango Delta, Botswana. This troop has been the subject of long-term study since Hamilton began work in the 1970s, and the environment and study population have been described in numerous publications (Hamilton et al., 1976; Hamilton, 1982; Busse, 1982; Cheney et al., 1995, 1996; Cheney and Seyfarth, 1997; Silk et al., 1999; Cheney et al., 2004; Cheney et al., this volume). Demographic data have been continually updated, so ages of all juveniles and infants and most adults are known as are maternal kin relationships. Features of the ecology and population dynamics critical to the current study include a high degree of seasonality and high population density. There are two elements that contribute to a complex seasonality. First, the annual distribution of precipitation varies, leading to distinctive wet and dry seasons along with a yearly flood in the dry season resulting from rainfall at the headwaters of the Okavango River in the highlands of southern Angola. Second, historical records suggest that rainfall fluctuates on an 18-year cycle (Thomas and Shaw, 1991), driving a succession of alternating wet and drought decades. Such a cyclical pattern directly and indirectly affects the population dynamics of a species such as the baboon as a result of variation in environmental extremes, food availability, and the species distribution and abundance of the biotic community. The size of this troop has been reported at a low of 43 and a high of 73 over an 8-year period (Bulger and Hamilton, 1987). During the present study, troop numbers oscillated at around 75 individuals, placing it at the upper end of observed density and suggesting intense within-group competition (Sterck et al., 1997; Bulger and Hamilton, 1987).

Female social relationships in the study group are organized in a strict linear dominance hierarchy with related females occupying adjacent ranks (Silk et al., 1999). For the purposes of this study, females were divided into high, medium, and low ranks. Repeated measures of mass for the infants and juveniles were obtained using a platform scale (Ohaus model D10L-M; see Johnson, 2003).

### 2.2. Data analysis

A multipronged analytical strategy was used to examine the effects of maternal rank, maternal age, mother's presence, and season of birth on offspring achieved mass-for-age. In the first set of analyses, the dependent variable was derived from residual data from separate regression analyses of mass-on-age for both male and female offspring. A dichotomous transformation was performed and the variable was coded as "negative residual = 1" and "positive residual = 0." Because the dependent variable is dichotomous and most likely follows a binomial distribution, the appropriate analytical technique is logistic regression. The second set of analyses focuses on the month of birth effect as it relates to seasonality and maternal rank. In the third set of analyses, the immature age range in which maternal age has the greatest effect on growth is estimated using piecewise ordinary least square (OLS) regression combined with a LOWESS smooth.

The application of logistic regression to biological data is comparable to ordinary least square OLS regression; however, the maximum likelihood method of least squares, which produces linear likelihood equations in an OLS regression, requires modification to deal with nonlinear likelihood equations (Hosmer and Lemeshow, 1989). The assumption in OLS regression is that the expected value of $y$ is a linear function of $x$, the relationship is linear across the values of $x$, and the error (an observation's deviation from the conditional mean) follows a normal distribution (Trexler and Travis, 1993). Alternatively, the explicit assumption in logistic regression is that the error follows a binomial distribution (Agresti, 1990) and this is addressed through the use of the logit transformation. The dependent variable is transformed as an odds' ratio, with the natural log of the odds' ratio as the logit transformation (Trexler and Travis, 1993). One of the advantages to this approach, analogous to the benefits of OLS regression, is that the logit is linear, it may be continuous, and it may range from $-\infty$ to $+\infty$ (Hosmer and Lemeshow, 1989). The independent variables in a logistic regression may be either discrete or continuous.

In order to test for the independent effects of birth seasonality, each environmental variable was independently evaluated, controlling for aspects of maternal condition such as rank and age. Model building is an essential part of logistic regression (Hosmer and Lemeshow, 1989). Consequently, a hierarchy of models was assessed in order to determine the important parameters affecting infant and juvenile growth. I compared the predictive power of models using different sets of variables; variables failing to improve the model as measured by the likelihood ratio test were excluded (Diggle et al., 1994).

Separate logistic models were constructed for males and females. For each, I report the odds' ratio, 95 percent confidence interval, and $p$-value for each variable. The odds' ratio is a measure of association and it estimates, in this case, how much more likely or unlikely it is for a mass to be below the curve among individuals of different maternal rank, maternal age, maternal presence, and season of birth. An odds' ratio of one indicates that the probability of an event occurring, having a negative mass-for-age residual, is equal to the probability of the opposite or having a positive mass-for-age residual. An odds ratio greater than one indicates how much more likely an event is to occur based on the value of the independent variable and an odds' ratio less than 1 indicates how much less likely an event is to occur (Hosmer and Lemeshow, 1989).

In order to determine the range for which maternal age exacts a negative effect on immature growth, piecewise OLS regression analyses, contrasting young and old mothers, were conducted separately for males and females. All analyses were conducted using SPSS 12.0.

## 3. RESULTS

Results from multivariate logistic regression analyses demonstrate that the female offspring of low-ranking females are approximately 20 times more likely to be below the mass-for-age growth curve than the offspring of high-ranking females (see Table 1). The female offspring of medium-ranking females are approximately 4 times more likely than the female offspring of high-ranking females to be below the mass-for-age growth curve. Maternal death poses significant risks for low achieved mass-for-age. Controlling for maternal rank, the female offspring of deceased mothers are approximately 8 times more likely to be below the mass-for-age curve than female offspring with living mothers.

Season of birth, as it affects rainfall, is a significant risk factor for animals low in terms of achieved mass-for-age. However, rainfall's effect varies across levels of maternal rank. Specifically, these analyses indicate a significant interaction effect between the amount of rainfall during the month of an offspring's birth and maternal rank. On the other hand, there is no significant effect of rainfall on the probability that immatures of high maternal rank will be below the mass-for-age curve. Thus, immatures of low maternal rank are 1.6 times more

**Table 1.** Risk factors for the probability of a mass measurement being below the average mass-for-age curve. Multivariate logistic regression analyses of repeated mass measurements on maternal and environmental characteristics were conducted separately for each sex. The sample size refers to the number of repeated measures for 22 immature females and 18 immature males. The dependent variable is a dichotomous transformation of the residuals of an OLS regression of mass on age for that sex. The dependent variable was coded as "negative residual = 1" and "positive residual = 0." Reference groups for the categorical risk factors (independent variables) are "high rank" and "mother alive." Results are reported as odds' ratio followed by upper and lower 95 percent confidence intervals within the parentheses and $p$-value

| Risk factors | Females ($N = 213$) | Males ($N = 150$) |
|---|---|---|
| Low rank | 20.03 (5.15,77.91) $p < 0.0001$ | 0.04 (0.01,0.27) $p = 0.001$ |
| Medium rank | 3.93 (1.20,12.80) $p = 0.023$ | 0.09 (0.02,0.54) $p =$ n.s. |
| Rainfall | N/A | 0.47 (0.27,0.84) $p = 0.01$ |
| Low rank* rainfall | 1.60 (1.06,2.42) $p = 0.025$ | 0.33 (0.02,5.64) $p =$ n.s. |
| Medium rank* rainfall | 2.13 (1.39,3.28) $p = 0.001$ | 18.33 (3.37,99.52) $p = 0.001$ |
| Mother dead | 8.51 (2.70,26.84) $p < 0.0001$ | N/A |

likely to be below the mass-for-age curve with each millimeter increase in rainfall during the month of their birth. Immatures of medium maternal rank are approximately 2 times more likely to be below the mass-for-age curve with each millimeter increase in rainfall during the month of their birth.

A parallel multivariate logistic regression analysis for males shows that maternal rank is a significant risk factor for low achieved mass-for-age, distinguishing the immatures of low-ranking females from those of high-ranking females (see Table 1). However, in contrast to the effect of maternal rank on female immature growth, male offspring of low-ranking mothers are significantly *less* likely to be below the mass-for-age growth curve. A small sample size precludes testing for effects of mother absence for males. The inclusion of rainfall in the model changes the coefficients for maternal rank, as it did for females, and the interaction term significantly improves the model. Male offspring of medium maternal rank are approximately 18 times more likely to be below the mass-for-age curve with each millimeter increase in rainfall. The immatures of high maternal rank did not show lower residuals as the amount of rainfall at the time of birth increased. A larger sample size is required to see if immatures of low-ranking mothers also would be more likely to be below the mass-for-age curve as rainfall increased. In the current sample, all the male immatures of low maternal rank were born during months with little to no rain.

Further multivariate logistic regression analyses indicate that maternal age at the time of the offspring's mass measurement is a significant risk factor for low achieved mass-for-age (see Table 2). Immature females are 1.2 times more likely to be below the mass-for-age growth curve with each additional year of maternal age. This effect is significant controlling for maternal rank. However, the effect of maternal age is consistent across ranks, as indicated by a nonsignificant interaction term. Controlling for maternal rank and maternal

**Table 2.** Risk factors for the probability of a mass measurement being below the average mass-for-age curve. These multivariate logistic regressions are similar to those reported in Table 1 except that in this analysis "maternal age" is substituted for the independent variable "mother dead." For further information see Table 1

| Risk factors | Females ($N = 203$) | Males ($N = 99$) |
|---|---|---|
| Low rank | 58.36 (6.80,501.21) $p < 0.0001$ | 0.59 (0.09,3.79) $p$ = n.s. |
| Medium rank | 25.84 (3.21,207.73) $p = 0.002$ | 2.75 (0.52,14.55) $p$ = n.s. |
| Rainfall | 1.59 (1.16,2.19) $p = 0.004$ | 2.78 (1.36,5.66) $p = 0.005$ |
| Maternal age (years) | 1.19 (1.06,1.34) $p = 0.003$ | 2.04 (1.44,2.90) $p < 0.0001$ |

age, season of birth remains a significant risk factor for immature female low achieved mass-for-age. Immature females are 1.5 times more likely to be below the mass-for-age curve with each millimeter increase in rainfall. Controlling for variation in season and maternal age, female immatures of low maternal rank are approximately 63 times more likely to have a mass measurement below the mass-for-age curve while female immatures of medium maternal rank are approximately 27 times more likely to have a mass measurement below the mass-for-age curve.

The effect of maternal rank on male growth is not significant when rainfall and maternal age are controlled. Immature males are 2 times more likely to be below the mass-for-age curve as maternal age increases. As the amount of rainfall at the time of birth increases, immature males are almost 3 times as likely to be below the mass-for-age curve.

Scatter plots of maternal age in years on the residual of mass-for-age of female offspring indicate that the negative effect of maternal age on immature growth intensifies at approximately 14 years of age (see Figure 1). There is no apparent negative effect of maternal age on offspring growth for younger

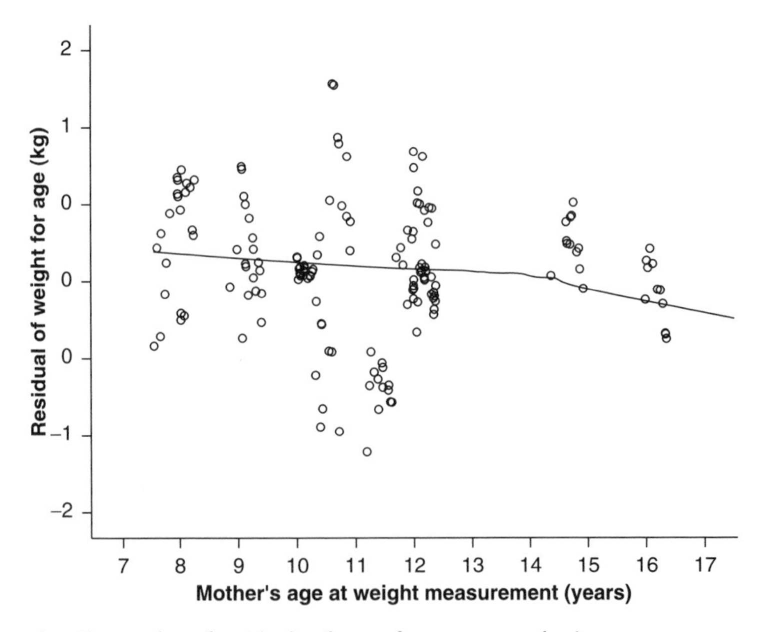


**Figure 1.** Scatterplot of residuals of mass-for-age on mother's age at mass measurement for females. Each marker represents one mass observation. A lowess curve was fit to the data. For comparable OLS regression analysis see Table 3.

mothers. A similar pattern is observed for males, with mothers between 12 and 14 years of age having a negative effect on achieved mass-for-age of male offspring (see Figure 2).

Two multiple OLS regression analyses of offspring mass by offspring age and maternal age were conducted by dividing the sample into younger mothers (age 6–12.99 years) and older mothers (13 and above). Controlling for age of the offspring produces no significant effect of maternal age on offspring mass for the immatures of young mothers (see Table 3 for males). However, maternal age is a significant predictor of female offspring mass for mothers 13 years and older (see Table 4).

## 4. DISCUSSION

This study demonstrates that maternal age, maternal rank, maternal absence, and variation in rainfall at the time of birth all have significant independent

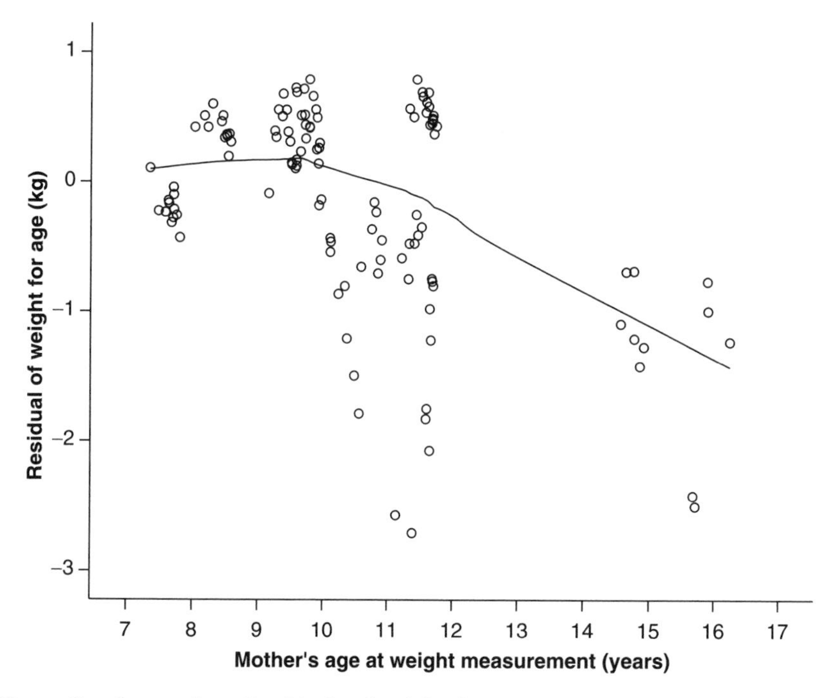


**Figure 2.** Scatterplot of residuals of weight-for-age on mother's age at mass measurement for males. Each marker represents one mass observation. A lowess curve was fit to the data. For comparable OLS regression analysis see Table 4.

**Table 3.** Three separate OLS regressions measuring the effect of maternal age in younger and older mothers on mass of male offspring. Standardized betas are adjusted controlling for age of the offspring. Young mothers are all females aged 6–12.99. Older mothers range in age from 13–18. Mothers of all ages are grouped together for the last analysis. There is no significant effect of parity when it is included in the model

| Males | Std beta | *p*-value |
|---|---|---|
| Young mothers | – 0.018 | n.s. |
| Older mothers | – 0.013 | n.s. |
| All mothers | – 0.046 | 0.008 |

**Table 4.** Three separate OLS regressions conducted separately for younger mothers, older mothers, and all mothers together to measure the effect of maternal age on mass of female offspring. Standardized betas are adjusted controlling for age of the offspring. Age categories follow those listed for the mothers of male offspring. There is no significant effect of parity when it is included in the model

| Females | Std beta | *p*-value |
|---|---|---|
| Young mothers | – 0.080 | n.s. |
| Older mothers | – 0.319 | <0.0001 |
| All mothers | – 0.121 | 0.005 |

effects on the growth of female chacma baboons. The relationship between maternal age and offspring growth is nonlinear and similar for males and females. This is consistent with expectations based on reproductive value and investment, coupled with the effects of reproductive senescence. However, the mediating variables remain unclear. Fairbanks and McGuire (1995) propose that maternal age extremes are related to poor condition and such females exhibit higher rates of maternal rejection in an attempt to preserve maternal reproductive capacity for the future. In this population, as maternal age increases, the length of lactation (estimated as the age of offspring at maternal resumption of cycling) increases and plateaus. Other studies have also demonstrated that younger mothers devote less time to lactation (Wilson et al., 1988). In contrast to the observed effect on immature growth, maternal age affects birth mass in common marmosets, but not postnatal growth rates (Tardiff and Bales, 2004).

An alternative explanation for lower growth rates among the youngest mothers relates to the cost of primiparity. Although many studies have demonstrated a cost to primiparity (Sade, 1990; Cheney et al., 2004), and younger mothers are obviously more likely to be primiparous, the present

ıan, C. M., 1990, Consistency in maternal behavior within families of free-ranging
esus monkeys: An extension of the concept of maternal style, *Am. J. Primatol.*
:159–169.

n, G. R. and Silk, J. B., 2002, Reconsidering the null hypothesis: Is maternal
k associated with birth sex ratio in primate groups? *Proc. Natl. Acad. Sci.*
11252–11255.

, J. and Hamilton, W. J., 1987, Rank and density correlates of inclusive fitness
sures in a natural chacma baboon (*Papio ursinus*) troop, *Int. J. Primatol.*
5–650.

C., 1982, Leopard and lion predation upon chacma baboons living in the
emi Wildlife Reserve, *Botswana Notes Rec.* **12**:15–21.

D. L. and Seyfarth R. M., 1997, Reconciliatory grunts by dominant female
ons influence victims behavior, *Anim. Behav.* **54**:409–418.

D. L., Seyfarth, R. M., and Silk, J. B., 1995, The role of grunts in reconcil-
pponents and facilitating interactions among adult female baboons, *Anim.*
**50**:249–257.

D. L., Seyfarth, R. M., and Palombit, R., 1996, The function and mecha-
underlying baboon contact barks, *Anim. Behav.* **52**:507–518.

D. L., Seyfarth, R. M., Fischer, J., Beehner, J., Bergman, T., Johnson, S. E.,
n, D. M., Palombit, R. A., Rendall, D., and Silk, J. B., 2004, Factors
g reproduction and mortality among baboons in the Okavango Delta,
na, *Int. J. Primatol.* **25**:401–428.

M. S. and Janson, C. H., 2000, When will the stork arrive? Patterns of birth
ity in neotropical primates, *Am. J. Primatol.* **50**:109–130.

J., Liang, K. Y., and Zeger, S. L., 1994, *Analysis of Longitudinal Data*,
University Press, Oxford.

L. A. and McGuire, M. T., 1995, Maternal condition and the quality of
care in vervet monkeys, *Behaviour* **132**:734–754.

V. J., 1982, Baboon sleeping site preferences and relationships to primate
patterns, *Am. J. Primatol.* **3**:41–53.

V. J., Buskirk, R. E., and Buskirk, W. H., 1976, Defense of space and
by chacma (*Papio ursinus*) baboon troops in an African desert and swamp,
:1264–1272.

. J., Buskirk, R. E., and Buskirk, W. H., 1978, Omnivory and utilization
ources by chacma baboons, *Papio ursinus*, Am. Nat. *112:911–924.*

M. and Louw, G. J., 1990, Height and weight differences among South
an school children born in various months of the year, *Am. J. Hum. Biol.*

I. and, Louw, G. J., 1993, Further studies on the month-of-birth effect
e: Rural schoolchildren and an animal model; *Am. J. Phys. Anthropol.*
.



analyses find no significant effect of parity on immature growth. Further study is merited to determine the senescent features of reproduction or behavior that contribute to the negative effect of maternal age on offspring growth. As the effect is consistent for males and females, it may be related to the energetics of aging through a compromise in energy transfer in utero or energy balance that promotes a change in behavior such as time devoted to infant carrying. One avenue for future research is to determine the offspring ages most sensitive to the effects of maternal age on growth.

These results suggest that the benefits to an infant of high maternal rank resist environmental perturbation. These benefits accrue in terms of energy transfer during gestation and lactation, as well as greater access to independently acquired resources. Month of birth has documented impacts on later height and mass in humans, but the mechanisms involved are unclear (Shephard et al., 1979; Henneberg and Louw, 1990, 1993; Weber et al., 1998; Banegas et al., 2001; Koscinski et al., 2004). Birth month may have more profound effects in groups of high-socioeconomic status than in low-status groups (Henneberg and Louw, 1993, Banegas et al., 2001). However, in the current study the growth of offspring of high-ranking mothers did not suffer if they were born in a month with higher rainfall, whereas the growth of offspring of low- and medium-ranking mothers was compromised if they were born in a month with higher rainfall. It is uncertain whether or not this effect of season of birth on growth stems from lower resource levels *in utero* or the seasonal availability of weaning foods. Birth seasonality can also contribute to maternal condition by influencing the distribution and abundance of resources at the time of conception, birth, and through the period of lactation (cf. Barrett et al., this volume). Independent of maternal condition, immature growth is further influenced by season of birth, as it dictates the availability of appropriate weaning foods (Altmann, 1980). Long-term data from the Okavango population indicate that high-ranking females exhibit slightly less reproductive seasonality than low-ranking females (Cheney et al., this volume). Reduced seasonality may reflect a higher probability of conception during a time of lower plant productivity, especially in the dry season (Cheney et al., 2004). Alternatively, or in addition, low- and medium-ranking females may under some circumstances facultatively adjust the timing of reproduction to maximize immature growth, but at a cost to reproductive rate.

The effects of maternal rank on male growth are more problematic to interpret. Immature males of low maternal rank are significantly less likely to be below the mass-for-age curve compared to the immatures of high maternal rank, at least when variation in maternal age remains uncontrolled. Although a previous study found no significant relationship between maternal rank and male growth, the regression of average residual on maternal rank (Johnson, 2003) suggests that the immatures from the lowest-ranking mothers may be slightly heavier than those from medium- and high-ranking mothers. The diminished impacts of maternal rank on male growth may denote reduced maternal influence. Alternatively, the stronger effect of maternal age and rainfall on male versus female growth may swamp the effects of maternal rank. It has been argued that variation in growth trajectories by sex is the result of variation in the selective pressures facing males and females (Plavcan, 2001). This study delineates both the shared and unique variables that affect male and female growth. It also suggests that the effects of maternal characteristics in the context of environmental conditions act differently on male and female immature baboon growth.

Since the Trivers–Willard hypothesis was first presented, researchers have examined biases in birth and secondary sex ratios among nonhuman primates (Altmann, 1980; Silk et al., 1981; Simpson and Simpson, 1982; Brown and Silk, 2002). Because maternal condition is heavily influenced by rank in baboons, it is a valuable commodity related to reproductive success that can be transmitted from mother to daughter (Altmann, 1980). Consistent with Leimar's model (1996), high-ranking females may be investing less in their sons than low-ranking mothers. In a previous study, the weaning mass of male offspring from high-ranking mothers was lower than for females of high-ranking mothers and the offspring of both sexes from low-ranking mothers (Johnson, 2003). If juvenile males face greater competitive pressure with size growth (Jarman, 1983; Leigh, 1995), then accelerating male growth may not represent the best strategy in a dangerous environment where early transfer may be associated with a high risk of mortality.

This study demonstrates that the maternal characteristics of rank and age have independent effects on the achieved mass-for-age in chacma baboons. Although offspring may be most vulnerable to the loss of a mother during the period of lactation, this study shows that the mother's presence significantly influences an offspring's achieved mass-for-age during the postweaning period (i.e., while offspring are foraging independently). Offspring of older mothers experience a negative effect on growth, independent of rank. These findings contribute to our understanding of the complex dynamics relating maternal and offspring life history characteristics and the intermediate variables that



affect trajectories of growth and development. The
maternal condition is itself composed of a suite of trai
ing functional effects on both maternal fitness and
decomposing this construct into its component ele
understanding of the synergistic relationship amon
environments and variation in life history traits.

## ACKNOWLEDGMENT

I would like to thank the Office of the Presider
sion to conduct this research in the Moremi Nat
thank Dorothy Cheney and Robert Seyfarth as
Wildlife and National Parks for permission to
Okavango field site. Funding was provided by
Dissertation Improvement Grant and an LSB
Grant, as well as a National Science Foundation
Robert Seyfarth. Many thanks to John Bock fo
lection. Indispensable research assistance was
thank John Bock, Steve Leigh, Larissa Swede
for their immensely helpful comments that
omissions and errors are strictly my responsib

## REFEREN

Agresti, A., 1990, *Categorical Data Analysis*,
Altmann, J., 1980, *Baboon Mothers and* Cambridge.
Altmann, J. and Samuels, A., 1992, Cost baboons, *Beh. Ecol. Sociobiol.* **29**:391–398
Altmann, J. and Alberts, S., 2003, Variabil life-history perspective in baboons, *Am.*
Banegas, J. R., Rodriguez-Artalejo, F., Gra Fisac, J. L., 2001, Month of birth and *Hum. Biol.* **28**:15–20.
Barton, R. A., Byrne, R. W., and Whiten, social structure in baboons, *Behav. Ec*
Berman, C. M., 1988, Maternal conditi ranging rhesus monkeys: An eleven-y



Berr
rh
22
Brow
ran
98
Bulge
mea
8:6
Busse,
Mor
Cheney
babo
Cheney
ing o
*Behav*
Cheney,
nisms
Cheney,
Kitche
affectin
Botswa
DiBitetti,
seasonal
Diggle, P.
Oxford
Fairbanks,
maternal
Hamilton,
grouping
Hamilton,
resources
*Ecology* 57
Hamilton, W
of food re
Henneberg,
African urb
2:227–233
Henneberg, M
on body siz
91:235–24

Hosmer, D. W. and Lemeshow, S., 1989, *Applied Logistic Regression*, John Wiley and Sons, New York.

Janson, C. H. and van Schaik, C. P., 1993, Ecological risk aversion in juvenile primates: Slow and steady wins the race, in: *Juvenile Primates*, M. E., Pereira and L. A. Fairbanks eds., Oxford University Press, Oxford, pp. 57–74.

Jarman, P. J., 1983, Mating system and sexual dimorphism in large terrestrial mammalian herbivores, *Biol. Rev.* **58**:485–520.

Johnson, S. E., 2003, Life history and the competitive environment: Trajectories of growth, maturation, and reproductive output among chacma baboons; *Am. J. Phys. Anthropol.* **120**: 83–98.

Johnson, S. E. and Bock, J., 2004, Trade-offs in skill acquisition and time allocation in chacma baboons. *Hum. Nat.* **15**:45–62.

Koscinski, K., Krenz-Niedbala, M., and Kozlowska-Rajewicz, A., 2004, Month-of-birth effect on height and weight of Polish children, *Am. J. Hum. Biol.* **16**:31–42.

Lancaster, J. B. and Lee, R. B., 1965, The annual reproductive cycle in monkeys and apes, in: *Primate Behavior: Field Studies of Monkeys and Apes*, I. Devore, ed., Holt, Rinehart, and Winston, New York, pp. 486–513.

Lee, P. C., 1999, Comparative ecology of postnatal growth and weaning among haplorhine primates, in: *Comparative Primate Socioecology*, P. C. Lee, ed., Cambridge University Press, Cambridge, pp. 111–136.

Lee, P. C. and Kappeler, P. M., 2003, Socioecological correlates of phenotypic plasticity of primate life histories, in: *Primate Life histories and Socioecology*, P. M. Kappeler and M. E. Pereira, eds., University of Chicago Press, Chicago, pp. 41–65.

Lee, P. C, Majluf, P, and Gordon, I. J., 1991, Growth, weaning and maternal investment from a comparative perspective, *J. Zool.* **225**:99–114.

Leigh, S. R., 1995, Socioecology and the ontogeny of sexual size dimorphism in anthropoid primates, *Am. J. Phys. Anthropol.* 97:339–356.

Leimar, O., 1996, Life-history analysis of the trivers and willard sex-ratio problem, *Behav. Ecol.* 7:316–325.

Lycett, J. E., Henzi, S. P, and Barrett, L., 1998, Maternal investment in mountain baboons and the hypothesis of reduced care, *Behav. Ecol. Sociobiol.* **42**:49–56.

Pereira, M. E. and Leigh, S. R., 2003, Modes of primate development, in: *Primate Life histories and Socioecology*, P. M. Kappeler, and M. E. Pereira, eds., The University of Chicago Press, Chicago, pp. 149–176.

Plavcan, M. J., 2001, Sexual dimorphism in primate evolution, *Yearbk. Phys. Anthropol.* **44**:25–53.

Roberts, S. B, Cole, T. J., and Coward, W. A., 1985, Lactational performance in relation to energy intake in the baboon, *Am. J. Clin. Nutr.* **41**:1270–1276.

Roff, D., 1992, *The Evolution of Life Histories*, Chapman and Hall, London.

Ross, K., 1989, *Okavango: Jewel of the Kalahari*, BBC Books, London.

Ross, C. and MacLarnon, A., 1995, Ecological and social correlates of maternal expenditure on infant growth in haplorhine primates, in: C. R. Pryce, R. D. Martin, and D. Skuse, eds., *Motherhood in Human and Nonhuman Primates*, Karger, Basel, pp. 37–46.

Sade, D., 1990, Intrapopulation variation in life-history parameters, in: C. J. DeRousseau, ed., *Primate Life History and Evolution*, Wiley-Liss, New York, pp. 181–194.

Schino, G., Cozzolino, R., and Troisi, A., 1999, Social rank and sex-biased maternal investment in captive Japanese macaques: Behavioural and reproductive data, *Folia Primatol.* **70**:254–263.

Shephard, R. J., Lavallee, H., Jequier, J. C., LaBarre, R., Volle, M., and Rajic, M., 1979, Season of birth and variations in stature, body mass, and performance, *Hum. Biol.* **51**:299–316.

Silk, J. B., 1983, Local resource competition and facultative adjustment of sex ratios in relation to competitive abilities, *Am. Nat.* **121**:56–66.

Silk, J. B, Clark-Wheatley, C. B., Rodman, P. S., and Samuels, A., 1981, Differential reproductive success and facultative adjustment of sex ratios among captive female bonnet macaques (*Macaca radiata*), *Anim. Behav.* **29**:1106–1120.

Silk, J. B, Seyfarth, R. M., and Cheney, D. L., 1999, The structure of social relationships among female savanna baboons in Moremi Reserve, Botswana; *Behaviour* **136**:679–703.

Simpson, M. J. A. and Simpson, A. E., 1982, Birth sex ratios and social rank in rhesus monkey mothers, *Nature* **300**:440–441.

Sterck, E. H. M., Watts, D. P., and van Schaik, C. P., 1997, The evolution of female social relationships in nonhuman-primates, *Behav. Ecol. Sociobiol.* **41**:291–309.

Takahata, Y., Koyama, N., Huffman, M. A., Norikoshi, K., and Suzuki, H., 1995, Are daughters more costly to produce for Japanese macaques? Sex of the offspring and subsequent interbirth intervals, *Primates* **36**:571–574.

Tardif, S. D., and Bales, K. L., 2004, Relations among birth condition, maternal condition, and postnatal growth in captive marmoset monkeys (*Callithrix jacchus*), *Am. J. Primatol.* **62**:83–94.

Tardif, S. D., Power, M., Oftedal, O. T., Power, R. A., and Layne, D. G., 2002, Lactation, maternal behavior and infant growth in common marmoset monkeys (*Callithrix jacchus*): Effects of maternal size and litter size. *Behav. Ecol. Sociobiol.* **51**:17–25.

Thomas, D. S. G. and Shaw, P. A., 1991, *The Kalahari Environment*, Cambridge University Press, Cambridge.

Trexler, J. C. and Travis, J., 1993, Nontraditional regression analyses, *Ecol.* **74**: 1629–1637.

Trivers, R. L. and Willard, D. E., 1973, Natural selection of parental ability to vary the sex ratio of offspring, *Science* **179**:90–92.

Weber, G. W., Prossinger, H., and Seidler, H., 1998, Height depends on month of birth, *Nature* **391**:754–755.

Wilson, M. E., Walker, M. L., Pope, N. S., and Gordon, T. P., 1988, Prolonged lactation and infertility in adolescent rhesus monkeys, *Biol. Reprod.* **38**:163–174.

CHAPTER NINE

# Whose Life Is It Anyway? Maternal Investment, Developmental Trajectories, and Life History Strategies in Baboons

*Louise Barrett, S. Peter Henzi, and John E. Lycett*


## CHAPTER SUMMARY

Levels of investment, interbirth intervals, and offspring survival vary across populations of savannah baboons in relation to habitat quality. In this chapter, we use detailed data from two populations of South African chacma baboons (De Hoop in the Western Cape and the Drakensberg Mountains in Kwa-Zulu Natal) to investigate the constraints on offspring ability to embark on independent foraging trajectories. These analyses reveal that both offspring independence and probability of survival are contingent on both timing of birth and habitat predictability. At De Hoop, where births are nonseasonal but the availability of



**Louise Barrett** • **John E. Lycett** • School of Biological Sciences, University of Liverpool, UK **Louise Barrett** • **S. Peter Henzi** • **John E. Lycett** • Behavioral Ecology Research Group, University of KwaZulu-Natal, Durban South Africa **S. Peter Henzi** • Department of Psychology, University of Central Lancashire, UK

weaning foods is contingent on rainfall, infants are at greater risk of making expensive "mistakes," (i.e., embarking on independent trajectories that cannot always be sustained). This contrasts with the Drakensberg, where the seasonal nature of births and food resources constrain infants more tightly with respect to achieving independence, but at the same time reduce their mortality risk. We extend these analyses to other *Papio* populations across Africa, revealing that the relationship between parental investment and offspring survival is a complex one, reflecting both the ease with which offspring can assume an independent lifestyle and the balance between intrinsic and extrinsic sources of mortality.

## 1. INTRODUCTION

Among primates, the relationship between life history parameters and both the social and ecological environment is receiving renewed attention (Kappeler and Pereira, 2003). This resurgence of interest represents a much more ambitious project than previous research efforts: Rather than merely describing differences in life history patterns and categorizing them along a crude fast–slow continuum, researchers are now investigating in some depth the developmental process in relation to life historical patterning. These analyses reveal that, within and across species, it is possible to show fast development in certain areas and retardation in others (Pereira and Leigh, 2003; Leigh, 2004; Leigh and Bernstein, this volume) and that these patterns can be related to a species' socioecological context (Leigh and Bernstein, this volume). There is also a greater focus on the degree of phenotypic plasticity shown within and between taxa (Lee and Kappeler, 2003), as researchers aim to discover the degree of flexibility in the timing of life historical events and the classic trade-off between the number and quality of offspring. Finally, the causes and consequences of mortality are also receiving greater attention. Predation risk, for example, has been shown to interact with longevity in an evolutionarily significant manner (Janson, 2002). This adds to the body of knowledge already available showing that the point at which mortality falls in the lifespan can dictate life history parameters (Promislow and Harvey, 1990). Furthermore, the source of mortality, in particular whether it is "care dependent" or "intrinsic" (i.e., mortality related to the cost of reproduction) versus "care independent" or "extrinsic" (i.e., mortality over which individual control cannot be exerted) can determine the rate at which offspring are produced and the level of care provided (Pennington and Harpending, 1988).

Using a simple mathematical model, Pennington and Harpending (1988) demonstrated that when sources of care-independent mortality were high relative to care-dependent sources, the optimal amount of parental investment was reduced. Thus, when the probability of survival to adulthood is largely determined by external forces of mortality, females will maximize their fitness by investing less in each individual offspring but producing more of them. These analyses therefore suggest that the optimum amount of investment may not be a simple function of habitat quality, but may instead reflect quite complex interactions between maternal condition, infant growth rates, and the sources of mortality to which individuals are exposed.

In this chapter, we wish to probe the relationship between environmental conditions, sources of mortality, and reproductive flexibility in some detail, reviewing data from two populations of chacma baboons (*Papio hamadryas ursinus*) in South Africa, as well as making use of cross-population data from baboons throughout Africa. Specifically, we wish to take an "infant's eye view," focusing on the efficiency with which offspring make the transition to nutritional independence, as opposed to the more traditional focus on the investment costs to parents. In doing so, we hope to show that the relationship between parental investment and offspring survival is a complex one, reflecting both the ease with which offspring can assume an independent lifestyle and the balance between intrinsic and extrinsic sources of mortality.

## 2. GROWING UP THE BATESON WAY

As our starting point, we take Bateson's 1994 study, which argues that, in the case of mammals, the tantrums and weaning conflicts shown by young animals do not reflect an underlying genetic conflict of interests between parents, as suggested by Trivers (1974), but rather form part of a dynamic process toward offspring independence in which both parent and offspring monitor each other's condition closely. In this view, tantrums and other forms of offspring solicitation behavior are means by which young animals signal to the mother that they require more investment than they are currently receiving if they are to survive (see also Godfray, 1991), rather than a way of manipulating parents to invest more care than is optimal.

Most importantly from our point of view, Bateson (1994) suggested that mammalian offspring reduce their nutritional dependence on their mothers spontaneously and without the need for maternal prompting as a result of several factors. First, a reduction in infant dependence occurs through the increasing inadequacy of maternal milk to satisfy their energetic needs as they grow. In addition, selection has favored quick and efficient acquisition of essential skills, like foraging, thus speeding the process of becoming an independently functioning adult (see also Lee et al., 1991; Bowman and Lee, 1995). Such a process requires monitoring on the part of the mother to ensure that remedial action can be taken should the infant be unable to sustain independence. This is the most crucial part of Bateson's (1994) argument, and represents his most dramatic departure from the parent–offspring conflict model of Trivers' (1974), in that Bateson dispenses with the fundamental assumption that parents and offspring differ with regard to investment optima. He argues instead that, over a wide range of conditions, parents and offspring converge on the ideal level of investment that the offspring should receive (see also Altmann, 1980).

This argument makes sense because the decision to become independent must ultimately lie with the infant and not the mother. Mothers cannot "know," in either a proximate or ultimate sense, when infants are ready to become independent. While it is true that certain proximate cues may indicate to the mother that the infant is potentially capable of feeding independently (e.g., the presence of teeth leading to increasing suckling discomfort to the mother), the ability of an infant to actually begin feeding independently is dependent on the availability of the right kinds of foods. As J. Altmann (1980) and S. A. Altmann (1998) have suggested, "weaning foods" are those that are easy to harvest, ingest, and digest, and include foods like flowers and young grass blades. Because it is the interaction of infant competence with food availability that determines when infants can begin foraging, mothers can only recognize that an infant is ready for weaning once the infant initiates the process. By the same token, mothers cannot prevent infants from embarking on an independent foraging trajectory once they have decided to do so, hence the need for careful monitoring. This also means that, while infants will be drawn into independent feeding by the availability of suitable weaning foods, their ability to continue on such a trajectory will depend upon the certainty of the environment. The availability of suitable foods must persist for long enough to sustain infants through the period when they are first learning to forage efficiently. If this is not the case, then the infant will be forced to return to the mother in an attempt to make up for the shortfall induced by its independent foraging efforts. This may require the offspring to signal to the mother, because even if there is monitoring of offspring condition, mothers will have imperfect knowledge of the state of their offspring (see Godfray, 1991).

## 2.1. Chacma Baboons as a Test Case

As an illustration of this point, we investigate two populations of South African chacma baboons (*Papio hamadryas ursinus*). At De Hoop, in the Western Cape, baboons live in a seasonal winter-rainfall environment, but births are distributed randomly across the year ($r=-0.284$, $n=12$, ns: Figure 1a). This

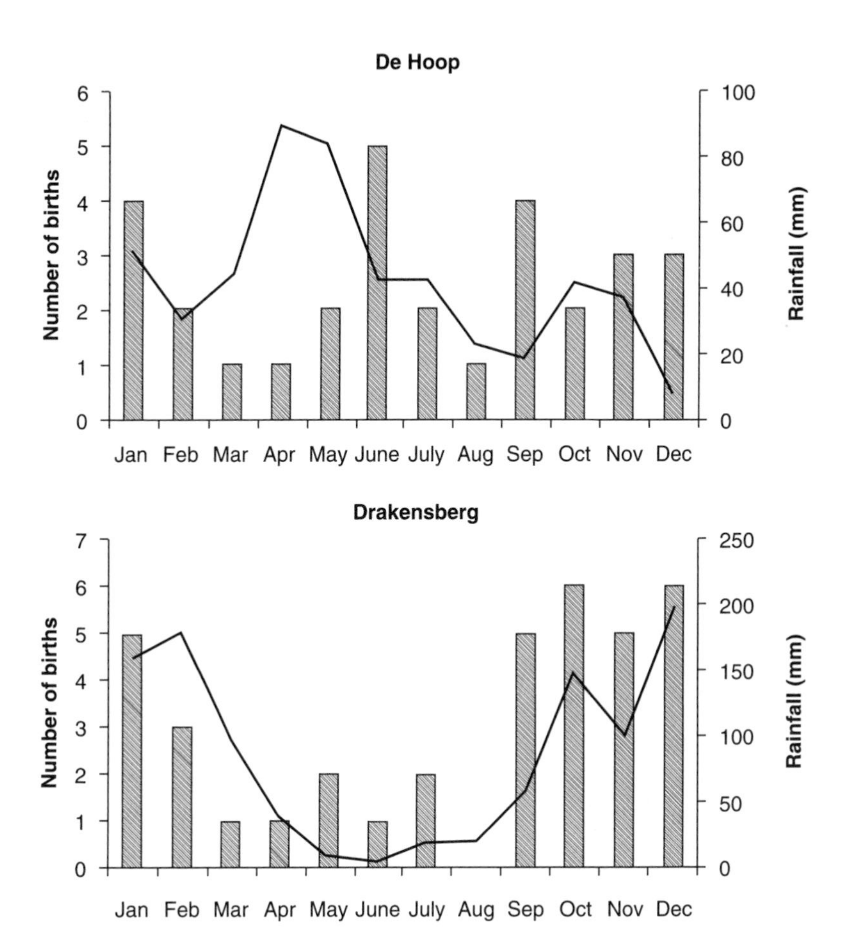


**Figure 1.** Relationship between rainfall (in mm; solid line) and number of births per month (bars) for (a) De Hoop, Western Cape (30 births across 14 females between 1996 and 2000: $r=-0.284$, $n=12$, ns) and (b) Drakensberg, Kwa-Zulu Natal (37 births across 28 females over the period 1982–1984, 1991–1995: $r=0.712$, $n=12$, $p=0.009$: Lycett et al., 1998, 1999).

means that infant vulnerability to perturbation of their developmental trajectories varies according to the time of year at which they are born (Barrett and Henzi, 2000). The austral summer months (December–March) are high-risk periods for weanlings because the availability of above-ground forage declines considerably (due to a combination of high temperatures and low rainfall), and infants lack the motor skills and coordination to utilize the subterranean food items upon which adults rely during this time (Figure 2; Barrett and Henzi, 2000). Any infant that attempts to become nutritionally independent during this period therefore runs the risk of being unable to fulfill its energetic requirements by its own means. The availability of weaning foods is highest, and the risk of infant starvation lowest, during the austral winter (June–September) after the onset of the winter rains (Hill, 1999).

By contrast, our other study population in the Drakensberg Mountains, Kwa-Zulu Natal, presents a very different pattern. This is a high elevation, summer-rainfall, montane grassland habitat. The quantity, nature, and distribution of food in these montane grasslands, combined with high thermal demand (mean minimum grass temperature is below freezing for 9 months of the year (Henzi et al., 1992)) increases the amount of time devoted to feeding by these baboons, relative to populations studied elsewhere (Whiten et al., 1987; Henzi et al., 1997). Therefore baboons in the Drakensberg allot, on average, 68 percent of the day to foraging, compared to an average of 39.8 percent of the day

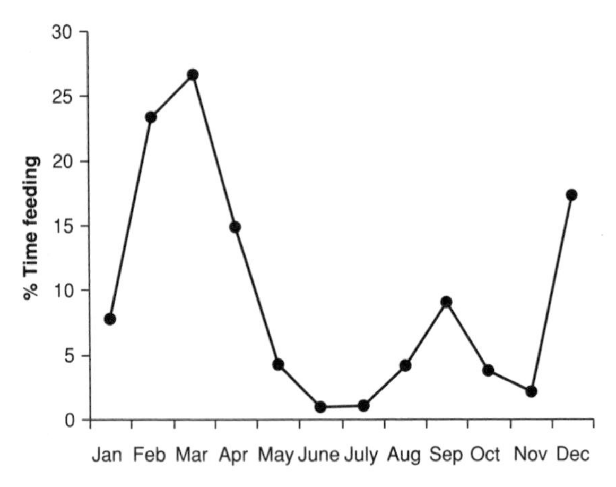


**Figure 2.** Percentage of time spent feeding on subterranean food items across the year for the De Hoop population.

at De Hoop, despite an overall similarity (with the exception of fruits) in the amount of time feeding on particular food types and comparable levels of protein in these food types (Table 1). The inherent harshness of the climate means that baboons in the Drakensberg have to work much harder to sustain themselves than those at De Hoop. This leaves comparatively little scope for variation in the ability of females to prepare for, and cope with, maternity. Even nonlactating females in the Drakensberg feed at levels higher than those predicted by Altmann's (1980) model of maternal time budgets, which means that lactating females simply have no choice but to use body reserves to sustain milk production. In the Drakensberg, these factors result in births that are markedly seasonal ($r = 0.712$, $n = 12$, $p = 0.009$: Figure 1b) (Lycett et al., 1999), due to a clustering of conceptions following the period of increased food availability. This in turn means that infants, which are potentially capable of independent foraging, are faced with the austral winter when suitable weaning foods are simply nonexistent. Consequently, as infants are not presented with any opportunity to forage independently, they continue to rely on their mothers as their sole source of nutrition. Thus, while infants at De Hoop achieve independence, on average, at around 9–10 months of age, those in the Drakensberg continue to suckle well into their second year of life (Figure 3: Barrett and Henzi, 2000; Lycett et al., 1998). It is also possible that, in addition to a lack of suitable foods, offspring in the mountains may grow more slowly due to the nutritional constraints on the mother and the consequent poor maternal condition in this harsh climate.

When Drakensberg infants finally do accelerate their commitment to independent foraging at around 10–12 months of age, they are both physically stronger by virtue by being 5–6 months older than infants foraging at the

**Table 1.** Details of diet composition and protein content for De Hoop and Drakensberg populations

| | Fruit | Leaves | Flowers | Corms/bulbs/tubers |
|---|---|---|---|---|
| *% Protein*[a] | | | | |
| De Hoop | 5.4 | 11.0 | 8.4 | 6.3 |
| Drakensberg | 8.1 | 9.9 | 10.8 | 4.8 |
| *% Feeding time* | | | | |
| De Hoop | 35 | 13 | 5 | 42 |
| Drakensberg | 3 | 26 | 14 | 53 |

[a]Data are % dry weight. Data for De Hoop taken from Hill (1999) and data for the Drakensberg taken from Byrne et al. (1992).

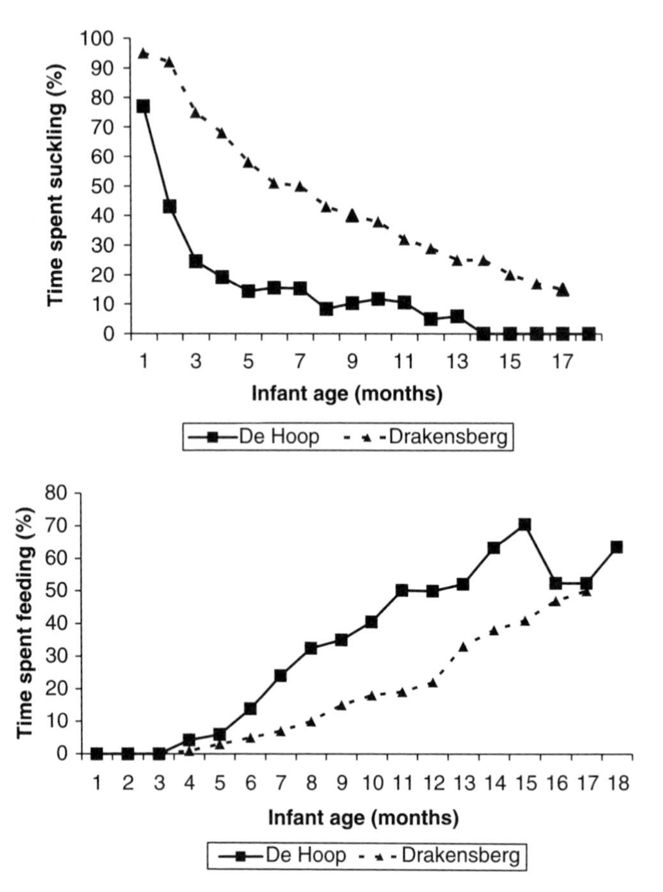


**Figure 3.** Comparison of (a) infant-suckling trajectories and (b) infant-independent feeding trajectories over time for De Hoop and Drakensberg.

same levels at De Hoop and are assured of a summer's worth of easily accessible foods. Consequently, interbirth intervals (IBI) also differ markedly; while the average IBI for De Hoop is 21 months, females in the Drakensberg experience an average IBI of 38.5 months (Figure 4). This unusually long IBI is forced on mothers by the strictures of the habitat. The end of their commitment to their current offspring occurs too close to winter for them to regain sufficient condition until the end of the next summer (Lycett et al., 1999), thereby lowering effective fecundability (Holman and Wood, 2001).

These contrasting conditions lead to interesting differences in the dynamics of the weaning process both within and between populations. At De Hoop,

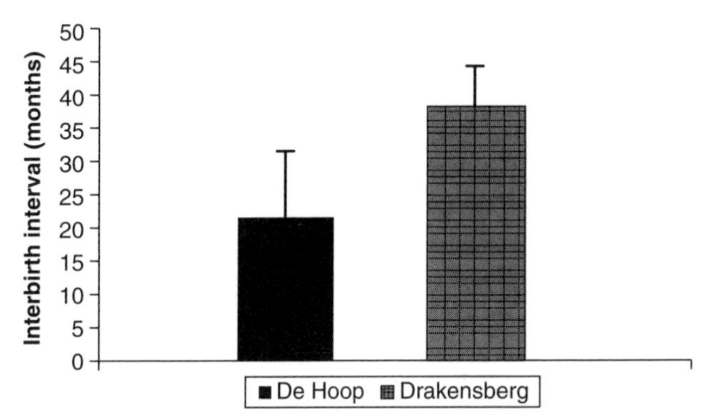


**Figure 4.** Mean interbirth intervals for De Hoop and Drakensberg populations. Interbirth intervals calculated using data for live births only where the infant survived to 12 months of age.

the so-called "weaning tantrums" were seen only among infants that were "tempted" into independence during periods when resource availability did not permit independent foraging to be sustained; in total, only three from a cohort of 10 infants showed any form of tantrum behavior (Barrett and Henzi, 2000). Two of these infants showed high-intensity tantrum behavior during the austral summer. (We defined high-intensity tantrum behavior as repeated attempts at the nipple with loud volume and intensity of vocalizations, plus frenzied behavior by the infants, including hurling themselves to the ground, jumping up and down on their mothers and, on several occasions, attempting to bite their mothers.) These infants had begun foraging at the end of winter when some suitable foods were still available, but, as summer approached, the disappearance of easily obtained food prevented continued independence. Moreover, these tantrums did indeed appear to function as "signals of need," as suggested by Godfray (1991) and Bateson (1994), because they were associated with an increase, rather than a decrease, in levels of parental investment (infants that showed high rates of tantrum behavior also displayed increased levels of suckling (Barrett and Henzi, 2000)). The third infant did not show tantrum behavior during the critical summer months, but did during another crisis period of epidemic disease in the study troop (Barrett and Henzi, 1998). This infant's mother became severely ill and spent most of her time lying on her stomach, preventing her infant's access to the nipple. Although the mother was unable to respond to the infant's signals, the infant did not show high-intensity tantrum behavior of the kind defined above and would return to independent feeding when its demands were not met (Barrett and Henzi, 2000). In contrast, where the timing of birth was such that infants were able to make a smooth transition from suckling to independent foraging, they did so without any evidence of tantrum behavior and, by around 9–10 months of age on average, were feeding independently (over 90 percent of time budget; Barrett and Henzi, 2000).

In the same way, the five Drakensberg infants for which detailed samples are available, although making their transition to independent foraging much later than those at De Hoop, did so without any tantrum behavior whatsoever (Lycett et al., 1998). Infants were not drawn into embarking on a foraging trajectory that was inherently unsustainable by the availability of ephemeral weaning foods and so did not leave their mothers early to begin independently foraging. Consequently, a situation in which infants were forced to return to their mothers for renewed investment simply did not arise, and there was never any need for infants to signal need.

## 3. COMPARISONS WITH OTHER BABOON POPULATIONS

### 3.1. Habitat Unpredictability and Survivorship

Taken together, the results from De Hoop and the Drakensberg suggest that some aspect of habitat "unpredictability" or "uncertainty" is of critical importance in determining the interaction between infant independence and levels of parental resource allocation over time, and that this in turn may have a negative impact on infant survivorship. In habitats where births are nonseasonal and levels of resource availability vary through the year unpredictably, infants may more frequently embark upon unsustainable foraging trajectories and be forced subsequently to return to their mothers, which themselves may be unable to support these renewed demands due to a decline in their own quality as resources diminish. It is important to note that better-quality habitats (in terms of primary productivity) are not necessarily more predictable habitats, such that the relationship between survival and predictability reduces to one between habitat quality and survival. This is because better-quality habitats are those in which female reproduction is less constrained by resource availability, with the result that females will be less likely to show any form of birth peak or seasonality in offspring production. Normal fluctuations in resource availability are then more likely to have a negative impact on offspring survival for those infants born at times of the year the transition to independence occurs under conditions that are not sustainable, as is seen at De Hoop.

We investigate these ideas further by examining the relationship between offspring survival and productivity across baboon populations for Africa as a whole. Analyses focus on comparing a measure of habitat productivity to demographic parameters. Specifically, we employ an index of habitat productivity and uncertainty, $P > 2t$, or the number of months where precipitation (in mm) is greater than twice the mean annual temperature (in °C) (le Houréau, 1984). Under ideal circumstances, plant evapotranspiration would be used to measure plant productivity (Rosensweig, 1968; Leith and Box, 1972). However, in the absence of data to compute evapotranspiration, le Houréau (1984) has suggested that $P > 2t$, can be used as an index of productivity. The drier and less productive a habitat, the lower the value of $P > 2t$. Williams (1997) has shown that this index is directly related to potential evapotranspiration ($r^2 = 0.99$, $p < 0.0001$) and gives results comparable to those obtained using these more complex measures. In addition, $P > 2t$ serves our purposes well here because it gives some feel for habitat seasonality, as well as productivity; $P > 2t$ is negatively correlated with an index of rainfall diversity (DIV), where high values of DIV indicate more seasonal rainfall (Williams, 1997). Thus, rainfall diversity across the year decreases as $P > 2t$ increases ($r^2 = 0.67$, $p < 0.0001$: Williams, 1997). The final advantage of $P > 2t$ is that because it only requires a simple count of the number of months in the year where rainfall exceeds a certain threshold value (i.e., the value of $2t$), it is therefore very easy to calculate from data that are usually collected as a matter of course in primate ecological studies and which are widely available.

We analyze data from 10 *Papio* populations across east and southern Africa (Table 2). Following our argument above, we suggest that, in habitats where $P > 2t$ is high, infant survivorship will be actually lower than in less productive, more seasonal habitats. This may seem counterintuitive but, to reiterate, increased productivity will tend to produce births that are more randomly distributed across the year, due to relaxed constraints on females' ability to conceive. This increases the probability that some offspring will encounter conditions that are not conducive to sustained foraging independence. Mothers in habitats where infants can potentially make "mistakes" with respect to the timing of independent foraging may also be less able to rectify these

**Table 2.** Details of interbirth intervals, survival, and habitat data for 10 *Papio* spp. populations

| Population | Inter-birth interval between live births (months) | $P$ (survival) | $P>2t$[a] | Predation risk index[b] | Source |
|---|---|---|---|---|---|
| Amboseli | 21.6 | 0.51 | 7 | 3 | Altmann, 1980 |
| Chololo | 26.9 | 0.38 | 10 | 2 | Kenyatta, 1995 |
| De Hoop | 21.8 | 0.75 | 5 | 2[c] | L. Barrett and S. P. Henzi, unpublished |
| Drakensberg | 38.5 | 0.93 | 6 | 1 | Lycett et al., 1998 |
| ErerGota | 24.0 | 0.74 | 5 | 2 | Sigg et al., 1982 |
| Gilgil | 26.5 | 0.50 | 12 | 2 | Nicolson, 1982 |
| Gombe | 25.0 | 0.63 | 7 | 3 | Collins et al., 1984 |
| Mikumi | 21.0 | 0.72 | 6 | 3 | R. Rhine, pers. commun. |
| Mkuzi | 20.1 | 0.81 | 6 | 3 | S. P. Henzi, unpublished |
| Moremi | 24.6 | 0.55 | 5 | 3 | Cheney et al., this volume |

[a]Values for $P>2t$ obtained from rainfall and temperature data given in Williams (1997), except for De Hoop, Mkuzi, and the Drakensberg where values were calculated from our own unpublished data and Moremi where the value was calculated from data presented in Cheney et al. (this volume).
[b]Predation risk index taken from Hill and Lee (1998). 1 = low, 2 = medium, 3 = high risk of predation.
[c]Although leopard and other standard predators are absent at De Hoop, deaths by snakebite are commonplace ($n=3$ confirmed deaths; 3 suspected) and can be considered as equivalent to deaths by predation as an extrinsic source of mortality.

mistakes because they will also be experiencing nutritional stress and, possibly, the reduction in milk production as the infant reduces suckling demand.

As predicted, offspring survivorship to 24 months is negatively related to $P>2t$ ($r^2=0.410$, $F_{1,8}=5.55$, $p=0.046$: Figure 5), suggesting that infants survive better where they are more likely to sustain themselves completely independently during the critical weaning period, even though these are habitats that could be considered as relatively low quality. This in itself may reflect strong past selection against infants born at the "wrong" time of year such that, in these habitats, births are timed so that offspring survival probabilities are enhanced. Dunbar et al. (2002) have recently made a similar point with respect to the timing of birth in gelada baboons. Although geladas are not seasonal breeders, most females in the Sankaber population gave birth after the main wet season from October to December, then faced peak lactation during the dry season when resource availability was low. Dunbar et al. argued that, although mothers that gave birth after December would reach

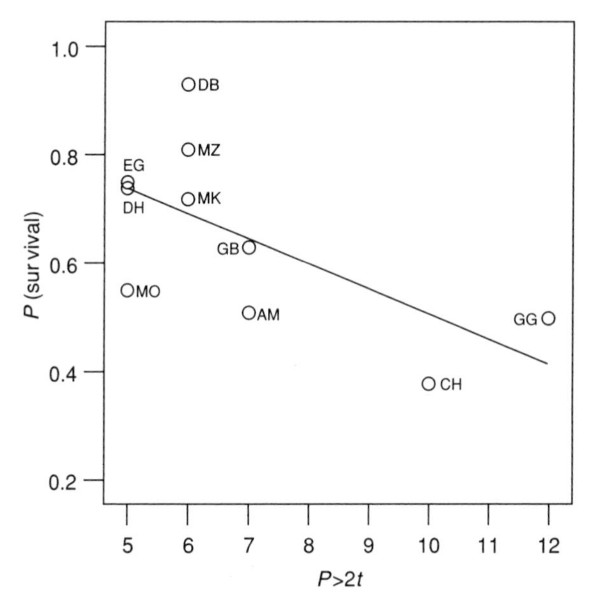


**Figure 5.** Relationship between $P>2t$ (see text for details) plotted and infant survivorship to 2 years for 10 *Papio* populations ($r^2 = 0.410$, $F_{1,8} = 5.55$, $p = 0.046$). Details of the populations are given in Table 2. Two-letter labels are as follows: Am = Amboseli; CH = Chololo; DB = Drakensberg; DH = De Hoop; EG = Erer Gota; GB = Gombe; GG = Gilgil; MK = Mikumi; MZ = Mkuzi).

peak lactation during the next wet season and therefore face less nutritional stress, their infants would be at a strong disadvantage because of the thermoregulatory costs associated with spending the wet season in their natal coats. Mortality for infants that spend the whole of the wet season in their natal coats is three times higher than it is for those born during the dry season (Dunbar, 1980), constituting a severe selection pressure on reproduction that is timed too late in the year (Dunbar et al., 2002). Another point to note is that these thermoregulatory costs constitute an extrinsic source of mortality, highlighting the interaction of intrinsic and extrinsic sources of mortality with respect to the timing of reproductive events.

### 3.2. Survival and Investment Patterns

Although the relationship we have uncovered is indirect, it does appear that, across populations, infant survival is linked to the timing of birth relative to periods of suitable nutritional conditions. Detailed data from De Hoop and the Drakensberg show how offspring and mothers are able to display behavioral flexibility with respect to the onset of independent feeding and the occurrence of, and response to, "weaning conflicts." Our interpretation is that increased probabilities of infant mortality result when this flexibility induces offspring "mistakes" that require remedial action, which may or may not be forthcoming. This, then, is a form of intrinsic mortality that mothers can potentially avoid by increasing investment in offspring if they can. Interbirth intervals are therefore predicted to be longer on average in such populations because some offspring will require extra maternal investment if they are to survive. This means that longer interbirth intervals should be associated with lower survival probabilities for infants across populations.

Again, we can investigate this using data from our 10 *Papio* populations. However, interbirth intervals may depend on overall habitat quality independent of the manner in which the habitat affects infant mortality risk. In resource-poor habitats, females may require longer lactational periods to produce offspring of equivalent size and weight to those born in better-quality habitats. They may also take more time to recover condition following lactation, irrespective of whether these infants require extra investment. To test this, we plotted the regression between $P > 2t$ and IBI for our sample of 10 *Papio* populations. As Figure 6a shows, there is no significant relationship between these variables ($r^2 = 0.026$, $F_{1,8} = 0.21$, $p = 0.658$). This is rather surprising, given that resource availability is known to have a profound effect on female reproductive performance (Whitten, 1983; Barton and Whiten, 1993). However, inspection of the graph shows that the Drakensberg population is an outlier. Removing the Drakensberg population from the analysis produces a marginally significant positive relationship between $P > 2t$ and IBI for the remaining nine populations ($r^2 = 0.421$, $F_{1,7} = 5.10$, $p = 0.059$: Figure 6b), but does not affect the relationship between offspring survival and $P > 2t$ ($r^2 = 0.486$, $F_{1,7} = 6.61$, $p = 0.037$).

Contrary to expectation, however, the relationship between IBI and $P > 2t$ is positive, rather than negative: IBI increases as habitat productivity improves. This means that when IBI and offspring survival are plotted against one another in all populations except the Drakensberg, a negative relationship is produced as predicted ($r^2 = 0.537$, $F_{1,7} = 8.12$, $p = 0.025$: Figure 7a). However, this relationship does not reflect poorer survival and greater nutritional stress

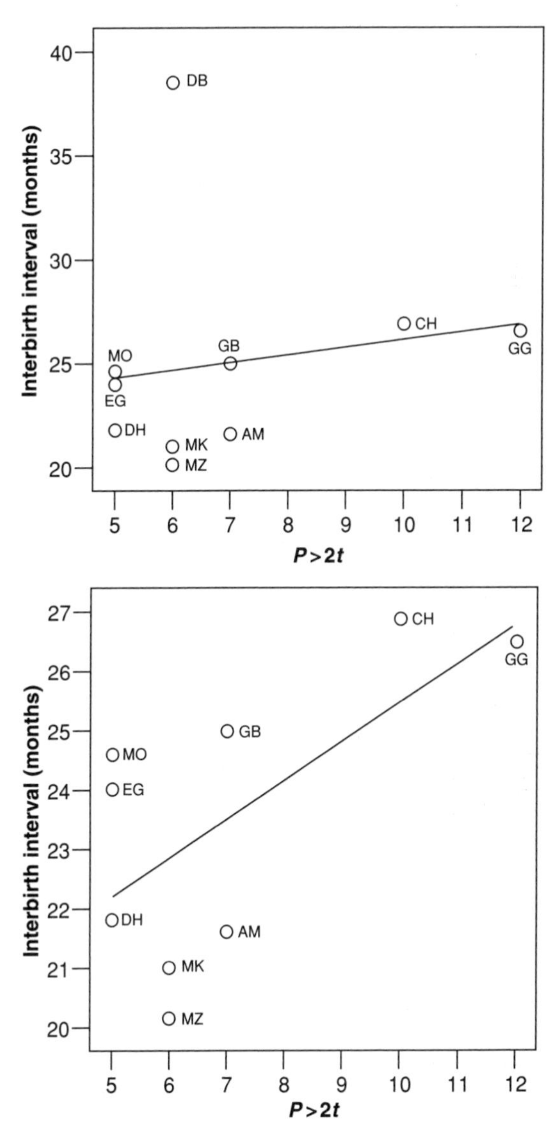


**Figure 6.** Relationship between $P > 2t$ and interbirth interval for *Papio* populations when (a) Drakensberg population is included ($r^2 = 0.026$, $F_{1,8} = 0.21$, $p = 0.658$) and (b) Drakensberg population is excluded ($r^2 = 0.421$, $F_{1,7} = 5.10$, $p = 0.059$). Labels as for Figure 5.

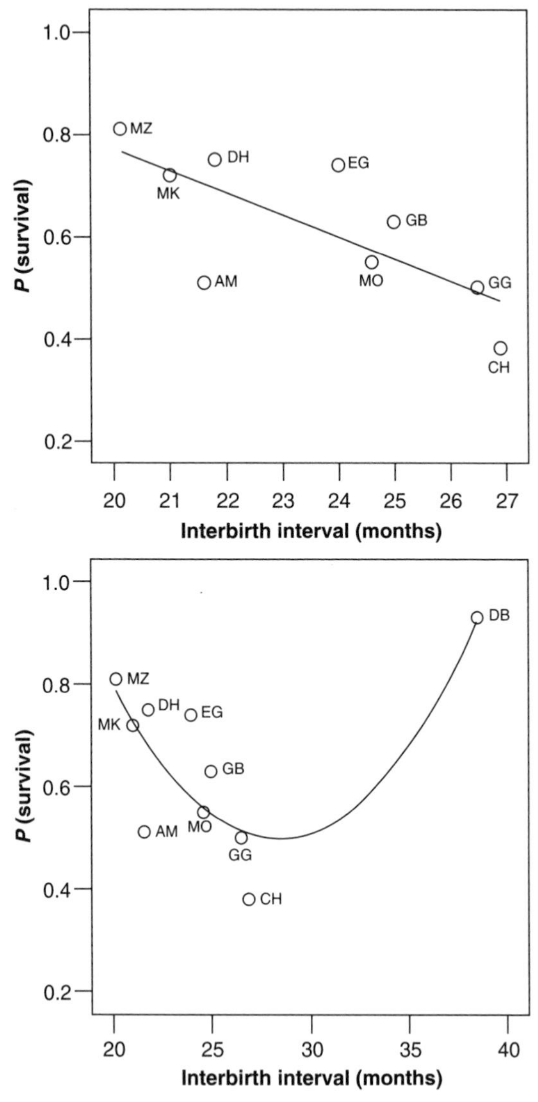


**Figure 7.** Relationship between interbirth interval and infant survivorship for *Papio* populations when (a) Drakensberg population is excluded ($r^2 = 0.537$, $F_{1,7} = 8.12$, $p = 0.025$) and (b) Drakensberg population is included ($r^2 = 0.651$, $F_{2,7} = 6.54$, $p = 0.025$). Labels as for Figure 5.

in poorer, more seasonal habitats because the relationships between IBI, survival, and habitat quality ($P > 2t$) go in the opposite direction to that predicted by such a hypothesis. Instead, it appears that longer IBIs and poorer survivorship are linked to situations where the transition to independence is more risky for the offspring. An interesting test case for this interpretation is provided by the changes that have occurred in the Amboseli population following home range shifts in the late 1980s and early 1990s. The new home ranges contained a relatively high density of *Acacia xanthophloea*, which are important food and refuge species for baboons. This shift produced an increase in the proportion of infant surviving from 0.51 to 0.71 and, as predicted by the relationship above, was accompanied by a reduction in interbirth interval from 21.6 months to 19.2 months (Altmann and Alberts, 2003). Substitution of these new data for Amboseli into the regression analysis above gives rise to an even stronger negative relationship between survival and IBI ($r^2 = 0.679$, $F_{1,7} = 14.78$, $p = 0.006$). In other words, then, the plasticity that females show in any given population is facultative and driven by the needs of the individual offspring during a particular reproductive event. Similarly, offspring mortality is the result of a contingency between the timing of birth and environmental conditions: A habitat will not always be risky for all infants all the time, but will depend on their particular circumstances.

### 3.2.1. *The Drakensberg Baboons: the Exception that Proves the Rule?*

A further complication is added to the situation by the fact that including the Drakensberg population in a plot of IBI versus survival generates a significant quadratic relationship ($r^2 = 0.651$, $F_{2,7} = 6.54$, $p = 0.025$: Figure 7b. The relationship remains equivalent if the new data for Amboseli are used: $r^2 = 0.621$, $F_{2,7} = 5.74$, $p = 0.033$). While one could argue that the Drakensberg baboons should simply be ignored as outliers, representing an odd population, not quite behaving as it should, in some sense this would miss the point. While it is true that the Drakensberg population is an outlier, this is not due to measurement error, which is the usual assumption made when excluding data points: Interbirth intervals were not recorded inaccurately or collected during a short time period that was unrepresentative of the true situation. The data come from several years of study and several troops, and if females in the mountains are taking twice as long as all other baboons and producing infants that survive better, then this needs to be explained. The Drakensberg population may thus be the exception that proves the rule with regard to understanding female baboon investment patterns. In support of this, gelada baboons living under similar circumstances to those of the Drakensberg population also show extremely long IBIs (36.5 months) and very high offspring survival (0.95; Ohsawa and Dunbar, 1984). As one might expect, using the best fit equations generated by both the linear and quadratic regressions to predict infant survival for gelada shows that the quadratic fit provides a more accurate estimate of offspring survival than the linear equation (linear fit: $P$(survival) $= 1.59 - 0.042[\text{IBI}] = 0.06$; quadratic fit: $3.88 - 0.24[\text{IBI}] + 0.004[\text{IBI}]^2 = 0.42$), although it substantially underestimates the actual value.

So, how to explain the Drakensberg pattern? Our detailed comparison of De Hoop and the Drakensberg revealed that infants in the Drakensberg were prevented from embarking on independent foraging trajectories due to a lack of appropriate weaning foods, so that they remained dependent on their mothers for much longer than average. This greater dependence reduces the mother's reproductive rate substantially, but enhances offspring survival. At first glance, this would seem to be an example of mothers making the "best of a bad job," rather than maximizing their reproductive success. If, on the other hand, we consider the potential impact of extrinsic, as well as intrinsic, sources of mortality on investment decisions, a more interesting picture emerges.

### 3.2.2. *Including extrinsic mortality*

Lycett et al. (1998) showed that IBIs decrease under conditions where sources of extrinsic mortality (in the form of predation) are high. This relationship also holds for the data presented here ($r_s = -0.880$, $n = 9$, $p = 0.001$: Figure 8), even though there is no relationship between extrinsic mortality and $P > 2t$ (including Drakensberg: $r_s = -0.014$, $n = 10$, $p = 0.970$; excluding Drakensberg: $r_s = -0.089$, $n = 9$, $p = 0.820$). Thus, the relationship between IBI and extrinsic mortality is independent of habitat quality and suggests that, under conditions where chances of surviving to adulthood are reduced, mothers attempt to produce offspring at or close to their maximum rate in order to increase the chances of producing at least some surviving offspring. Further support for this comes from the comparative analyses of Hill (1999). In baboon populations with virtually no risk of predation, Hill (1999) measured high variance in IBI. For populations with high predation risk, however, variance in IBI was low. Hill (1999) suggested that this was because females in high-risk

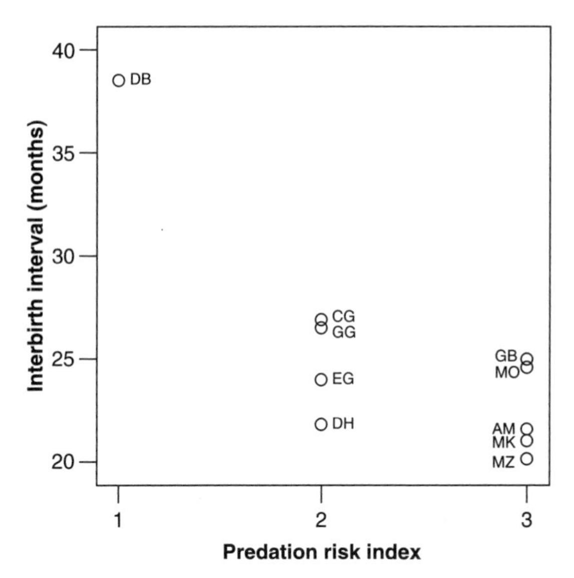


**Figure 8.** Relationship between interbirth interval and predation risk index (Hill and Lee, 1998) for 10 *Papio* baboon populations ($r_s = -0.880$, $n = 9$, $p = 0.001$). Labels as for Figure 5.

populations were constrained to reproduce at close to their maximal rate in order to alleviate the effects of predation-based mortality on lifetime reproductive output, whereas females in low-risk populations could show greater flexibility of investment and adjust levels of investment to minimize the impact of intrinsic mortality.

It is relevant to note here that predation is only one of several forms of extrinsic mortality that should be considered in these kinds of analysis and, of these, disease is probably the most significant. Detailed analysis of data from the De Hoop population reveals that, after an initial early reduction in survivorship (partly explained by infanticide (Henzi and Barrett, 2003; see also Cheney et al., this volume)), survivorship continues to drop quite substantially during the first 5 years of life (Figure 9). Almost all of this mortality can be attributed to disease. We also know that disease plays a large role in controlling the population at De Hoop as a whole (Barrett and Henzi, 1998). Ideally, this should be factored into our analyses. Unfortunately, information on the impact of disease on mortality and morbidity is not known for most populations, and the crude assessment above is all we currently have.

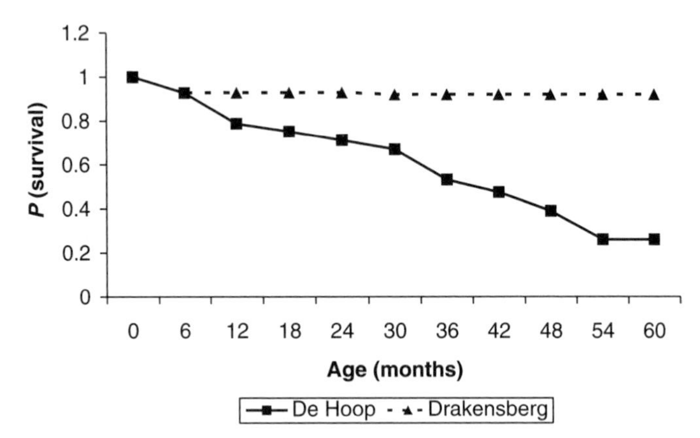


**Figure 9.** Probability of offspring survival against offspring age for De Hoop ($n$ = 34 individuals of known birth date) and the Drakensberg (S. P. Henzi and J.E. Lycett, unpublished).

## 3.3. Investment Patterns and the Interaction of Intrinsic and Extrinsic Mortality

### *3.3.1. De Hoop and Drakensberg*

Given the nature of the relationship between IBI and extrinsic mortality, we suggest that the surprising quadratic relationship between survival and interbirth intervals arises as a consequence of the interaction between intrinsic and extrinsic sources of mortality (Promislow and Harvey, 1990; Pennington and Harpending, 1988). To put this in concrete terms, under conditions where an infant baboon's survival depends on the level of nutritional investment it receives, it pays a mother to invest such care in her offspring because this will increase survivorship. Thus, if an infant is tempted into independence but subsequently forced to return to its mother for extra nutritional investment, it will pay the mother to provide it, thus increasing her IBI.

However, where survival to reproductive maturity is more strongly influenced by extrinsic sources of mortality (e.g., predation, disease), mothers should invest less in each offspring in order to increase overall offspring production and increase the chances of any individual offspring surviving to reproduce. Again, the De Hoop and Drakensberg populations provide an excellent contrast in this respect. At De Hoop, offspring have a 75 percent chance of surviving until the age of 2, but then have only a 26 percent chance of surviving until age 5 (Figure 9). By contrast, in the Drakensberg, infants have a 93 percent chance of surviving to age 2, and this remains constant across the first 5 years of life (Figure 9). Thus, the marginal value of extra units of care is greater for Drakensberg mothers than for those at De Hoop (Lycett et al., 1998).

We therefore suggest that the curvilinear nature of the relationship between IBI and survivorship can be explained as follows: Where infant survival is high and IBI is short, mothers have been selected to invest less due to high extrinsic mortality risk while, at the same time, infants are more likely to be able to cope independently with foraging demand, even though overall habitat quality is relatively low (as demonstrated by the relationship between survival and $P>2t$). Our hypothesis is that this results from selection acting to ensure that births tend to coincide with periods favorable to infant survival, perhaps in conjunction with a clustering of conceptions as females respond to seasonal variability in resources.

When infant survivorship declines, this reflects the improvement of habitat conditions so that births are more likely to be randomly distributed in time, with the result that a proportion of infants will embark on unsustainable foraging trajectories and require more contingent investment from their mothers. Under such conditions, mortality thus becomes dependent on intrinsic, as well as extrinsic, factors. Consequently, mothers should respond flexibly and increase investment in infants that are in need; hence inter-birth intervals will be longer on average. However, survivorship to 2 years is nevertheless reduced because infants are more likely to make expensive "mistakes" that extra investment cannot always remediate, as is seen at De Hoop.

When sources of care-independent mortality fall away altogether and maternal investment alone becomes the key to infant survival, as seen in the Drakensberg, then females should invest heavily in their offspring and the marginal value of such care is very high because the probability of losing an infant to extrinsic causes is low. Consequently, there is no simple relationship between IBI and infant survival because this relationship hinges on the relative balance of intrinsic and extrinsic mortality, which, in turn, reflects a complex relationship between habitat quality, predictability, and the probability of infants experiencing harsh conditions during weaning.

### 3.3.2. *Other sites*

The situation is likely to be even more complicated, given that we have not considered the possibility that sources of extrinsic mortality may also show seasonal patterns that could affect both the timing and success of reproductive events. For example, among the Moremi baboons, predation is the most important source of mortality for this population: Practically all adult and juvenile deaths could be attributed to predation (Cheney et al., this volume). In addition, an important source of infant mortality was infanticide, which, like predation, is a source of extrinsic mortality: Females cannot prevent infanticide either alone or in alliance with kin (Cheney et al., this volume). Interestingly, mortality from both predation and infanticide is highest during the period of peak flooding (July–September) in the delta, possibly due to the increased predictability of troop movements during this period. Moreover, mortality was highest among infants born during the second quarter of the year (April–June), which is the time when fewest infants overall are born. Infants born in the second quarter of the year will be most vulnerable to infanticide (i.e., between the ages of 3–6 months) at precisely the time when the risk of infanticide is at its highest. Higher mortality among these infants may partly explain why most births occur in the months following July and August, because one would expect strong selection against females producing infants during the second quarter of the year. This seasonal extrinsic pressure acts in conjunction with a clustering of conceptions around March–April in response to seasonal variation in habitat productivity, which could be viewed as a response to sources of intrinsic mortality. Thus, at Moremi, seasonal variability in high extrinsic mortality risk reinforces the pressures exerted on mothers to reproduce seasonally and lower intrinsic risk. This contrasts with the situation in geladas, where timing births to lower intrinsic risk has the effect of increasing extrinsic risk (Dunbar et al., 2002). Consequently, variations in extrinsic and intrinsic mortality risk through the year can be expected to exert selection pressures on the relative timing of births across the year (as demonstrated by the gelada, Drakensberg, and Moremi baboons) as well as on the level of investment a mother should provide in a given habitat. Identifying potential conflicts between the timing of birth relative to seasonal variation in intrinsic and extrinsic sources of mortality and integrating them into models of optimal investment would seem to be the logical next step to take in studies of maternal investment and life history strategies.

## 4. CONCLUSIONS

At present, the hypotheses proposed by this study remain tentative, requiring more data for thorough tests. In particular, more and better data are required on extrinsic sources of mortality (with more attention paid to disease) because this seems central to a better understanding of variation in maternal investment strategies across populations and species. Recent work by Janson (2002), demonstrating how a life-historical perspective can explain some of the "puzzles" of group living, also points the way for theoretical explorations of this issue. Moreover, it needs to be established whether extrinsic or "care-independent" mortality is indeed independent of the level of investment an infant receives. It seems possible, in principle, for high investment to increase resistance to disease or reduce the likelihood of being taken by a predator and for these effects to last well into adulthood. Such an enterprise requires that, as suggested by Pereira and Leigh, we need to "describe not only 'average' juveniles and adult males and females but also individuals' responses to varied circumstances throughout development" (2002:150). In this manner, we will be able to determine the benefits of high investment and whether it is possible to compensate for low investment together with the consequences of this for an individual's own life history trajectory. Understanding the success of baboons as a species means understanding how females are able to produce surviving offspring in the face of mortality risks that are often severe. We hope that, by highlighting what we regard as the important factors influencing infant nutritional independence and survival, we have set out an agenda for the work that might be done to improve and extend our understanding of the ecological context of baboon life histories.

## ACKNOWLEDGMENTS

We would like to thank Steve Leigh and Larissa Swedell for inviting us to participate in their AAPA symposium and to contribute a chapter to this book. We also thank the other contributors for interesting discussions of these and other issues, and three anonymous reviewers, plus Larissa and Steve, for very helpful comments on an earlier draft of this chapter. Our research was funded by grants from the National Research Foundation (South Africa), the Centre for Science Development (South Africa), and the University of Liverpool Research Development Fund. LB was supported by a Leverhulme Research Fellowship during the writing of this chapter. We are grateful to the Natal Parks Board and the Cape Nature Conservation for permission to conduct research in the Drakensberg and at De Hoop.

## REFERENCES

Altmann, J., 1980, *Baboon Mothers and Infants*, Harvard University Press, Cambridge, MA.

Altmann, S. A., 1998, *Foraging for Survival: Yearling Baboons in Africa*, University of Chicago Press, Chicago.

Altmann, J. and Alberts, S. C., 2003, Variability in reproductive success viewed from a life-history perspective, *Am. J. Hum. Biol.* **15**:401–409.

Barrett, L. and Henzi, S. P., 1998, Epidemic deaths in a chacma baboon population, *S. Afr. J. Sci.* **94**:441.

Barrett, L. and Henzi, S. P., 2000, Are baboon infants Sir Phillip Sydney's offspring? *Ethology* **106**:645–658.

Barton, R. A. and Whiten, A., 1993, Feeding competition among female olive baboons, *Papio anubis*, *Behav. Ecol. Sociobiol.* **38**:321–329.

Bateson, P., 1994, The dynamics of parent–offspring conflict relationships in mammals, *Trends Ecol. Evol.* **9**:399–403.

Bowman, J. and Lee, P. C., 1995, Growth and threshold weaning weights among captive rhesus macaques, *Am. J. Phys. Anthropol.* **96**:159–175.

Byrne, R. W., Whiten, A., Henzi, S. P., and McCulloch, F. M., 1992, Nutritional constraints on mountain baboons (*Papio ursinus*): Implications for baboon socioecology, *Behav. Ecol. Sociobiol.* **33**:233–246.

Collins, D. A., Busse, C. D., and Goodall, J., 1984, Infanticide in two populations of savanna baboons, in: *Infanticide: Comparative and Evolutionary Perspectives*, G. Hausfater and S.B. Hrdy, eds., Aldine, New York, pp. 193–215.

Dunbar, R. I. M., 1980, Demographic and life history variables of a population of gelada baboons (*Theropithecus gelada*), *J. Anim. Ecol.* **49**:485–506.

Dunbar, R. I. M., Hannah-Stewart, L., and Dunbar, P., 2002, Forage quality and the costs of lactation for female gelada baboons, *Anim. Behav.* **64**:801–805.

Godfray, H. C., 1991, Signalling of need by offspring to their parents, *Nature* **352**:328–330.

Henzi, S. P. and Barrett, L., 2003, Evolutionary ecology, sexual conflict and behavioral differentiation among baboon populations, *Evol. Anthropol.* **12**:217–230.

Henzi, S. P., Byrne, R. W., and Whiten, A., 1992, Patterns of movement by baboons in the Drakensberg mountains: Primary responses to the environment, *Int. J. Primatol.* **13**:601–629.

Henzi, S. P., Lycett, J. E., Weingrill, T., Byrne, R. W., and Whiten, A., 1997, The effect of troop size on travel and foraging in mountain baboons, *S. Afr. J. Sci.* **93**:333–335.

Hill, R. A., 1999, The ecological and demographic determinants of time budgets in baboons. PhD Dissertation, University of Liverpool.

Hill, R. A., Lee, P. C., 1998, Predation risk as an influence on group size in Cercopithecoid primates, *J. Zool. Lond.* **245**:447–456.

Holman, D. J. and Wood, J. W., 2001, Pregnancy loss and fecundability in women, in: *Reproductive Ecology and Human Evolution*, P. T. Ellison, ed., Aldine De Gruyter, New York, pp. 15–38.

le Houréau, H. N., 1984, Rain use efficiency: A unifying concept in arid-land ecology, *J. Arid Environ.* 7:213–247.

Janson, C. H., 2002, Puzzles, predation and primates: Using life-history to understand selection pressures, in: *Primate Life Histories and Socioecology*, P. M. Kappeler, and M. E. Pereira, eds., Chicago University Press, Chicago, pp. 103–131.

Kappeler, P. M. and Pereira, M. E., eds., 2003, *Primate Life Histories and Socioecology*, University of Chicago Press, Chicago.

Kenyatta C. G., 1995, Ecological and social constraints on maternal investment strategies. Ph.D. Dissertation, University College, London.

Lee, P. C. and Kappeler, P. M., 2003, Socioecological correlates of phenotypic plasticity of primate life histories, in: *Primate Life Histories and Socioecology*, P. M. Kappeler and M. E. Pereira, eds., Chicago University Press, Chicago, pp. 41–65.

Lee, P. C., Majluf, P., and Gordon, I. J., 1991, Growth, weaning and maternal investment from a comparative perspective, *J. Zool. Lond.* **225**:99–114.

Leigh, S. R., 2004, Brain growth, life history, and cognition in primate and human evolution, *Am. J. Primatol.* **62**:139–164.

Leith, H. and Box, E. O., 1972, Evapotranspiration and primary production: C.W. Thornthwaite memorial model, *Publ. Cimatol.* **25**:37–46.

Lycett, J. E., Henzi, S. P., and Barrett, L., 1998, Maternal investment in mountain baboons and the hypothesis of reduced care, *Behav. Ecol. Sociobiol.* **42**:49–56.

Lycett, J. E., Weingrill, T., and Henzi, S. P., 1999, Birth patterns in the Drakensberg Mountain baboons (*Papio cynocephalus ursinus*), *S. Afr. J. Sci.* **95**:354–356.

Nicolson, N., 1982, Weaning and the development of independence in olive baboons. PhD Dissertation, Harvard University.

Ohsawa, H. and Dunbar, R. I. M., 1984, Variations in the demographic structure and dynamics of gelada baboon populations, *Behav. Ecol. and Sociobiol.* **15**:231–240.

Pennington, R. and Harpending, H., 1988, Fitness and fertility among Kalahari !Kung. *Am. J. Phys. Anthropol.* 77:303–319.

Pereira, M. E. and Leigh, S. R., 2003, Modes of primate development, in: *Primate Life Histories and Socioecology*, P. M. Kappeler, and M. E. Pereira, eds., University of Chicago Press, Chicago, pp. 149–176.

Promislow D. E. L. and Harvey P. H., 1990, Living fast and dying young: A comparative analysis of life-history variation among mammals, *J. Zool. Lond.* **220**:417–437.

Rosensweig, M. L., 1968, Net primary productivity of terrestrial communities: Prediction from climatological data, *Am. Nat.* **102**:67–74.

Sigg, H., Stolba, A., Abegglen, J.-J., and Dasser, V., 1982, Life history of hamadryas baboons: Physical development, infant mortality, reproductive parameters and family relationships, *Primates* **23**:473–487

Trivers, R., 1974, Parent–offspring conflict, *Am. Zool.* **14**:249–264.

Whitten, P., 1983, Diet and dominance among female vervet monkeys (*Cercopithecus aethiops*), *Am. J. Primatol.* **5**:139–159.

Whiten, A., Byrne, R. W., and Henzi, S. P., 1987, The behavioral ecology of mountain baboons, *Int. J. Primatol.* **8**:367–388.

Williams, D. K., 1997, Primate socioecology: development of a conceptual model for the early hominids. PhD Dissertation, University of London.

CHAPTER TEN

# Ontogeny, Life History, and Maternal Investment in Baboons

*Steven R. Leigh and Robin M. Bernstein*


## CHAPTER SUMMARY

This chapter compares the ontogeny of *Papio* baboons to other papionin primates through a theoretical perspective that prioritizes ontogeny in the study of life history. This viewpoint anticipates that life history variables are dissociable, or capable of responding to selection independent of one another. The result is diversity in how primate life histories unfold. *Papio* baboons provide excellent evidence for this view of life history, illustrating a mode of life history with clear ties to female reproduction. Specifically, relative to other papionins, life history in *Papio* baboons involves tightly coordinated patterns of development for somatic variables, including body mass, skeletal dimensions, and dental eruption. Growth hormones in *Papio* baboons are highly intercorrelated. However, brain growth follows a distinct pattern from other systems, ceasing very early in *Papio* baboons.

This life history mode reflects heavy metabolic burdens on baboon mothers to produce "high-quality" offspring that can cope with intense selection during early postnatal development. Brain growth is dissociated from


**Steven R. Leigh** • Department of Anthropology, University of Illinois, Urbana, IL, USA **Robin M. Bernstein** • Department of Anthropology, George Washington University, Washington, D.C., USA

development of other somatic systems, inducing high maternal gestational costs, but possibly reflecting the neural capabilities to survive the infant period. These costs appear to have selectively favored an integrated pattern of somatic, dental, and hormonal development, along with large female adult size. Ties between reproduction and life history are integral to understanding baboon evolution.

## 1. INTRODUCTION

Our understanding of primate life history variation and evolution relies on broad-scale interspecific analyses (Harvey et al., 1987; Ross, 1988, 1998). These studies provide a robust understanding of primate life history, generally finding that a continuum, ranging from "fast" to "slow," reasonably describes life history patterns (Ross and Jones, 1999). However, recent advances point to deficiencies with regard to how well this notion addresses basic questions about primate variation and evolution, particularly with regard to the relations between fitness components and life history. In general, analyses of life history in relation to behavior (Garber and Leigh, 1997; Janson and van Schaik, 1993), diet (S. A. Altmann, 1998; Godfrey et al., 2003; Leigh, 1994), demography (Altmann and Alberts, 2003; DeRousseau, 1990; Johnson, 2003; Sade, 1990; Stucki et al., 1991), endocrinology (Finch and Rose, 1995), and morphology (Pereira and Leigh, 2002; Leigh, 2004) have generated insights that are incompatible with hypotheses shaped by the concept of a "fast versus slow" continuum. The inherent limitations of comparative methods typically used in life history analyses compound these emerging theoretical difficulties (Altmann and Alberts, 2003; Martin, 2002). At a more specific level, broad-scale interspecific analyses may not provide the kinds of close-grained information necessary for assessing the relations of reproductive adaptations or strategies to the scheduling of life history.

Moving beyond traditional theoretical perspectives on primate life histories requires attention to ontogenetic processes (Shea, 1990). Consequently, the present study seeks to advance our understanding of life history evolution by investigating links between life history, reproduction, and morphological and hormonal ontogeny in baboons (*Papio*) and other papionins (*Macaca, Mandrillus, Cercocebus, Lophocebus*). We conduct these analyses within a theoretical framework predicting that life history "modes" are fundamental to primate life history evolution (Leigh and Blomquist, 2006; Pereira and Leigh,

2002). The concept of a life history mode contrasts with a simple "fast versus slow" continuum by stipulating that primate life history adaptations can be understood in terms of qualitatively and quantitatively different ways of reaching maturation and adulthood, ultimately driving life history variation in primates. More explicitly, this concept predicts that primate life history does not merely reflect a sliding scale of "faster" or "slower" pathways to maturation and reproduction. While differences in age at maturation are, of course, vitally important (Charnov and Berrigan, 1993), the concept of a life history mode suggests that adaptations involving rates of growth (Janson and van Schaik, 1993; Godfrey et al., 2003; Leigh, 1994) and sequences of development (Smith, 2002; Watts, 1985, 1990) contribute significantly to the structure of primate life histories. Rates of growth and sequences of development can respond to evolutionary forces, potentially leading to an array of ontogenetic configurations that vary both by sex and species. These life history variables are influenced either independently or together by factors that modify ontogenetic programs, including physiological mechanisms such as hormones. The ways in which morphological and hormonal ontogenies are structured comprise life history modes, and these modes may vary predictably among species. In effect, life history modes reflect alternative ways of "assembling" and integrating morphological and behavioral systems during development: They comprise different pathways to adulthood, representing evolutionary responses to selection throughout the life course.

Baboons present outstanding opportunities to investigate the concept of a life history mode as well as the relations of life history modes to reproductive strategies mainly because a remarkable body of research has concentrated directly on baboons during ontogeny (J. Altmann, 1980, 1983; S. A. Altmann, 1998; J. Altmann and Alberts, 1987, 2003; J. Altmann et al., 1977, 1978, 1993; Bercovitch and Strum, 1993; Johnson, 2003, this volume; Moses et al., 1992; Pereira and Altmann, 1985; Strum, 1991). This research has myriad implications for understanding baboon life history adaptations in social and ecological contexts, and provides insight into life history evolution more generally. Baboons are especially important in this regard, given Stuart Altmann's findings that selection encountered by yearling baboons virtually determines lifetime reproductive success (1998). This pioneering research presents opportunities to assess life history theory, increases our understanding of how juvenile periods evolve, and evaluates the relations between reproductive strategies and ontogeny. Consequently, the present study hypothesizes that baboons have evolved a distinctive life history mode relative to other papionin primates. Compared to closely related species, we expect that the baboon life history mode involves major prenatal maternal investment in offspring brain growth, a coordinated pattern of morphological and hormonal ontogeny, high but consistent growth rates, and, finally, large body size. The timing of maturation may be intertwined with aspects of this life history mode, both in baboons and in other papionins. We can note that the attributes of infancy are poorly known in other papionins, precluding more definitive expectations for other species. However, we anticipate that baboons are exceptional in their highly eclectic foraging style relative to these other species.

## 2. MATERIALS AND METHODS

### 2.1. Materials

Defining and understanding the baboon life history mode requires detailed study of numerous developing systems. Rates, time spans, and sequences of ontogeny must be analyzed, with attention to variation both within and among behavioral, anatomical, and hormonal systems. Consequently, this analysis explores links between life histories and reproductive strategies by analyzing the ontogeny of specific anatomical and physiological features, including brains, teeth, body mass, somatometrics, and hormones.

Details on data sources must, regrettably, be cast aside in the interest of brevity. However, data are derived from a variety of sources, including literature, museum-curated wild-shot specimens, and live captive primates (Table 1; see also Bernstein, 2004; Buchanan, 2006; Leigh, 2004; Leigh et al., 2003). The majority of live captive data were collected during an ongoing comparative study of ontogeny in papionin primates (see Leigh et al., 2003; Leigh, 2006). When possible, we investigate three or more papionin species, so as to avoid certain inherent limitations of two-species comparative studies (Garland and Adolf, 1994). However, two-species comparisons are, by necessity, conducted in some cases, particularly for hormonal analyses. Our most detailed comparisons rely on longitudinal somatometric and hormonal data for captive baboons (*Papio hamadryas*, Southwest Foundation for Biomedical Research) and sooty mangabeys (*Cercocebus atys*, Yerkes Regional Primate Research Center). These data represent measurements obtained every 6 months from core groups of 20 baboons and 20 sooty mangabeys

**Table 1.** Data type, taxa represented, and data sources

| Data type | Species | Source | Comments |
|---|---|---|---|
| Prenatal brain mass growth | *Papio hamadryas* | Hendrickx and Houston, 1971 | All necropsy, captive |
| | *Papio hamadryas* | Tame et al., 1998 | |
| | *Macaca mulatta* | Cheek, 1975 | |
| Postnatal brain growth, cranial capacity | *P. hamadryas*, *M. mulatta*, *Cercocebus* sp., *Lophocebus* sp. | This study; Leigh, 2004 | Cranial capacities of wild-shot specimens |
| Postnatal brain mass growth | *P. hamadryas* | Mahaney et al., 1993a,b | Necropsy, captive |
| Postnatal relative brain size growth | *P. hamadryas*, *M. mulatta*, *Mandrillus* sp., *Cercocebus* sp., *Lophocebus* sp. | Leigh et al., 2003 | Cranial measures of wild-shot specimens |
| Age at first birth | *P. hamadryas* | Williams-Blangero and Blangero, 1995 | Captive |
| | *P. hamadryas* | Bercovitch and Strum, 1993; Cheney et al., this volume; Sigg et al., 1982; Altmann et al., 1981 | Wild |
| | *M. mulatta* | Bercovitch and Berard, 1993; Sade, 1990 | Provisioned |
| | *Cercocebus atys* | This study | Captive |
| | *M. sphinx* | Setchell et al., 2002 | Provisioned |
| Dental development | *P. hamadryas* | Bernstein et al., 2000 | All captive or provisioned |
| | *Cercocebus atys* | Bernstein et al., 2000 | |
| | *M. mulatta* | Cheverud, 1981 | |
| Mass ontogeny | *P. hamadryas* | This study | |
| | *C. atys* | This study | |
| | *M. mulatta* | Leigh, 1992 | |
| Body length | *P. hamadryas*, *C. atys* | This study | |
| | *M. sphinx* | Setchell et al., 2001 | Crown-rump length |
| Hormones analyzed | *P. hamadryas*, *C. atys* | This study | IGF-I, IGFBP-3, DHEAS, estradiol, radioimmunoassay |

over a 5-year time span (1997–2002 for baboons, 1998–2003 for mangabeys) (Table 2). Core group individuals were replaced as needed (usually upon death or transfer of the animal), and both sexes are equally represented in the core groups. Additional cross-sectional data were collected from

**Table 2.** Life history and morphometric variables for papionin primates

| Species | Age at first birth (years) | Adult size (kg) | Adult size at first birth (%) | Adult length at first birth (%) | Sample source |
|---|---|---|---|---|---|
| *Papio h. anubis* | 6.32[a] | 15.96 | 89 | ~100 | SFBR |
| *Papio h. anubis* | 6.92[b] | 15.2[c] | – | – | Noncaptive |
| *Papio h. ursinus* | 6.75[d] | | | | Noncaptive |
| *Papio h. hamadryas* | 6.1[e] | | | | Noncaptive |
| *Papio h. cynocephalus* | 6[f] | | | | Noncaptive |
| *Papio h. papio* | – | 14.3[g] | – | – | Captive |
| *Macaca mulatta* | 4.1[h] | 8.37[9] | 77 | – | Captive, free ranging |
| *Cercocebus* | 4.88 | 7.90 | 79 | 90 | Yerkes, captive |
| *Mandrillus*[i] | 4.63 | 9.91 | 74 | 94 | Free ranging |
| *Mandrillus* | | 16.4[9] | | | Captive |

[a]Williams-Blangero and Blangero (1995). [b]Bercovitch and Strum (1993). [d]Cheney et al. (this volume). [e]Sigg et al. (1981). [f]Altmann et al. (1981). [h]Bercovitch and Berard (1993). [i]Mandrill values from Setchell et al. (2001, 2002). weighted average of "founder" and "colony born" animals (12.8 and 9.1 kg, respectively). [c]Asymptotic parameter for nongarbage raiding animals from Strum (1991). [g]Leigh (1992). All other values estimated from the current study.

other animals (Leigh, 2006). Sex, age, subspecies, and general condition of each animal were recorded. Most of the baboons are olive baboons (*P. h. anubis* > 75 percent), with most of the remainder representing olive–yellow hybrids. Unless otherwise specified, baboon subspecies are combined in these analyses in order to maximize samples. Dental analyses are based on visual inspection of teeth.

Serum samples for hormone assays were obtained from core group animals (both species) at the time of measurement. These enable hormone analyses focusing on insulin-like growth factor I (IGF-I), IGF-binding protein-3 (IGFBP-3), 17β-estradiol (E2, or estradiol), testosterone, and dehydroepiandrosterone sulfate (DHEA-S). IGF-I is active in postnatal growth and exerts potent mitogenic action on the cells of connective tissues, organs, and bones (Liu and LeRoith, 1999). IGFBP-3 (Yu et al., 1999) is the principal carrier of IGF-I and IGF-II in serum (binding 90–96 percent of these growth factors), often traveling in a ternary complexed form with ALS (acid-labile subunit). IGFBP-3 actions in vivo include enhancing IGF-I actions in bone, muscle, and visceral growth (Baxter, 2000). Both testosterone and estrogen appear to have a stimulatory effect on IGFBP-3, possibly through indirect effects on GH secretion (Hall et al., 1999; Pazos et al., 2000). Estradiol plays an essential role in the control of development of secondary sexual characteristics in most female vertebrates, and in the regulation of female reproductive function (Bentley, 1998). Dehydroepiandrosterone sulfate (DHEA-S) is the sulfate of another steroid (DHEA), produced by the adrenal cortex, and can function as a major precursor of testosterone and estradiol in peripheral tissues (Hadley, 2000).

### 2.2. Methods

Patterns of brain growth are assessed by regression techniques. First, we describe both prenatal and postnatal absolute brain growth by nonparametric loess regression (Leigh, 2004, 2006). To assess postnatal brain ontogeny, we use summary statistics for baboon necropsy data (reported by Mahaney et al., 1993a,b) and measurements of cranial capacity from wild-shot specimens (Leigh, 2004; Pereira and Leigh, 2002). Second, relative growth trajectories compare allometric growth among papionin genera. Here, we present reduced major axis regressions of neurocranial size plotted against facial size (Leigh et al., 2003). Dental comparisons between baboons and sooty mangabeys are based on calculated average completed eruption age for each tooth represented by a 52-stage scale (Bernstein et al., 2000).

Analyses of somatic growth concentrate on a large data set compiled from baboons and mangabeys (Leigh, 2006). Mass (in kg) is estimated directly from weight measures of both core- and noncore-group animals. Additional mass data for baboons were provided by Mahaney (Mahaney et al., 1993a). Mass data for macaques are derived from captive colonies (Leigh, 1992). The remaining somatometric dimension, body length (vertex to ventral base of tail), was measured to the nearest millimeter. All data are treated cross-sectionally, including the multiple observations from core group animals. The vast majority of noncore group baboons were measured only once, but most noncore group mangabeys contribute longitudinal observations. Treating longitudinal data cross-sectionally poses problems in significance testing but does not generally affect estimates of central tendencies (Leigh, 1992). Formal investigations of longitudinal data are pending. Statistics for age at first birth in several papionins are derived from literature sources (Table 1). Data for mangabeys are calculated from colony records for 12 births during a period when females in the colony were at risk of pregnancy. Since several sources represent captive data, management practices may influence these estimates. We limit analyses to female patterns of ontogeny.

We mainly employ nonparametric regression techniques to describe ontogenetic variation (Efron and Tibshirani, 1991; Leigh, 1992; Mahaney et al., 1993a). Loess regression has been used extensively for analyses of primate growth (Leigh, 1992, 1996, 2001, 2004, 2006; Leigh and Park, 1998). This approach estimates a locally weighted regression line by successively analyzing small segments or "windows" of a bivariate data scatter (Efron and Tibshirani, 1991). Age at growth cessation is estimated visually, following earlier procedures (Leigh, 1992).

Hormone analyses utilize radioimmunoassay methods. To dissociate the IGF from its binding proteins (IGFBPs), serum samples (500 μL) are chromatographed in 0.2 M formic acid on a 0.9 × 100-cm column containing Sephadex G-50 beads (Pharmacia Fine Chemicals, Piscataway, NJ). IGF-I, estradiol, and DHEA-S are measured by radioimmunoassay. IGFBP-3 is measured using immunoradiometric assay. Changes in hormone levels with age were analyzed using protocols used for somatometric data, particularly nonparametric regression. Correlations among hormones and between hormones and measures of size (body mass and body length) are analyzed to assess the degree of hormonal integration, among hormones and between hormones and morphology. All statistical analyses are performed using Systat (version 9.01) statistical software (Wilkinson, 1999).

## 3. RESULTS

### 3.1. Brain Ontogeny

Nonparametric loess regressions illustrate that both baboons and macaques begin brain growth at about the same gestational age (Figure 1a). Macaque brains are marginally larger early in development, but are exceeded after baboons initiate a brain growth spurt at about 120 gestational days, or the beginning of the third trimester. This leads to neonates with large brains compared to macaques. Adjusting for average female adult size (Table 2) indicates that fetal brain sizes as a percentage of adult female body mass are smaller in baboons than macaques (Figure 1b).

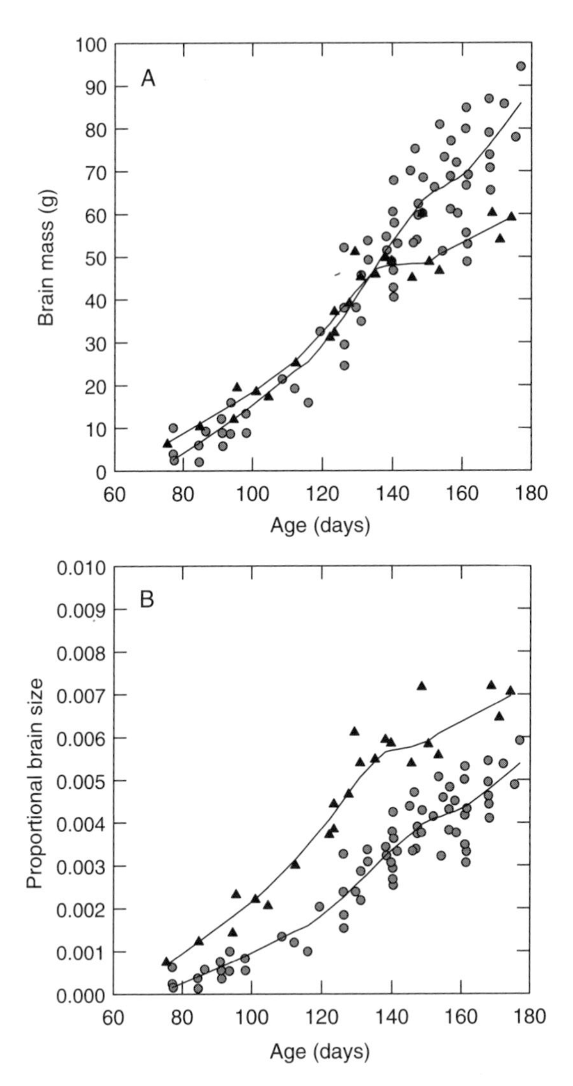


**Figure 1.** (A) Absolute brain growth in baboons (circles) and rhesus macaques (triangles). (B) Fetal brain size divided by average female body size for baboons (circles) and rhesus macaques (triangles).

Allometric comparisons of postnatal relative brain size ontogeny further demonstrate that baboons are born with brains that are large relative to the remainder of the skull, as illustrated by separation of regression lines (Figure 2a). It is important to note that mandrills, which reach adult body

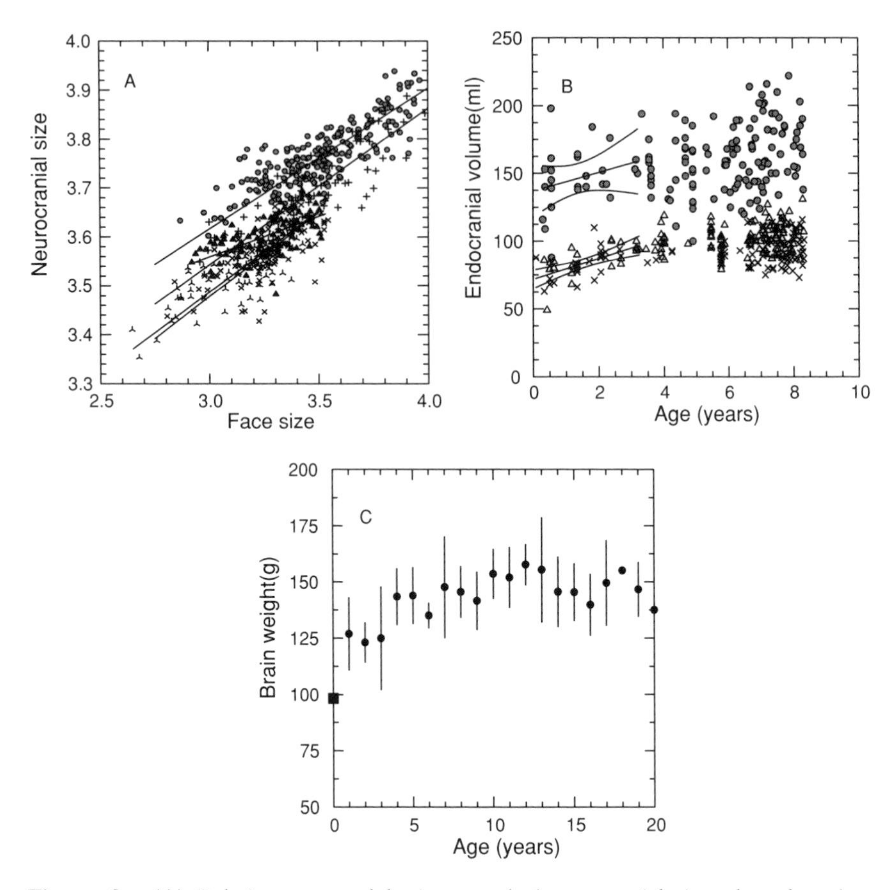


**Figure 2.** (A) Relative postnatal brain growth (neurocranial size plotted against facial size) for papionins. *Papio* is represented by filled circles, rhesus macaques by filled triangles, mandrills by +, *Lophocebus* by inverted "Y" symbols, and *Cercocebus* by "X" symbols. (B) Endocranial volume against estimated age for baboons (filled circles) and a combined sample of *Lophocebus* (X) and *Cercocebus* (triangles). Lines represent least squares regressions with 95 percent confidence intervals. (C) Postnatal brain size in female baboons from Mahaney et al., 1993a,b. Circles represent means, and bars denote standard deviations.

and brain sizes comparable to baboons, have smaller relative brain sizes early in life. Postnatal relative brain growth rates are low in baboons.

Absolute postnatal brain growth trajectories are difficult to interpret, given reliance on either wild-shot data of estimated chronological age or necropsy data. Wild-shot baboons show no statistically significant change in the relation between brain size and age during the first 3 years of postnatal life (Figure 2b; Pereira and Leigh, 2002). However, we find statistically significant postnatal brain size increases in some other papionins, notably in mangabeys (*Cercocebus* and *Lophocebus*). Samples for mandrills are small and difficult to interpret. Unfortunately, necropsy data for baboons are ambiguous (Figure 2c). Data are sparse for young specimens, but summary statistics are consistent with an interpretation of limited postnatal changes in brain size.

In summary, baboons have relatively and absolutely large brains that increase size dramatically during later gestation. At birth, some may fall within the adult size range. In contrast, other papionins appear to complete brain growth later in the postnatal period, with mangabeys possibly extending brain growth well into the postnatal period.

### 3.2. Age at First Birth

Age at first birth in the Southwest Foundation baboon colony averages 6.3 years (Williams-Blangero and Blangero, 1995), while noncaptive olive baboons achieve first birth at 6.9 years, on average (Bercovitch and Strum, 1993), chacmas at 6.75 (Cheney et al., this volume), and hamadryas at 6.1. Mangabeys first give birth at an average age of 4.9 years (Table 2). A *t*-test indicates that means for captive baboons and mangabeys are significantly different ($p < 0.05$), despite the small mangabey sample size. First birth in Cayo Santiago rhesus macaques (*Macaca mulatta*) occurs at an average of 4.1 years (Bercovitch and Berard, 1993; Sade, 1990). Free ranging mandrill females average 4.63 years for age at first birth, ranging between 3.29 and 6.14 ($N = 19$, Setchell et al., 2002). Baboons show an absolutely later age at first birth than these papionins. Literature reported values for noncaptive animals are consistent with these values (Ross and Jones, 1999). We urge caution in interpreting these results, given the variety of data sources (Table 1).

### 3.3. Dental Development

Patterns of tooth eruption show some distinctions between baboons and mangabeys (Figure 3). This and several subsequent plots show average ages

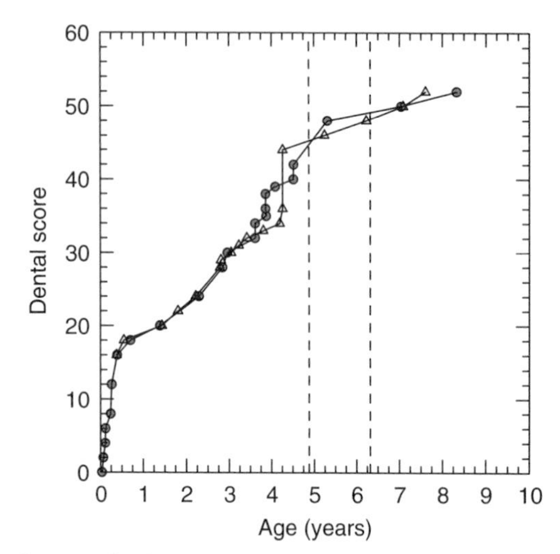


**Figure 3.** Dental scores for baboons (filled circles) and sooty mangabeys (triangles). Mean ages at first birth are designated by the dashed lines.

at first birth, so that results can be referenced to this important life history milestone. Female dental scores show identical progressions between the two species until about 3.75 years of age. At this point, the baboon premolar complex erupts very suddenly, with several teeth emerging in close succession. Mangabeys erupt premolars later, simultaneously erupting several teeth. Consequently, baboon dental development is accelerated relative to age at first birth in comparison to mangabeys. Analyses of macaques show that by age at first reproduction, eruption of adult teeth is nearly complete (with the exception of M3s) by 4 years of age (Cheverud, 1981).

### 3.4. Somatic Ontogeny

#### *3.4.1. Body Mass*

We find substantial similarities in the timing of body mass growth among baboons, mangabeys, and rhesus macaques, with all species reaching adult size between 6 and 7 years of age (Figure 4a–c). However, baboons grow at faster rates to become larger than either of the other two species. Female baboons older than 6 years of age average nearly twice the size of rhesus macaques and mangabeys (Table 2). Average age at first birth occurs when baboons are close to adult size (89 percent of adult value), but macaques, mangabeys, and mandrills are relatively much smaller (Table 2).

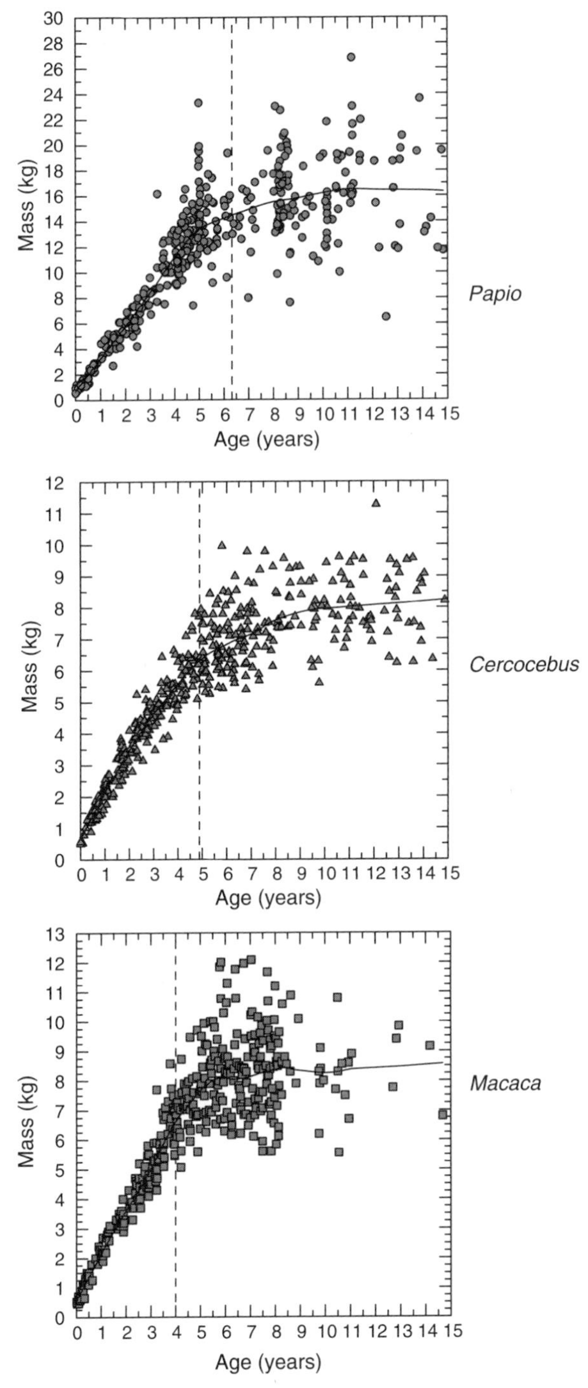


**Figure 4.** Body mass plotted against age for papionin primates. Growth curves are calculated with loess regressions, dashed lines denote average ages at first birth.

### *3.4.2 Body Length*

Baboons increase body length at higher rates than mangabeys and actually may reach adult body length slightly before mangabeys (Figure 5; Table 2). Mangabeys can be expected to attain adult length well after average age at first birth. By average age at first birth, mangabeys have reached 90 percent

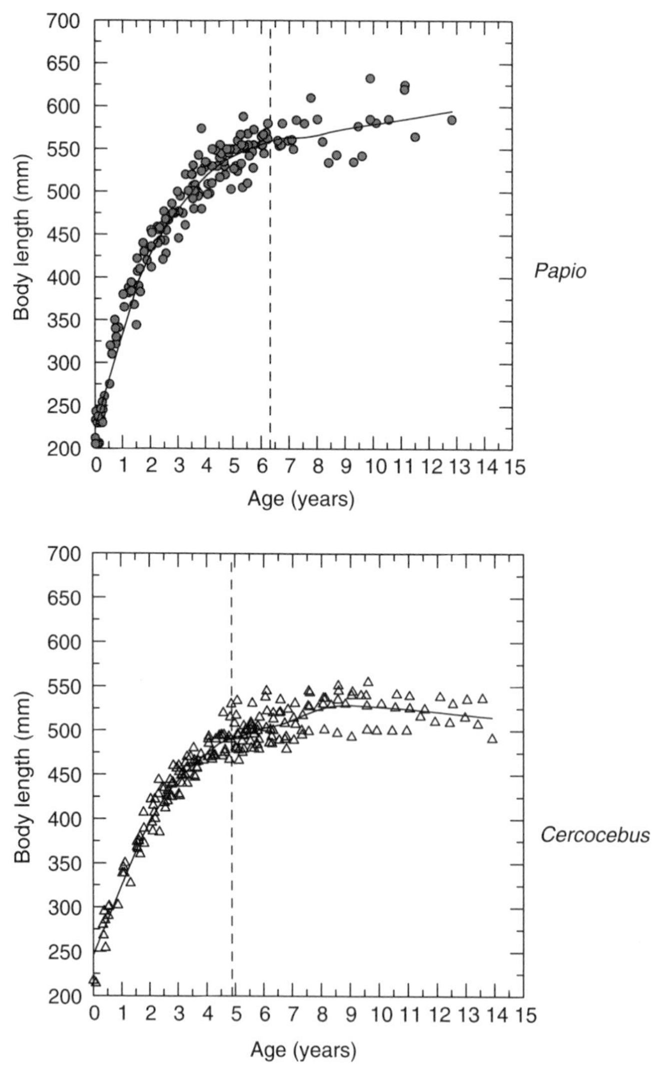


**Figure 5.** Body length growth in baboons and sooty mangabeys estimated by loess regression. Dashed lines represent age at first birth.

of adult size (523 mm), but baboon body length growth is virtually complete (adults average 571 mm). On a comparable measure of body length, mandrill females fall between these values (Table 2) (Setchell et al., 2001). It should be noted that mangabeys have long trunks relative to mass when compared to baboons. Although this study examines only body length growth, previous research indicates that baboons show few differences in the ages at which skeletal elements cease growth compared to macaques (Leigh, 2006). Specifically, overall skeletal growth appears to be better synchronized in baboons than in macaques (and possibly mangabeys). For example, macaque trunk dimensions apparently cease growth up to 2.5 years later than limbs (Cheverud et al., 1992; Turnquist and Kessler, 1989), with macaque species generally growing slower for longer periods of time than baboons. These patterns may be related to different management conditions, but this is unlikely because the macaques differ most in these terms. Thus, these patterns seem to represent species differences between baboons and macaques, implying that baboon skeletal growth is highly integrated. Preliminary analyses suggest that mangabeys show greater levels of growth rate variation than baboons. Males in particular show prominent growth spurts in numerous skeletal dimensions that are not apparent in baboons (Buchanan, 2006).

### 3.5. Hormonal Ontogeny

Hormonal analyses are restricted to baboons and mangabeys. Age-related changes in baboon hormone profiles are characterized by more regularity than those of mangabeys. For example, age-related changes in IGF-I in baboons are steadier than in mangabeys (Figure 6a and b). Estradiol is generally uncorrelated with age in each species, with young pregnant mangabeys showing high levels. In addition, serum concentrations for both hormones are notably lower in baboons than in mangabeys. IGFBP-3 levels are correlated with age in each species (Pearson $r = 0.64$ for baboons, $r = 0.63$ for mangabeys), although mangabey values are overall slightly higher than baboons.

Female baboons show strong, statistically significant correlations between IGF-I, IGFBP-3, and body mass and length (Table 3). Female mangabeys show significant correlations between IGFBP-3, mass, and length; additionally, estradiol and mass show a modest correlation. For both species, DHEA-S showed negative, nonsignificant correlations with both body mass and length and these values are therefore not shown.

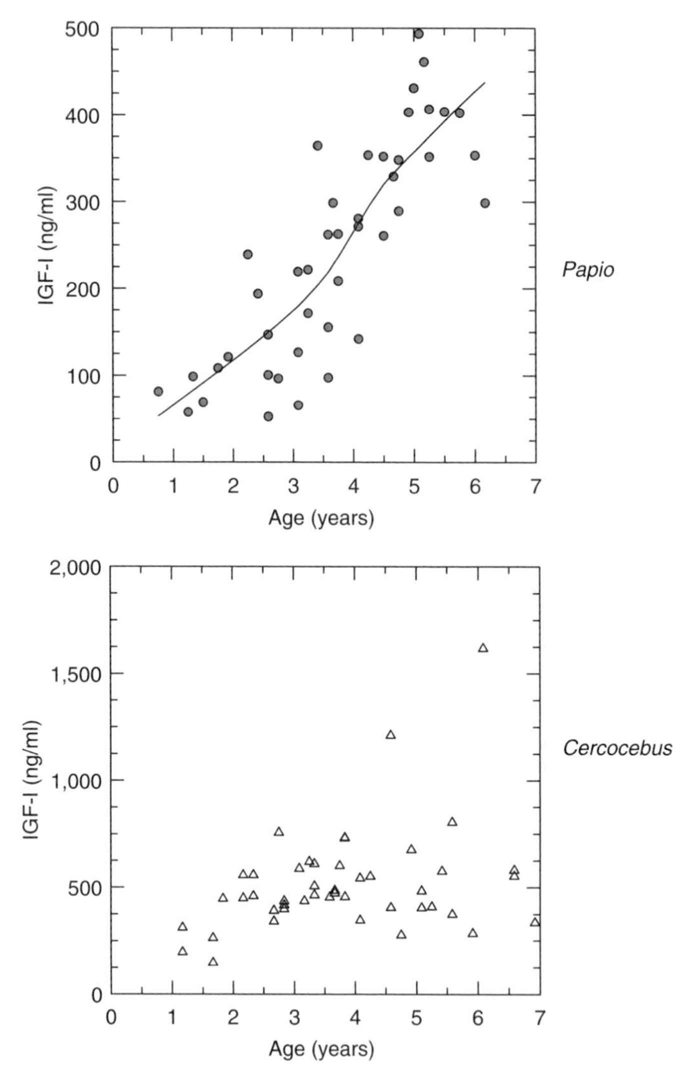


**Figure 6.** Hormone concentrations for IGF-I plotted against age for baboons and sooty mangabeys. The *Papio* regression line is calculated by loess regression.

Differences between species are evident when considering the correlations among the hormones themselves. Female baboons show strong and statistically significant correlations between IGF-I and IGFBP-3, and estradiol and IGFBP-3 (Table 4). Although not statistically significant, the positive correlation between estradiol and DHEA-S in female baboons is moderately strong, while correlations between DHEA-S and all other hormones (and all

**Table 3.** Correlations among hormones and size measures

| | IGF-I | | IGFBP-3 | | Estradiol | |
|---|---|---|---|---|---|---|
| | Papio | Cerco | Papio | Cerco | Papio | Cerco |
| Weight | 0.841 | 0.593 | 0.696 | 0.658 | 0.367 | 0.507 |
| Crown rump length | 0.834 | 0.483 | 0.685 | 0.655 | 0.319 | 0.393 |

**Table 4.** Pearson correlations among hormones for *Papio*, below the diagonal, and *Cercocebus*, above the diagonal

| | IGF-I | IGFBP-3 | Estradiol | DHEA-S |
|---|---|---|---|---|
| IGF-I | – | 0.028 | 0.195 | –0.201 |
| IGFBP-3 | 0.767 | – | 0.141 | –0.78 |
| Estradiol | 0.106 | 0.561 | – | –0.201 |
| DHEA-S | –0.162 | –0.134 | 0.333 | – |

hormones for mangabeys) are negative. Mangabeys do not show any significant correlations among any hormones.

## 4. DISCUSSION

### 4.1. Life History

These findings have important implications for understanding baboon ontogeny and maternal investment, as well as primate life history in a more general sense. Comparisons among papionins reveal that the level of integration among developing systems may vary during ontogeny. Furthermore, these results point to a complex and distinctive life history mode in baboons. This mode is characterized by brain and tooth ontogeny that is complete relatively early, high somatic growth rates, large size, and, probably, deferred age at first reproduction. Brain size, dental eruption, body mass, and somatic dimensions present a range of variations in ontogeny when compared across species. Female baboons show consistently high correlations between IGF-I, IGFBP-3, and size variables, and hormones appear to be integrated. Taken together, these findings indicate high degrees of diversity both among and within developing systems.

Our results have implications at general and specific levels. At the broadest level, papionin ontogenetic variation is consistent with recently established theoretical advances in developmental biology (Raff, 1996). Specifically, dissociability, or the capacity of developing systems to respond independently to selection, is now recognized as a major source of evolutionary diversity (Gould, 1977; Needham, 1933; Raff, 1996). Dissociability leads to modularity in ontogeny. Our comparative investigations reveal a prominent role for modularity in the organization of papionin ontogeny and life history. The presence of modularity is notable for at least three reasons. First, theoretical concepts in developmental biology have traditionally focused on the earliest phases of ontogeny (Raff, 1996). Modularity during postnatal periods implies that ideas about dissociability apply to all phases of ontogeny, including life histories. Second, the traditional view of life history as a "fast versus slow" continuum is incompatible with the concept of modularity. This continuum requires high levels of ontogenetic integration such that life history variables comprise a "suite" (Alberts and Altmann, 2002) of inter-related traits, running on either fast or slow trajectories with linkages to body size. Differing time scales of growth among variables recorded by our study compromise this assumption. Third, standard life history theory anticipates that selection on either maturation age or the duration of body size growth drives size differences among species (Charnov and Berrigan, 1993). Our analyses demonstrate that size, age at first reproduction, and age at body size growth cessation can be decoupled: Differing sizes may be reached in comparable time periods (see also Garber and Leigh, 1997; Leigh, 1992; Leigh and Terranova, 1998; Watts, 1990).

These basic points strongly imply that life history theory must accommodate both patterns of morphological ontogeny and the division of the life course into phases. We suggest that the concept of a life history mode accomplishes this goal by recognizing phases and components of life history (Figure 7). Two phases of life history traditionally have been recognized, including ontogenetic and reproductive periods. Classic life history theory focuses on explaining the allocation of the life course to these phases (Charnov and Berrigan, 1993; Cole, 1954), but life history components have received little theoretical attention. Components are, however, critical because they represent targets of selection on morphological, hormonal, or behavioral attributes during these life history phases. The rate and timing of ontogeny for various components or traits may differ considerably, and components can respond to selection in a modular fashion. Our "components" can be fitness components (Altmann and Alberts, 2003; Hughes and Burleson, 2000), although we use this term in a broader sense, referring to

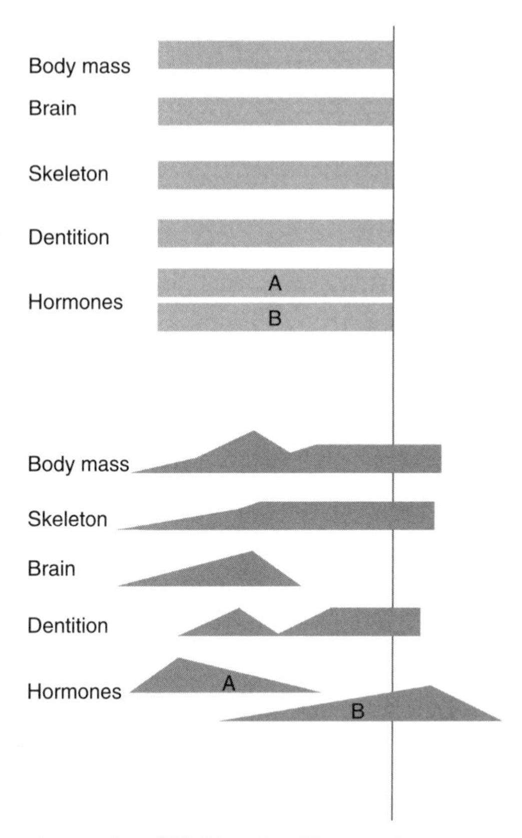


**Figure 7.** Alternative modes of life histories. Top panel represents a species in which patterns of ontogeny are well synchronized, and growth is complete by age at maturation (vertical line). Growth rates and periods are consistent among developing components. The bottom panel portrays a species in which components are dissociated, with some components growing past age at maturation. Vertical changes qualitatively represent changes in growth rates or changes in hormone concentrations.

the skeletal system, the brain, body mass, and hormones. The ways that components grow through different phases comprise life history modes. We suggest that attention to life history modes undermines a traditional "fast versus slow" continuum.

Analyses of baboons illustrate key features of life history modes, facilitating conceptualization of life history phases and components. First, the temporal organization of ontogeny varies among components, producing variable levels of integration among components. In baboons, somatic growth and dental eruption cease at about the time of average age at first birth, but the period of brain growth is well separated from the reproductive phase. Consequently, the brain growth period overlaps minimally with the growth periods of other components, but all components grow over similar time frames in other papionins. Investigations of baboons indicate that selection on individual components may have limited consequences for other components. Second, diversity in temporal patterning of ontogeny is complemented by variable levels of integration within each developing component. Components in baboons appear to be internally well integrated, particularly hormones and skeletal growth (Leigh, 2006). Baboons contrast with mangabeys and possibly macaques (Watts, 1990) in this regard, given that lower levels of integration are apparent in these taxa. Third, phases are distinct in baboons, with a clear switch between growth and reproduction, possibly stemming from hormonal integration. Taken together, these dimensions of variation indicate that components behave in a modular fashion, with interspecific differences in life history modes produced by differing configurations of components and phases.

Consideration of brain growth patterns in relation to other components provides direct insight into these processes. The case of brain size growth is especially significant given the emphasis accorded on the relations between brain size and life history (Allman and Hasenstaub, 1999; Deaner et al., 2002; Harvey et al., 1987; Sacher, 1959; Sacher and Staffeldt, 1974). Although much better data are needed for understanding patterns of brain growth, baboons seem to illustrate large increases in brain size over an abbreviated ontogenetic period. Baboon neonatal brain size is very large, even though gestation periods in papionins are strikingly uniform, ranging from 165 to 183 days across the clade (Table 5) (see also Martin and MacLarnon,

**Table 5.** Gestation lengths in papionin primates

| Taxon | Approximate gestation length (days) | Reference |
|---|---|---|
| *Macaca mulatta* | 165 | Cerroni et al., 2003 |
| *Macaca* | 180 | Ardito, 1976 |
| *Cercocebus* | 167 | Gordon et al., 1991 |
| *Mandrillus* | 168–176, 175 (mean, $n = 61$, range 168–183) | Carman, 1979; Setchell et al., 2001 |
| *Lophocebus* | 175 | DePutte, 1991 |
| *Papio* | 177 (mean, $n = 59$) | Hendrickx, 1967 |

1990). This finding suggests that evolutionary changes in brain size need not entail increases in gestation length (*contra* Sacher and Staffeldt, 1974), either through selection directly on gestation length, or as a result of more general selection for a "slow" life history. Postnatally, baboons seem to reach adult brain size very soon after birth, in contrast to mangabeys. Differences among papionins are apparent, despite a strong tendency for Old World monkeys to complete brain growth early in life (Leigh, 2004).

Comparative analyses of ontogeny suggest that baboons could be described as a "fast" species in terms of some aspects of morphological ontogeny by reaching large body and brain sizes during growth periods comparable to those of smaller papionins. Baboons are also legitimately considered a "slow" species in terms of age at first reproduction. However, consistency in the amount of time allocated to size growth among papionins precludes categorization of a particular species as "fast" or "slow." Furthermore, in baboons, the time period of brain growth and body size growth overlap minimally. Skeletal growth in baboons and, apparently, other papionins, occupies a shorter time frame than skeletal growth in macaques (cf. Cheverud et al., 1992; Turnquist and Kessler, 1989). Other papionins seem to reach adulthood very differently from baboons. For example, mangabeys can be described as "slow" in terms of brain growth, but "fast" in terms of age at first reproduction. Mangabey brain growth occupies a substantial portion of the postnatal period, attaining brain sizes absolutely and relatively smaller than in baboons. In addition, somatic growth rate variation is evident in mangabeys, but growth rates are generally lower than in baboons. These examples suggest that the concept of a life history mode, with attention to components and their allocation to different phases, helps account for life history variation. In baboons, patterns of brain growth, somatic growth, dental eruption, and levels of hormonal integration are related, but not directly through selection solely on the pace of life histories.

### 4.2. Baboon Life History Evolution and Reproduction

These findings, while providing general insights into primate life history theory, have specific implications for understanding the baboon life course. We define the baboon life history mode as one comprising discrete life history phases, with little overlap between the periods of ontogeny and reproduction. The brain follows an abbreviated time schedule, while maturation age appears to be deferred. However, anatomical and hormonal components appear to be relatively well integrated both internally and with one another. Mangabeys, and possibly other papionins, seem to show a contrasting mode, lacking clear separation among life history phases and a different pattern of relations among components. It is important to recognize that high levels of variability have been shown to characterize baboon life history attributes (Altmann and Alberts, 2003), so our inferences must be regarded in the context of species differences.

We hypothesize that the baboon life history mode is ultimately a response to severe selection during the juvenile period, based largely on S. Altmann's results illustrating major fitness consequences of foraging success in yearling baboons (1998). Selection seems to have favored heavy investment in offspring, especially production of absolutely and relatively large-brained infants. The large brain of infant baboons may, in turn, provide the necessary infrastructure for their highly eclectic foraging style (assuming that brain morphology relates to cognitive traits). Aspects of dental and somatic development may be organized around this heavy maternal metabolic investment. In other words, completion of female ontogeny is a precondition for this investment. Our model suggests that selection has favored large body size through growth at high rates, and completion of ontogeny prior to reproduction as a maternal reproductive strategy supporting infant brain growth costs. Thus we see both direct effects of selection on offspring—brain size and cognitive precocity—and indirect effects of this selection on females—a coordinated pattern of somatic development and discrete separation of life history phases (see also Leigh, 2004).

Attributes of the baboon life history mode are explicable in terms of maternal metabolic costs and resultant adaptations. For example, rapid dental and somatic ontogeny may represent a way to ensure that the costs of gestation can be met without the competing demands of maternal somatic growth. Accelerated dental eruption in baboons also ensures functional competence of adult dentition by age at first reproduction. High somatic growth rates in baboons may play a similar role, yielding large body size without trade-offs against reproductive phase duration. Large body size may also reduce the relative costs of fetal brain growth during gestation. Although detailed data exist for estimating reproductive costs in baboons (J. Altmann, 1980, 1983), such data are lacking for other papionins.

The relations between morphology, hormones, and life histories may be largely structured by the selective landscape for female baboons, particularly selection on maternal characteristics. Morphologically, baboon growth seems to be keyed to producing mothers that are capable of heavy investment in offspring, especially during late gestation and early postnatal life. Coordination among hormones may reflect selection on female baboons favoring clear distinctions between growth and reproductive periods. Intense selection on juveniles should be expected to result in a finely tuned system that controls growth and development. This model explains selection on both mothers and offspring. For female baboons, reproduction requires an individual that is hormonally and morphologically adult. Alternatively, mangabeys and other papionins apparently pay the costs of continued growth along with those of gestation and lactation. These other species appear not to face expensive trade-offs between growth and reproduction that may characterize baboons.

Accepting these hypotheses requires evaluation of numerous alternatives drawn from traditional life history theory. While more detailed tests are beyond the scope of this analysis, we anticipate that predictions from traditional theories would perform poorly in explaining baboon life history. First, variation in ontogenetic time scales of various components is not well accommodated by existing theoretical notions. For example, a "fast" life history should result in early cessation of brain and body growth, coupled with high coordination among developing systems and early reproductive maturation. Instead, we see "fast" life history species with early maturation such as mangabeys that may show relatively extended periods of brain and body ontogeny. Second, baboons unambiguously violate a fundamental assumption of traditional life history theory by reaching a body size larger than most other papionins absent a longer growth period. Thus, size differences need not be produced exclusively by variation in age at maturation, as assumed by some models (see Leigh, 2001). Growth rates respond adaptively to selection (Garber and Leigh, 1997; Godfrey et al., 2003; Janson and van Schaik, 1993; Leigh, 1994), combining with maturation age to produce size variation. Overall, standard life history theories fail to make predictions about growth rates, patterns of growth in components, and the allocation of component growth to different life history phases.

Despite our limited ability to appraise alternatives, we can suggest that the mangabey (*Cercocebus*) life history mode provides an instructive contrast to baboons. Most importantly, mangabeys blur separations among components and phases, apparently investing relatively little in prenatal brain growth, but growing brains well into the postnatal period (Leigh, 2004). Pregnancies in mangabeys are common prior to either completion of dental or somatic ontogeny. For example, during this study, Yerkes female mangabeys typically presented mixed deciduous/adult dentitions throughout their first pregnancy. In many cases, females had shed only deciduous incisors at the time of conception, actually erupting first molars during pregnancy. Obviously, female mangabeys can simultaneously meet the metabolic costs of both their own growth as well as costs of pregnancy despite incomplete adult dentition. The mangabey mode is characterized by production of a small-brained infant, prolonged brain growth into the postnatal period, indistinct life history phases, and loosely coordinated components. Unlike mangabeys, female baboons shoulder most of the costs of offspring brain growth, either during gestation or during a short postnatal brain growth period. Relative to body size, these costs may be comparatively low, but large size is also partly a product of maternal investment (Johnson, 2003, this volume), particularly when high growth rates occur. By age at first birth, baboons possess most of their adult teeth, and have completed the vast majority of body mass and length ontogeny. It should be noted that large brain size in young organisms can be a product of selection solely on size (Gould, 1971). This possibility has been considered for baboons, and while an evolutionary size increase in and of itself may be a factor, it cannot fully account for variation in brain size among papionins (Leigh et al., 2003). Finally, these results indicate that ties between brain ontogeny and life history involve both direct and indirect factors (see Leigh, 2004).

These analyses strongly imply that ontogenetic variability is fundamental to life history evolution and reproductive strategies. Ultimately, brain growth differences may relate to ecological and cognitive demands on these species. In baboons, the response to selection involves discretely separated life history phases along with coordinated development among certain components. Selection related to foraging in young animals has favored females that can support infant brain growth costs. Coordinated somatic ontogeny at high rates, followed by cessation of growth prior to age at first reproduction, may facilitate this adaptation. In contrast, other papionins may encounter a selective regime that favors very early first birth rather than production of "high-quality" offspring. Earlier age at first reproduction is possible because selection apparently has not favored large, rapidly developing brains in offspring, enabling female reproductive opportunities despite continued investment in maternal growth. Baboons follow a contrasting pathway, with an alternative pattern of growth among components and clear shifts between life history phases, possibly stemming from expensive trade-offs between growth and reproduction.

## 5. CONCLUSIONS

Investigations of ontogeny in papionin primates, with special emphasis on baboons, show that life histories in these species depart substantially from expectations of traditional life history theory. Specifically, information about growth and development suggests that predictions regarding "fast versus slow" life histories in primates cannot account for ontogenetic variability among taxa. We propose that the concept of a life history mode, which predicts that there are qualitatively and quantitatively different ways of reaching adulthood, better accounts for baboon and papionin life histories. The concept of a life history mode forces attention to phases and components of ontogeny. Most papionins seem to present a life history mode without clear separations of ontogenetic and reproductive phases. Developing systems, including the brain, skeletal dimensions, and hormones, appear to be only loosely coordinated. In contrast, the baboon life history mode involves discrete separation of ontogenetic and reproductive phases, with apparently high levels of integration among components. This life history mode may be a response to intense selection during the juvenile period involving both ontogeny and maternal reproductive strategies. The baboon life history mode may reveal expensive trade-offs between ontogeny and reproduction that are not faced by other papionins.

## ACKNOWLEDGMENTS

The authors thank volume contributors and Dr. Larissa Swedell for productive discussions, interactions, and support. Funding for this project was provided by NSF BNS-9707361, the Wenner-Gren Foundation for Anthropological Research, the LSB Leakey Foundation, and the University of Illinois. Dr. Michael Mahaney (Southwest Foundation for Biomedical Research and Southwest National Primate Research Center; Department of Genetics, Southwest Foundation for Biomedical Research) and M. Michelle Leland, D.V.M., (Department of Comparative Medicine, Southwest Foundation for Biomedical Research) kindly provided body mass data from their earlier study, with support from the National Institutes of Health:

HV053030, P01 HL028972, and P51 RR013986. Staff at the Southwest Foundation for Biomedical Research, including Drs. Karen Rice, K.D. Carey, Ms. Elaine Windhorst, Ms. Shannon Theriot, and others provided valuable assistance throughout the course of the study. Staff at the Yerkes Primate Center, including Dr. Tom Gordon, the late Dr. Harold McClure, Dr. Katherine Paul, Ms. Tracy Meeker and others aided in aspects of the study conducted at their institution. Ms. Meeker's contribution to ongoing research with the mangabey colony has been especially significant. The Yerkes Primate Center is supported by NIH RR-00165. Dr. Jean Turnquist and the Carribbean Primate Research Center Museum provided access to rhesus macaque skeletal collections, supported by the NIH Award RR-03640. Collaborations with Drs. Sharon Donovan and Marcia Monaco (University of Illinois) were instrumental in conducting hormonal analyses.

## REFERENCES

Alberts, S. C. and Altmann, J., 2002, Matrix models for primate life history analysis, in: *Primate Life Histories and Socioecology*, P. M. Kappeler and M. E. Pereira, eds., Unversity of Chicago Press, Chicago, pp. 66–102.

Allman, J. and Hasenstaub, A., 1999, Brains, maturation times, and parenting, *Neurobiol. Aging* **20**:447–454.

Altmann, J., 1980, *Baboon Mothers and Infants*, Harvard University Press, Cambridge, MA.

Altmann, J., 1983, Costs of reproduction in baboons (*Papio cynocephalus*), in: *Energetics: The Costs of Survival in Vertebrate*, W. P. Aspey, and S. I. Lustick, eds., Ohio State University Press, Columbus, OH, pp. 67–88.

Altmann, S. A., 1998, *Foraging for Survival*, University of Chicago Press, Chicago.

Altmann, J. and Alberts, S. C., 1987, Body mass and growth rates in a wild primate population, *Oecologia* **72**:15–20.

Altmann, J. and Alberts, S. C., 2003, Variability in reproductive success viewed from a life-history perspective in baboons, *Am. J. Hum. Biol.* **15**:401–409.

Altmann, J., Altmann, S. A., Hausfater, G., and McCuskey, S. A., 1977, Life history of yellow baboons: Physical development, reproductive parameters, and infant mortality, *Primates* **18**:315–330.

Altmann, J., Altmann, S. A., and Hausfater, G., 1978, Primate infant's effects on mother's future reproduction, *Science* **201**:1028–1029.

Altmann, J., Altmann, S. A., and Hausfater, G., 1981, Physical maturation and age estimates of yellow baboons, *Papio cynocephalus*, in Amboseli National Park, Kenya, *Am. J. Primatol.* **1**:389–399.

Altmann, J., Hausfater, G., Altmann, S. A., 1988, Determinates of reproductive success in savannah baboons (*papio cynocephalus*), in: *Reproductive Success: Studies of Individual Variation in Contrasting Breeding Systems*, T. H. Clutton-Brock, ed., University of Chicago Press, Chicago, pp. 403–418.

Altmann, J., Schoeller, D., Altmann, S. A., Muruthi, P. and Sapolsky, R. M., 1993, Body size and fatness of free-living baboons reflect food availability and activity levels, *Am. J. Primatol.* **30**:149–161.

Arditio, G., 1976, Check-list of the data on the gestation length of primates. *J. Hum. Evol.* **5**:213–222.

Baxter, R. C., 2000, Insulin-like growth factor (IGF)-binding proteins: interactions with IGFs and intrinsic bioactivities, *Am. J. Physiol. Endocrinol. Metab.* **278**: E967–E976.

Bentley, P., 1998, *Comparative Vertebrate Endocrinology, third edition*, Cambridge University Press, Cambridge.

Bercovitch, F. B. and Berard, J. D., 1993, Life history costs and consequences of rapid reproductive maturation in female macaques, *Beh. Ecol. Sociobiol.* **32**: 103–109.

Bercovitch, F. B. and Strum, S. C., 1993, Dominance rank, resource availability, and reproductive maturation in female savanna baboons, *Beh. Ecol. Sociobiol.* **33**:313–318.

Bernstein, R. M., Rowan, K., Leigh, S. R., 2000, New dental age standards for baboons (*Papio hamadryas anubus*) and mangabeys (*Cercocebus atys*). *Am. J. Phys. Anthropol.* **30**:107–107.

Bernstein, R. M., 2004, Hormones and Evolution in Papionin Primates. Ph.D. Dissertation, University of Illinois, Urbana, IL.

Buchanan, L. S., 2006, The evolution of primate growth spurts. Ph.D. Dissertation, University of Illinois, Urbana, IL (in preparation).

Carman, M., 1979, The gestation and rearing periods of the mandrill, *Mandrillus sphinx, Int. Zoo Yrbk.* **19**:159–160.

Cerroni, A. M., Tomlinson, G. A., Turnquist, J. E., Grynpas, M. D., 2003, Effect of parity on bone mineral density in female rhesus macaques from Cayo Santiago, *Am. J. Phys. Anthropol.* **121**:252–269.

Charnov, E. L. and Berrigan, D., 1993, Why do female primates have such long lifespans and so few babies? *Or* Life in the slow lane, *Ev. Anthropol.* **1**:191–194.

Cheek, D. B., 1975, Appendix II. Data on normal fetal and postnatal *Macaca mulatta*, in: *Fetal and Postnatal Cellular Growth: Hormones and Nutrition*, Cheek, D. B., ed., John Wiley & Sons, New York, pp. 521–534.

Cheverud, J. M., 1981, Epiphyseal union and dental eruption in *Macaca mulatta*, *Am. J. Phys. Anthropol.* **56**:157–167.

Cheverud, J. M., Dittus, W. P. J., and Wilson, P., 1992, Primate population studies at Polonnaruwa. III. Somatometric growth in a natural population of toque macaques (*Macaca sinica*), *J. Hum. Evol.* **23**:51–78.

Cole, L. C., 1954, The population consequences of life history phenomena, *Q. Rev. Biol.* **29**:103–137.

Deaner, R. O., Barton, R. A., and van Schaik, C. P., 2002, Primate brains and life histories: Renewing the connection, in: *Primate Life Histories and Socioecology*, P. M. Kappeler and M. E. Pereira, eds., Unversity of Chicago Press, Chicago, pp. 233–265.

Deputte, B. L., 1991, Reproductive parameters of captive grey-cheeked mangabeys, *Folia Primatol. (Basel)* **57**:57–69.

DeRousseau, C. J., 1990, Life history thinking in perspective, in: *Primate Life History and Evolution*, C. J. DeRousseau, ed., Alan R. Liss, New York, pp. 1–13.

Efron B. and Tibshirani R., 1991, Statistical data analysis in the computer age, *Science* **253**:390–395.

Finch, C. E. and Rose, M. R., 1995, Hormones and the physiological architecture of life history evolution, *Q. Rev. Biol.* **70**:1–52.

Garber, P. A. and Leigh, S. R., 1997, Ontogenetic variation in small-bodied New World Primates: Implications for patterns of reproduction and infant care, *Folia Primatol.* **68**:1–22.

Garland, T. and Adolf, S. C., 1994, Why not to do two species comparative studies: Limitations on inferring adaptation, *Physiol. Zool.* **67**:797–828.

Godfrey, L. R., Samonds, K. E., Jungers, W. L., Sutherland, M. R., and Irwin, M. T., 2003, Ontogenetic correlates of diet in Malagasy lemurs, *Am. J. Phys. Anthropol.* **123**:250–276

Gordon, T. P., Gust, D. A., Busse, C. D.,Wilson, M. E., 1991, Hormones and sexual behavior associated with postconception perineal swelling in the sooty mangabey (*cercocebus atys*), *Int. J. Primatol.* **12**:585–597.

Gould, S. J., 1977, *Ontogeny and Phylogeny*, Belknap Press, Cambridge, MA.

Gould, S. J., 1971, Geometric similarity in allometric growth: A contribution to the problem of scaling in the evolution of size, *Am. Nat.* **105**:113–136.

Hadley, M., 2000, *Endocrinology, fifth edition*, Prentice Hall, New Jersey.

Hall, K., Hilding, A., Thoren, M., 1999, Determinants of circulating insulin-like growth factor-I. J. *Endocrinol. Invest.* **22**:48–52.

Harvey, P. H., Martin, R. D., and Clutton-Brock, T. H., 1987, Life histories in comparative perspective, in: *Primate Societies*, B. B., Smuts, D. L., Cheney, R. M., Seyfarth, R. M., Wrangham, and T. T. Struhsaker, eds., University of Chicago Press, Chicago, pp. 181–196.

Hendrickx, A. G., 1967, Studies in the development of the baboon, in: *The Baboon in Medical Research, Volume II*, H. Vagtborg, ed., University of Texas Press, Austin, TX, pp. 283–307.

Hendrickx, A. G. and Houston, M. L., 1971, Appendix: Fetal growth, in: *Embryology of the Baboon*, A. G. Hendrickx, ed., University of Chicago Press, Chicago, pp. 173–196.

Hughes, K. A. and Burleson, M. H., 2000, Evolutionary causes of variation in fertility and other fitness traits, in: *Genetic Influences on Human Sexuality and Fertility*, J. L. Rodgers, D. C. Rowe, and W. Miller, eds., Kluwer Academic Press, Dordrecht, The Netherlands, pp. 7–34.

Janson, C. and van Schaik, C., 1993, Ecological risk aversion in juvenile primates: Slow and steady wins the race, in: *Juvenile Primates: Life History, Development and Behavior*, M. E. Pereira and L. A. Fairbanks, eds., Oxford University Press, New York, pp. 57–76.

Johnson, S. E., 2003, Life histories and the competitive environment: Trajectories of growth maturation and reproductive output among chacma baboons, *Am. J. Phys. Anthropol.* **120**:83–98.

Leigh, S. R., 1992, Patterns of variation in the ontogeny of primate body size dimorphism, *J. Hum. Evol.* **23**:27–50.

Leigh, S. R., 1994, Ontogenetic correlates of diet in anthropoid primates, *Am. J. Phys. Anthropol.* **99**:499–522.

Leigh, S. R., 1996, Evolution of human growth spurts, *Am. J. Phys. Anthropol.* **101**:455–474.

Leigh, S. R., 2001, Evolution of human growth, *Evol. Anthropol.* **10**:223–236.

Leigh, S. R., 2004, Brain growth, life histories, and cognition in primate and human evolution, *Am. J. Primatol.* **62**:139–164.

Leigh, S. R., 2006, Growth and development, in: *The Baboon in Biomedical Research*, J. L. VandeBerg, S. Williams-Blangero, and S. Tardiff, eds., Kluwer Academic Press, New York (in press).

Leigh, S. R. and Blomquist, G. E., 2006, Life history, in: *Primates in Perspective*, C. Campbell, A. Fuentes, K. C. MacKinnon, M. Panger, and S. Bearder, eds., Kluwer Academic/Plenum Press, New York (in press).

Leigh, S. R., and Park, P. B., 1998, Evolution of human growth prolongation, *Am. J. Phys. Anthropol.* **107**:331–350.

Leigh, S. R. and Terranova, C. J., 1998, Comparative perspectives on bimaturism, ontogeny, and dimorphism in lemurid primates, *Int. J. Primatol.* **19**:723–749.

Leigh, S. R., Shah, N., and Buchanan, L. S., 2003, Ontogeny and phylogeny in papionin primates, *J. Hum. Evol.* **45**:285–316.

Liu, J. and LeRoith, D., 1999, Insulin-like growth factor I is essential for postnatal growth in response to growth hormone, *Endocrinol.* **140**:5178–5184.

Mahaney, M. C., Leland, M. M., Williams-Blangero, S., and Marinez, Y. N., 1993a, Cross-sectional growth standards for captive baboons: I. Organ weight by chronological age, J. *Med. Primatol.* **22**:400–414.

Mahaney, M. C., Leland, M. M., Williams-Blangero, S., Marinez, Y. N., 1993b, Cross-sectional growth standards for captive baboons: II. Organ weight by body weight, *J. Med. Primatol.* **22**:415–427.

Martin, R. D., 2002, Foreward, in: *Primate Life Histories and Socioecology*, P. M. Kappeler and M. E. Pereira, eds., Unversity of Chicago Press, Chicago, pp. xi–xx.

Martin, R. D. and MacLarnon, A. M., 1990, Reproductive patterns in primates and other mammals: The dichotomy between altricial and precocial offspring, in: *Primate Life History and Evolution*, C. J. DeRousseau, ed., Alan R. Liss, New York, pp. 47–80.

Moses, L. E., Gale, L. C., and Altmann, J., 1992, Methods for analysis of unbalanced, longitudinal growth data, *Am. J. Primatol.* **28**:39–59.

Needham, J., 1933, On the dissociability of the fundamental processes of ontogenesis, *Biol. Rev.* **8**:180–223.

Pazos, F., Sanchez-Franco, F., Balsa, J., Escalada, J. Palacios, N., and Cacicedo, L., 2000, Mechanisms of reduced body growth in the pubertal feminized male rat: unbalanced estrogen and androgen action on the somatotropic axis. *Ped. Res.* 48: 96–103.

Pereira, M. E. and Altmann, J., 1985, Development of social behavior in free-living nonhuman primates, in: *Nonhuman Models for Human Growth and Development*, E. S. Watts, ed., Alan R. Liss, New York, pp. 217–309.

Pereira, M. E. and Leigh, S. R., 2002, Modes of primate development, in: *Primate Life Histories and Socioecology*, P. M. Kappeler and M. E. Pereira, eds., Unversity of Chicago Press, Chicago, pp. 149–176.

Raff, R., 1996, *The Shape of Life*, University of Chicago Press, Chicago.

Ross, C., 1988, The intrinsic rate of natural increase and reproductive effort in primates, *J. Zool. (Lond.)* **214**:199–219.

Ross C., 1998, Primate life histories, *Evol Anthropol.* **6**:54–63.

Ross, C. and Jones, K. E., 1999, Socioecology and the evolution of primate reproductive rates, in: *Comparative Primate Socioecology*, P. C. Lee, ed., Cambridge University Press, New York, pp. 73–110.

Sacher, G.A., 1959, The relation of life span to brain weight and body weight in mammals, in: *CIBA Foundation Colloquia on Aging. Vol. 5. The Lifespan of Animals*, G. E. W. Wolstenholme, and M. O'Connor, eds., Churchill, London, pp. 115–133.

Sacher, G. and Staffeldt, E., 1974, Relation of gestation time to brain weight for placental mammals: Implications for the theory of vertebrate growth, *Am. Nat.* **108**:593–615.

Sade, D. S., 1990, Intrapopulation variation in life-history parameters, in: *Primate Life History and Evolution*, C. J. DeRousseau, ed., Alan R. Liss, New York, pp. 181–194.

Setchell, J. M., Lee, P. C., Wickings, E. J., and Dixson, A. F., 2001, Growth and ontogeny of sexual size dimorphism in the mandrill (*Mandrillus sphinx*), *Am. J. Phys. Anthropol.* **115**:337–348.

Setchell, J. M., Lee, P. C., Wickings, E. J., and Dixson, A. F., 2001, Growth and ontogeny of sexual size dimorphism in the mandrill (*Mandrillus sphinx*), *Am. J. Phys. Anthropol.* **115**:349–360.

Setchell, J. M., Lee, P. C., Wickings, E. J., and Dixson, A. F., 2002, Reproductive parameters and maternal investment in mandrills (*Mandrillus sphinx*), *Int. J. Primatol.* **23**:51–68.

Shea, B. T., 1990, Dynamic morphology: Growth, life history, and ecology in primate evolution, in: *Primate Life History and Evolution*, C. J. DeRousseau, ed., Alan R. Liss, New York, pp. 325–352.

Sigg, H., Stolba, A., Abegglen, J.-J., Dasser, V., 1982, Life history of hamadryas baboons: Physical development, infant mortality, reproductive parameters and family relationships, *Primates* **23**:473–487.

Smith, K. K., 2002, Sequence heterochrony and the evolution of development, *J. Morphol.* **252**:82–97.

Strum, S. C., 1991, Weight and age in wild olive baboons, *Am. J. Primatol.* **25**:219–237.

Stucki, B. R., Dow, M. M., and Sade, D. S., 1991, Variance in intrinsic rates of growth among free-ranging rhesus monkey groups, *Am. J. Phys. Anthropol.* **84**:181–192.

Tame, J. D., Winter, J. A., Li, C., Jenkins, S., Giussani, D. A., Nathanielsz, P. W., 1998, Fetal growth in the baboon during the second half of pregnancy, *J. Med. Primatol.* **27**:234–239.

Turnquist, J. E., and Kessler, M. J., 1989, Free-ranging Cayo Santiago rhesus monkeys (*Macaca mulatta*): I. Body size, proportion, and allometry, *Am. J. Primatol.* **19**:1–13.

Watts, E. S., 1985, Adolescent growth and development of monkeys, apes and humans, in: *Nonhuman Primate Models of Human Growth and Development*, E. S. Watts, ed., Alan R. Liss, New York, pp. 41–66.

Watts, E. S., 1990, Evolutionary trends in primate growth and development, in: *Primate Life History and Evolution*, C. J. DeRousseau, ed., Alan R. Liss, New York, pp. 89–104.

Williams-Blangero, S. and Blangero, J., 1995, Heritability of age at first birth in captive olive baboons, *Am. J. Primatol.* **37**:233–240.

Wilkinson L. 1999. *Systat for Windows, Version 9.* Systat, Inc, Evanston, IL.

Yu, H., Mistry, J., Nicar, M., Khosrab, M., Diamandis, A., van Doom, J., and Juul, A., 1999, Insulin-like growth factors (IGF-I, free IGF-I and IGF-II) and insulin-like growth factor binding proteins (IGFBP-2, IGFBP-3, IGFBP-6, and ALS) in blood circulation, *J. Clin. Lab. Anal.* **13**:166–172.

CHAPTER ELEVEN

# Testicular Size, Developmental Trajectories, and Male Life History Strategies in Four Baboon Taxa

*Clifford J. Jolly and Jane E. Phillips-Conroy*

**CHAPTER SUMMARY**

Sociobiological theory predicts that natural selection via sperm competition will favor greater relative testicular size in adults of polyandrous species than in their monandrous relatives. We have previously shown that, among baboons of the Awash National Park, Ethiopia, "multimale" olive baboons have testes larger relative to total body mass than "one-male unit (OMU)" hamadryas, with most of the difference attributable to a late growth spurt in olives. In this chapter, we use a sample of yellow baboons captured in the Mikumi National Park, Tanzania, and Guinea baboons living in a captive colony to test the prediction that they will resemble olive and hamadryas baboons respectively, in relative, adult testicular size. Like olives, yellow

**Clifford J. Jolly** • Department of Anthropology, New York University, New York, NY, USA **Jane E. Phillips-Conroy** • Department of Anatomy & Neurobiology, Washington University School of Medicine and Department of Anthropology, Washington University, St. Louis, MO, USA

baboons develop large testes in a late growth spurt, while Guineas, like hamadryas, do not. Yellow baboons apparently have relatively smaller testes than olives at all ages, but this effect is probably an artifact of their long-limbed body build and disappears if a measure of trunk volume (rather than total body mass) is used as a proxy of functional body size. Previous work also showed a difference between olive and hamadryas baboons in juvenile testicular ontogeny, explicable in terms of dispersal rather than adult testicular size. Male hamadryas, which breed in their natal group, undergo testicular enlargement earlier than olives, perhaps reflecting general sexual precocity and/or opportunities to sneak copulations while "following" an OMU. Olive and yellow baboons, which have few mating opportunities before dispersal, have less developed testes as juveniles. Yellow baboons seem to be more extreme than olives in this respect, perhaps reflecting a lower propensity to disperse as juveniles, and thus fewer preadult mating opportunities. The few available data suggest that Guineas tend to resemble hamadryas in testicular ontogeny.

## 1. INTRODUCTION

The starting point of this study is the observation that testicular size in mammals can be related to social behavior, especially to polyandry and specifically to the intensity of sperm competition among males (Short, 1979; Harcourt et al., 1981; Harvey and Harcourt, 1984; Møller, 1988; Stockley and Purvis, 1993; Harcourt, 1997). Among baboons, the most obvious contrast with respect to these variables is between the putatively monandrous hamadryas baboon (*Papio hamadryas hamadryas*) and other populations of the species, especially East African olive and yellow baboons, in which polyandry is common. We have shown elsewhere (Jolly and Phillips-Conroy, 2003; Phillips-Conroy and Jolly, 2004) that in the Awash National Park, Ethiopia, where both forms occur parapatrically, adult hamadryas baboons have significantly smaller testes than adult olive baboons. In Awash, olive baboons are polygynous and polyandrous, and (in multimale troops, which are the majority) each female typically mates with several males during a single ovulatory cycle. Hamadryas baboons are polygynous but typically not polyandrous, with each female of reproductive age tightly bonded to a single adult male, usually her only mate during a given ovulatory cycle (see Swedell and Saunders, this volume). Scattered records of testicular size in olive and hamadryas baboons of unknown (but non-Awash) provenience (Kinsky, 1960; Kummer, 1968, 1995; Short, 1979; Harcourt et al., 1981; Abegglen, 1984; Harvey and Harcourt, 1984; Smuts, 1985; Møller, 1988; Stockley and Purvis, 1993; Harcourt, 1997) suggest that the contrast between the two taxa is widespread and probably universal.

Because our data included many immature animals, we were also able to document the developmental trajectories by which the adult values are attained (Jolly and Phillips-Conroy, 2003). These showed that the intertaxon difference in adult testicular size and body mass is due entirely to differential growth in the young adult to adult phase, between the emergence of the earliest third molar at about 80 months and the completion of M3 and canine emergence at about 100 months (Phillips-Conroy and Jolly, 1988). Until full adulthood there is little difference, age for age, in absolute or relative testicular size. Olive baboons then undergo a final, rapid increase in body mass and, especially, testicular size, while hamadryas acquire their spectacular mane and increase somewhat in body mass, but absolute testicular size shows no significant change, while relative testicular size actually decreases slightly.

The contrast in developmental trajectories between olive and hamadryas baboons can be explained in terms of their life histories: Male hamadryas remain in their natal troop and attain breeding status by attracting females to their one-male unit (OMU). Male olive baboons typically emigrate from their natal troop immediately after their early-adult growth spurt, then use both brawn and social skills to establish a position in the breeding hierarchy of an adoptive troop, where they typically experience sperm competition.

A second, less pronounced but still real difference between the taxa was seen in juvenile animals. Enlargement and descent of the testes apparently begins earlier, and proceeds faster, in juvenile (approximately 3.5–5 year old) hamadryas than in olive baboons. Our post hoc explanation for the difference is that the opportunity for surreptitious copulations by juveniles, and hence sperm competition with an adult male, may be greater in a hamadryas group, where even 2–3-year-old males typically attach themselves as followers to a one-male unit (see Swedell and Saunders, this volume). In olive baboon society, young males have few opportunities for fertile mating until after they emigrate and successfully join a new group.

In the present chapter we extend the study to cover two more samples, one of wild yellow baboons from northern Tanzania, and one of a captive colony of Guinea baboons. Field studies of this yellow baboon population

(Phillips-Conroy et al., 1988; Rogers and Kidd, 1993; Rhine et al., 2000) and others (Alberts and Altmann, 1995a,b, 2001) have described a social system much like that of Ethiopian olive baboons, with multimale groups, polyandrous mating, and almost universal male emigration before breeding. Accounts of wild Guinea baboons are much less complete (Dunbar and Nathan, 1972; Sharman, pers. commun.; Galat-Luong et al., this volume), but suggest a structure approaching that of hamadryas in some respects, with large, bisexual aggregations within which are clusters of animals apparently centered upon single adult males. Hamadryas-type herding behavior was not observed, however, and it is not known whether males are philopatric nor whether, like hamadryas, they exhibit precocious sexual interest. A similar substructuring, with mainly endogamous, OMU-like clusters, has been observed in the captive Guinea baboon colony at the Brookfield Zoo in Chicago (Boese, 1975; Maestripieri et al., 2005).

The developmental patterns represented by olive and hamadryas baboons differ in three distinct features: (a) absolute and relative testicular size in full adults (greater in olive baboons), (b) difference in testicular size between subadults and full adults (significant in olive baboons, not in hamadryas) and (c) timing of juvenile testicular growth (earlier in hamadryas). If social behavior accurately predicts the adult size and developmental trajectory of testes, we expect yellow baboons to resemble olive baboons in those features where the latter differ from hamadryas. Guinea baboons might or might not show some hamadryas-like features of testicular size and development, depending upon whether males are philopatric and the extent to which the apparent OMU-like clusters represent exclusive, monandrous mating groups.

## 2. MATERIALS AND METHODS

The hamadryas and olive baboon samples, from the Awash National Park hybrid zone, have been described previously (Phillips-Conroy and Jolly, 1981, 1986, 1988; Phillips-Conroy et al., 1991, 1992). The 78 male yellow baboons were among those live-trapped and released in the Mikumi National Park in 1984–1986 (Phillips-Conroy et al., 1987, 1988; Rogers and Kidd, 1993). The 16 Guinea baboons, measured in 1985, came from a colony housed, and mostly born and bred, at the Brookfield Zoo, Chicago. In all cases, body mass was measured on a spring scale accurate to approximately 250 g. Our experience indicates that testicular measurements are much more consistent if taken by a single observer. Length and breadth measurements were taken on the left testis by one of us (J. P.-C.), using a dial caliper, as previously described (Jolly and Phillips-Conroy, 2003). A standard variable proportional to testicular volume was calculated as $(\pi/6) \times$ testicular breadth$^2 \times$ testicular length (Glander et al., 1992; Jolly and Phillips-Conroy, 2003). Relative testicular size was calculated as testis volume/body mass. To assess and compare patterns of testicular growth, we assigned each animal to one of five age classes, based on dental eruption (Phillips-Conroy and Jolly, 1988). Age class 1, consisting of animals with deciduous teeth only, included too few individuals with measurable testes, and so was excluded from the analysis. Age classes 2–4 included animals with first, second, and third molars erupting, representing younger juveniles, older juveniles, and subadults, respectively. Age class 5 included fully adult animals with all teeth, including canines and third molars, fully erupted and in occlusion (Jolly and Phillips-Conroy, 2003). We calculated mean body mass, testicular volume, and relative testicular size for each age class and taxon, and compared adjacent age classes by nonparametric statistics. To retrieve some of the information lost by lumping growing animals into broad classes, we also calculated linear regressions of $\log_e$ testicular volume on $\log_e$ body mass, with all taxa combined, both overall and by age class, and retained the resulting residuals as variables for intertaxon comparison, especially in age classes 2, 3, and 4. As several of the variables are non-normally distributed, we used nonparametric statistical tests to compare samples, although means and standard deviations (SD) are shown as descriptors.

## 3. RESULTS

Table 1 includes mean and SD, for each of the samples, of the expression of testicular volume (DITEV), body mass (WTK), and testicular volume relative to body mass (TVREL). Table 2 shows the results of nonparametric tests comparing taxa in each age class, and Table 3 shows the results of similar tests of difference between adjacent age classes within taxa. Figure 1 depicts the profile of testicular growth in all four samples, obtained by plotting $\log_e$ testis volume against $\log_e$ body mass. The lines represent taxon-specific trajectories, produced by locally weighted regression smoothing using an iterative weighted least-squares' method (Lowess), with three iterations, set to fit 50 percent of points. This line-fitting option was selected as the most purely descriptive, making the fewest assumptions about the distribution of the data.

**Table 1.** Descriptive statistics for each taxon by age class

| Taxon | Age class | Testicular volume | | | Body mass | | | Relative testicular volume | | |
|---|---|---|---|---|---|---|---|---|---|---|
| | | *N* | Mean | SD | *N* | Mean | SD | *N* | Mean | SD |
| Olive | 2 | 16 | 1.72 | 0.88 | 16 | 7.14 | 1.48 | 16 | 0.25 | 0.12 |
| | 3 | 85 | 14.10 | 11.89 | 85 | 12.74 | 3.04 | 85 | 0.98 | 0.73 |
| | 4 | 31 | 33.35 | 12.20 | 31 | 17.64 | 2.44 | 31 | 1.88 | 0.58 |
| | 5 | 148 | 49.78 | 14.75 | 148 | 22.72 | 2.34 | 148 | 2.19 | 0.62 |
| Hamadryas | 2 | 6 | 1.25 | 0.64 | 6 | 6.10 | 1.76 | 6 | 0.20 | 0.09 |
| | 3 | 16 | 14.34 | 7.50 | 16 | 12.42 | 2.71 | 16 | 1.11 | 0.51 |
| | 4 | 11 | 26.00 | 7.06 | 11 | 17.72 | 3.15 | 11 | 1.47 | 0.30 |
| | 5 | 34 | 29.26 | 9.28 | 34 | 20.83 | 1.96 | 34 | 1.40 | 0.41 |
| Yellow | 2 | 5 | 0.51 | 0.31 | 5 | 5.62 | 1.07 | 5 | 0.10 | 0.08 |
| | 3 | 16 | 3.70 | 3.50 | 16 | 10.66 | 2.31 | 16 | 0.32 | 0.26 |
| | 4 | 19 | 20.68 | 6.87 | 19 | 18.46 | 2.75 | 19 | 1.11 | 0.32 |
| | 5 | 35 | 37.79 | 14.79 | 34 | 22.76 | 2.70 | 34 | 1.65 | 0.55 |
| Guinea | 2 | 2 | 0.57 | 0.53 | 2 | 4.66 | 0.96 | 2 | 0.11 | 0.09 |
| | 3 | 1 | 3.08 | – | 1 | 9.09 | – | 1 | 0.34 | – |
| | 4 | 6 | 30.99 | 14.42 | 6 | 19.39 | 3.53 | 6 | 1.56 | 0.56 |
| | 5 | 7 | 33.00 | 11.35 | 7 | 23.51 | 1.59 | 7 | 1.41 | 0.47 |

**Table 2.** Comparisons among taxa and within age classes (Mann–Whitney *U*-test)

| Taxon pairs | Age class | Testicular volume | Body mass | Residuals | Relative testicular volume |
|---|---|---|---|---|---|
| **Olive–Yellow** | 5 | *** | ns | *** | *** |
| | 4 | *** | ns | *** | *** |
| | 3 | *** | ** | ** | *** |
| | 2 | ** | 0.06 | ns | * |
| **Hamadryas–Yellow** | 5 | ** | ** | ns | 0.05 |
| | 4 | 0.09 | ns | * | ** |
| | 3 | *** | ns | *** | *** |
| | 2 | * | ns | ns | 0.05 |
| **Olive–Guinea** | 5 | ** | ns | ** | ** |
| | 4 | ns | ns | 0.05 | ns |
| | 3 | ns | ns | ns | ns |
| | 2 | ns | 0.05 | ns | ns |
| **Guinea–Hamadryas** | 5 | ns | ** | ns | ns |
| | 4 | ns | ns | ns | ns |
| | 3 | ns | ns | ns | ns |
| | 2 | ns | ns | ns | ns |
| **Yellow–Guinea** | 5 | ns | ns | ns | ns |
| | 4 | ns | ns | ns | ns |
| | 3 | ns | ns | ns | ns |
| | 2 | ns | ns | ns | ns |

*0.05 $< p <$ 0.01. **0.001 $< p <$ 0.01. ***$p <$ 0.001.

**Table 3.** Comparisons within taxa and between age classes (Mann–Whitney *U*-test)

| | Age class | Testicular volume | Body mass | Relative testicular volume | Residuals |
|---|---|---|---|---|---|
| **Olive** | 2–3 | *** | *** | *** | ns |
| | 3–4 | *** | *** | *** | ns |
| | 4–5 | *** | *** | ** | ns |
| **Hamadryas** | 2–3 | *** | *** | ** | ns |
| | 3–4 | *** | *** | * | ** |
| | 4–5 | ns | *** | ns | ** |
| **Guinea** | 2–3 | ns | ns | ns | ns |
| | 3–4 | ns | ns | ns | ns |
| | 4–5 | ns | * | ns | ns |
| **Yellow** | 2–3 | ** | *** | * | ns |
| | 3–4 | *** | *** | *** | ns |
| | 4–5 | *** | *** | *** | ns |

*0.05 < $p$ < 0.01. **0.001 < $p$ < 0.01. ***$p$ < 0.001.

The data from olive and hamadryas baboons, which we have previously described and compared in more detail (Jolly and Phillips-Conroy, 2003), provide contrasting, standard types for comparison with the new information from yellow and Guinea baboons. Comparing age class for age class, yellow baboons from Mikumi are very similar in body mass to olive baboons from

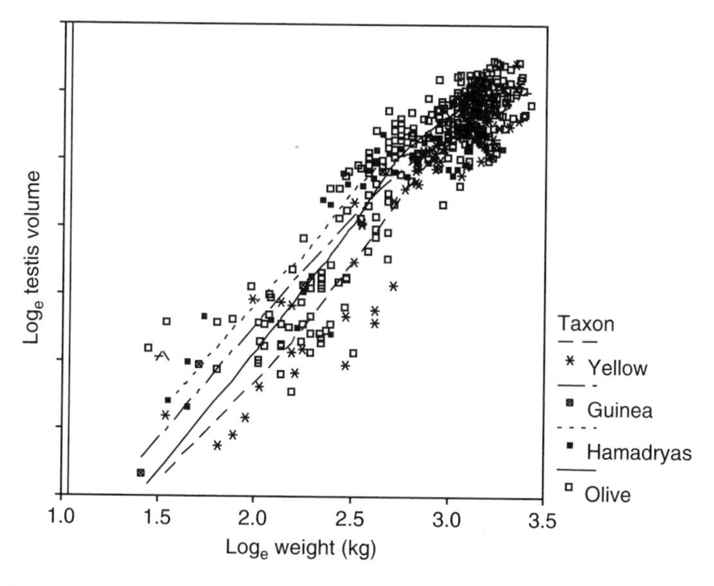


**Figure 1.** Log–log plot of testis volume and body mass.

Awash, but in all age classes they have significantly smaller testes, both absolutely and relative to body mass (Table 1). Body mass also resembles that of hamadryas in all age classes except full adults, where yellow baboons are distinctly heavier (Table 2). Absolute and relative testicular size is similar to that of hamadryas in age class 2, significantly smaller in age class 3, somewhat smaller in age class 4, and significantly greater in full adults. As in olive baboons, differences between all adjacent age classes in yellow baboons are highly significant ($p < 0.01$ or lower) in body mass and in absolute and relative testicular size (Table 3). However, the overall developmental trajectory in yellow baboons does differ from both the hamadryas and the olive baboon patterns. Figure 2 compares the three samples with respect to the difference in mean testicular volume between adjacent age classes, expressed as a percentage of the total increment from class 2 to class 5. The proportion of the total increment lying between classes 3 and 4 is remarkably constant, at 40.1, 41.6, and 45.5 percent in olive, hamadryas, and yellow baboons, respectively. In olive baboons, most of the remaining increment (34 percent) occurs late, between classes 4 and 5, while in hamadryas most of it occurs early, from class 2 to 3, as previously recorded (Jolly and Phillips-Conroy, 2003; Phillips-Conroy and Jolly, 2004). In yellow baboons, the developmental trajectory is an exaggerated version of that seen in olive baboons, with very little testicular growth occurring in the early stages and the largest single increment in absolute testicular size (45.9 percent) seen in the latest interval (4–5).

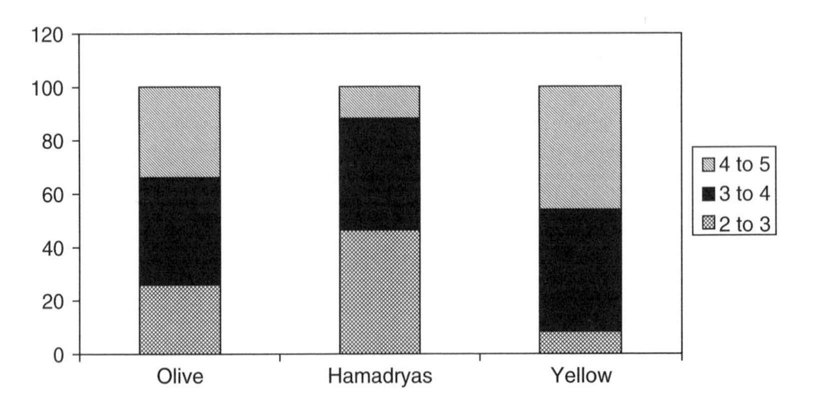


**Figure 2.** Mean testicular volume between adjacent age classes as a percentage of total growth increment between age classes 2 and 5.

A similar picture emerges from variation *within* age classes. Figure 3 shows values of the slope of the regression of $\log_e$ testis "volume" on $\log_e$ body mass, calculated for each age class, in the three taxa. A high slope value indicates that testicular volume is increasing rapidly relative to body mass within that age class. In hamadryas, the slope is already high in class 2, reflecting the early testicular growth, but yellow and olive baboons show little or no sign of growth in this class (that is, before the eruption of the earliest second molar, at about 4 years of age). In all three, the slope is greatest in class 3, corresponding to a period (from about 4 to about 6.5 years) when testicular growth outpaces somatic growth. Among subadults and adults, the slope declines in both olive and hamadryas baboons, as testicular and somatic growth keep pace with each other, although the rate of growth of both parameters is much faster in the case of olive baboons. In yellow baboons, by contrast, a testis-on-body-mass slope almost as high as that seen in age class 3 is maintained through age classes 4 and 5, so that in the largest individual adults testicular size converges on, and eventually reaches, an absolute value close to that seen in similarly sized olive baboons.

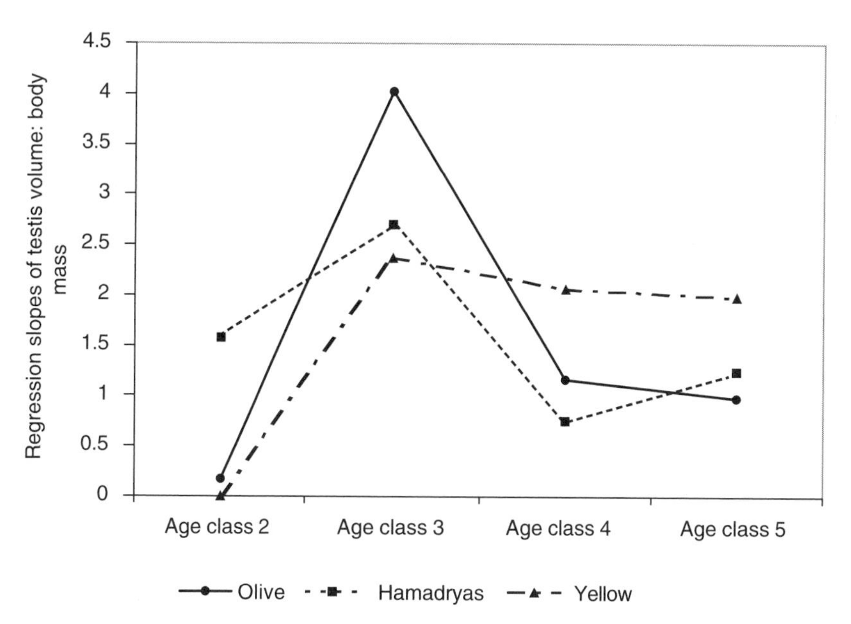


**Figure 3.** Comparison of regression slopes of log testis volume on body mass across age classes in olive, hamadryas, and yellow baboons.

The much smaller sample of Guinea baboons includes very few animals in the younger age classes, and thus is not amenable to such detailed analysis, but some significant results are apparent in comparisons of sub- and full adults with the monandrous and polyandrous standards provided by hamadryas and olive baboons. Guinea baboons of age classes 4 and 5 do not differ significantly from olive baboons in body mass, but those of age class 5 have significantly smaller testes, both absolutely and relatively (Table 2). The two oldest age classes do not differ from hamadryas in measures of absolute or relative testicular size, but full adults have greater body mass. Of successive age classes, only 4 and 5 have sample sizes adequate for statistical treatment. As in hamadryas, but unlike olive or yellow baboons, measures of testicular size (absolute and relative) do not distinguish subadults from full adults, but full adults are slightly but significantly heavier (M–W $U = 6$, $p = 0.03$, Table 3). This hamadryas-like pattern, with testes reaching maximum size in the subadult age class, and failing to match a modest increase in body size, is also apparent in the log–log plot. However, Guinea baboons of age class 4 have a higher slope of testis size regressed on body mass than do hamadryas ($1.73 \pm 0.85$ SE versus $0.76 \pm 0.34$ SE). Though these high SEs caution against over-interpretation, this result suggests that in Guinea baboons testicular growth is perhaps more active *within* age class 4, rather than being largely completed by the beginning of this interval, as in hamadryas. The same conclusion (again tentative) can be drawn from the more rightward position of the flex point, where testis size levels off at the adult value, in the Guinea baboon trajectory of testis growth (Figure 1). In the juvenile Guinea baboons, both absolute and relative testicular sizes are small, resembling yellow baboons, but the sample is very limited and this observation cannot be regarded as robust.

## 4. DISCUSSION

Yellow baboons of Mikumi, as represented in this sample, conform to the primary prediction of the sperm competition model, namely that, as a "multi-male" form with polyandrous mating, their adult testicular size, like that of olive baboons, is absolutely and relatively larger than that of the monandrous hamadryas baboon. Also, as predicted, they resemble olive baboons in exhibiting a powerful, late, spurt in growth, in both testis volume and body mass, corresponding to full eruption of the canine teeth. We interpret this as an adaptation to a reproductive strategy involving emigration from the natal group and agonistic competition for a place in the breeding hierarchy of an adoptive troop (Alberts and Altmann, 1995a,b, 2001).

However, the Mikumi yellow baboons also differ from olive baboons in two ways that were not predicted. The first is that, although there is no indication that sperm competition is less intense among yellow than among olive baboons, their full adult testicular size is significantly less, both absolutely and relative to body mass. We suggest that the explanation lies not in selection by sperm competition but in the use of body mass as a standard of comparison. Yellow baboons in general are often said to be long limbed, with a linear build. This impression is confirmed by metric data from the Mikumi sample (unpublished), which indicate that more of the body mass is concentrated in the limbs, and less in the trunk, in yellow baboons than in hamadryas or olive baboons. We obviously could not test this conjecture by weighing body parts separately. We therefore devised a test based upon volume, on the assumption that trunk volume would be similarly related to trunk mass in all baboons. We approximated trunk volume as a cylinder with circumference equal to the average of maximum and minimum trunk girths, and length equal to the distance from occiput to tail root. When values of this "volume" are used as standards of comparison for testicular growth, the log–log regression line for Mikumi lies closer to, but is still not coincident with, that for Awash olive baboons (Figure 4). However, comparison of residuals by age class shows that by this standard, yellow baboons in age class 5 are equal to olive baboons (Figure 5). It is evidently the relatively long (and thus proportionately heavy) limbs of yellow baboons that make them *appear* to have smaller testes, when total body mass is used as the standard. In other words, yellow baboons can be considered small baboons with heavy limbs. Whatever the explanation for the distinctive body build of yellow baboons, in which the trunk comprises a smaller proportion of total mass, it is unlikely to be related to selection for sperm competition. We therefore see no reason to doubt that, as predicted by their comparable rates of polyandrous mating, olive baboon and yellow baboons have functionally equivalent testicular size as full adults. Cases such as this also have more general implications for studies of allometry. They should remind us that in any study of relative size, allometry, or growth, results are likely to vary according to the dimension used as a standard of comparison, that total body mass may not always be the most appropriate choice for the independent variable in such studies, and that alternative standards should at least be investigated.

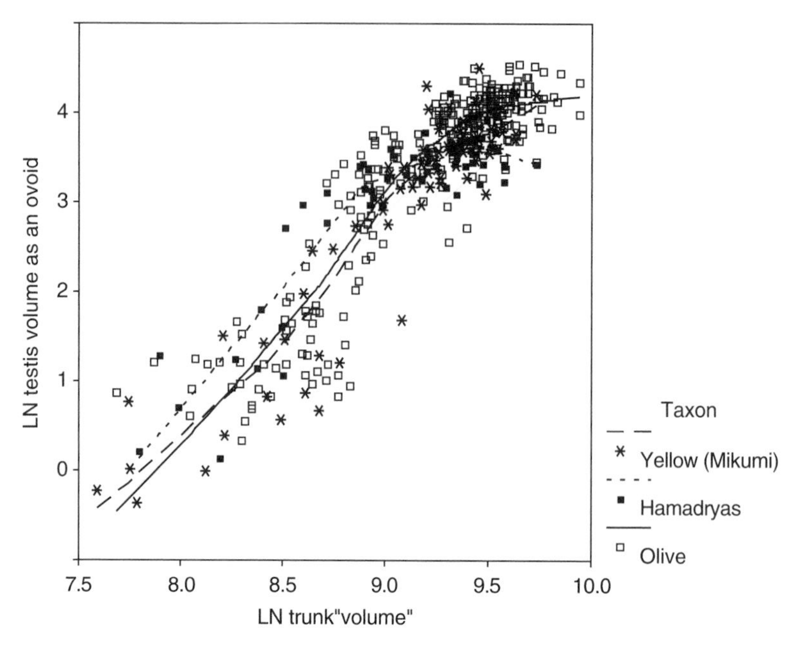


**Figure 4.** Log–log plot of testis volume and trunk volume for olive, hamadryas, and yellow baboons.

The other unexpected peculiarity of yellow baboons is that their late growth spurt is much more delayed, and more prolonged, than that experienced by olive baboons. This effect is seen when either body mass or trunk volume is used as the standard of comparison, and so is unlikely to be an artifact of the "heavy limbs" effect. In our previous comparison of olive and hamadryas baboons, we suggested that the earlier testicular enlargement seen in hamadryas might be a function of their precocious sexual development, and the occasional opportunities for surreptitious copulations afforded to larger juveniles by their role as OMU followers in hamadryas society (see Swedell and Saunders, this volume for further discussion of surreptitious copulations in hamadryas baboons). We contrasted this with the olive baboon situation, where opportunities for fertile copulation are minimal until after the male emigrates, normally in late preadulthood, and a greater return in fitness is therefore predicted if resources are concentrated in the immediate preadult phase of growth. By this logic, the growth trajectory of yellow baboons suggests that they should be even more extreme than olive baboons in terms of delaying reproductive effort until after full maturity. As males of both forms

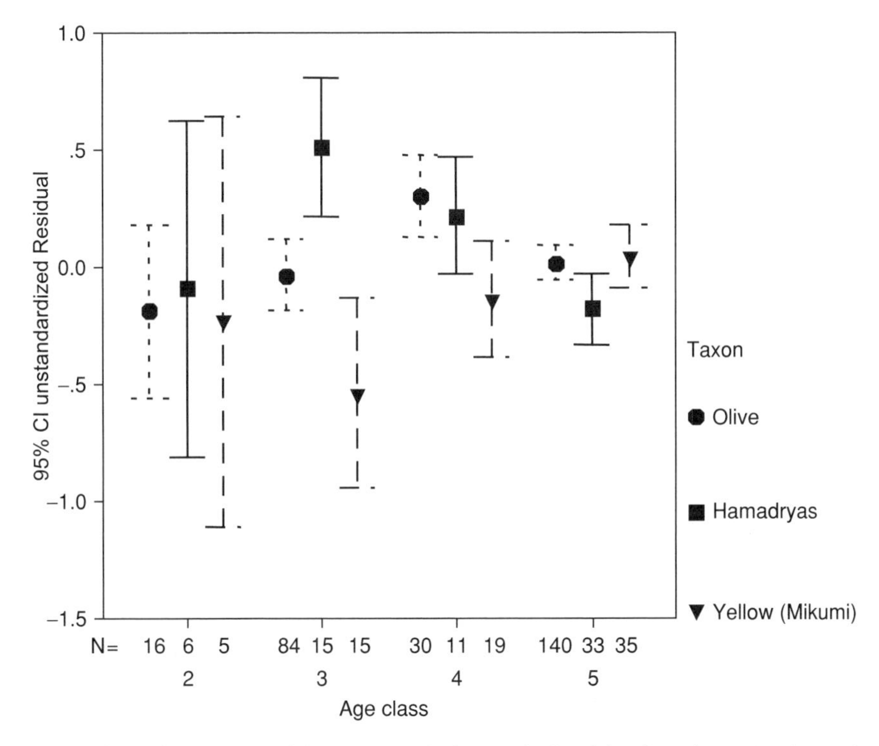


**Figure 5.** 95 percent confidence intervals for residuals of the data shown in Figure 4.

normally breed only after emigrating, a difference might arise if olive baboons more often violate the "savanna baboon" norm of remaining in their natal group until full maturity. While we are unaware of systematic comparative data bearing on this point, there is evidence to suggest that juvenile emigration, presumably affording a low but nonzero chance of fertile copulation, does indeed occur quite frequently among olive baboons, but is much less frequent in yellow baboons (Alberts and Altmann, 1995a,b, 2001).

Results obtained from the Guinea baboons must be interpreted with caution, as the sample of animals is relatively small and derived from a captive colony. One likely effect of captivity might be a somewhat higher mean body mass than seen in the wild. Indeed, the mean body mass of adult males in our sample (24 kg) is somewhat higher than expected from the fact that the Guinea baboon is cranially one of the smaller forms of the genus, comparable to hamadryas (Jolly, 1964). A rather lower mean adult male body mass, about 21 kg, would be expected from cranial size. In our data, mean relative testis size in adult Guinea baboons is very close to the mean of hamadryas (1.4 in both cases). Substituting an "estimated wild" mean body mass of 21 kg raises mean relative testis size, but only to a value of 1.5, still well below the olive baboon mean value of 2.2. In fact, for Guinea baboons to attain an olive-like mean relative testis size, they would need a mean body mass of only 15 kg, which is certainly unrealistically small. We can be confident, then, that, even allowing for added mass produced by rearing in captivity, mean adult testis size in Guinea baboons is much closer to hamadryas than to olive or yellow baboons, and is thus adapted to a mating system with little sperm competition among full adult males. As Guinea baboon troops are known to be large and multimale, this presumably means that each troop includes a substructure of endogamous, OMU-like or harem groups. Such an interpretation is supported by behavioral observations of both wild and captive Guinea baboons (Boese, 1975; Maestripieri et al., 2005; Galat-Luong et al., this volume).

Guinea baboons also clearly differ from olive and yellow baboons in their testicular developmental trajectory. This difference is equally unlikely to be produced by any potential systemic effect of zoo life on body mass. Like hamadryas, subadult Guinea baboons are equal to full adults in testicular size, although somewhat smaller in body mass. Our data suggest, however, that in Guinea baboons testicular growth may be more vigorous *within* age class 4 than it is in hamadryas. If real, and not an artifact of small sample size, this effect would represent a less extreme deviation from the olive–yellow pattern than is seen in hamadryas, but many more data points are needed to confirm it. The same is even more true for testicular development in juvenile Guinea baboons, represented in our sample by only three individuals. Their residuals to the log–log regression are not significantly different from either olive or hamadryas baboons, so the fact that the lowess regression line derived from them falls exactly between the olive baboon and hamadryas lines, though intriguing, cannot be regarded as highly meaningful.

Our findings point to a number of general conclusions about baboon evolution and testicular development. First, and most obviously, *adult* testicular size in olive, hamadryas, and yellow baboons clearly fits the general model, with much larger testes seen in the forms where rapid consort turnover and successional polyandry regularly occurs within single female reproductive cycles, and sperm competition is therefore very frequent. The adult testicular size of Guinea baboons seems to suggest that polyandry is as rare among them as in hamadryas, but this remains unconfirmed. Second, there is apparently some subtle diversity among the taxa in the timing of testicular enlargement, independent of adult testicular size. Hamadryas are the most precocious, and yellow baboons the least. Precocity seems to correlate with the incidence of opportunities for juvenile males to enter into sperm competition with adults via surreptitious copulations, but much more information is needed to test this proposition.

Our findings also have implications for the phylogeny of baboon social behavior. Guinea baboons share exclusively with hamadryas a suite of pelage traits that were presumably inherited from their common ancestor, and which are seen, in more derived form, as autapomorphies in hamadryas (see, e.g., Jolly, 1964, 1993). Our results suggest that the same may be true of characters pertaining to testicular size and development, as Guinea baboons are uniquely similar to, but less derived than, hamadryas in features that we interpret as adaptations to lower levels of polyandry. At the same time, evidence from the mitochondrial genome (Newman et al., 2004; Wildman et al., 2004; Burrell et al., unpublished) strongly suggests that the separation of Guinea baboons from hamadryas (and other northern hemisphere baboon lineages) is ancient, occurring shortly after the dispersal of a stem population from southern Africa. The most parsimonious inference is that the common *mitochondrial* ancestor of all the extant *Papio* baboons in the northern hemisphere had a generally Guinea-like morphology, and perhaps a social organization including some incipient OMU-associated behavior.

To test these propositions it will clearly be important to investigate testicular size and development in chacma baboons, which apparently exhibit forms of social behavior unlike any of the forms studied here (Henzi and Barrett, 2005) and which are close to the base of the extant baboon clade (Newman et al., 2004; Wildman et al., 2004). To our knowledge, the relevant data have yet to be gathered from large samples of wild animals.

Finally, it is also worth emphasizing that our findings offer no support for the "savanna versus hamadryas" dichotomy widely espoused by socioecologists (e.g., Pusey and Packer, 1987; Smuts et al., 1987; Barton, 2000). The four major taxa usually recognized among the "savanna baboons" are clearly diverse in this respect, as in many others (see Swedell and Leigh, this volume), as might be expected from their polyphyly (Newman et al., 2004). In fact, we should be cautious even about generalizing from single populations to the major taxa themselves, especially widespread and ecologically diverse ones such as olive or chacma baboons. The extant members of the genus *Papio* comprise many distinct allopatric populations (Jolly, 1993). It is uncertain how many of these are truly diagnosable, in the sense of the Phylogenetic Species Concept (Cracraft, 1989), but it is certainly more than the five traditionally recognized as full species (e.g., by Groves (2001) and Kingdon (1997)) or major subspecies. Thus, although our data represent four of the five divisions, they cannot be assumed to comprise an adequate representation of testicular variation in four-fifths of the populations of the genus, and more work will be needed to determine how widely our findings apply.

## ACKNOWLEDGMENTS

We thank Larissa Swedell and Steve Leigh for the invitation to participate in the symposium. Data used in this study were accumulated over many years with too many helpers to mention individually, but special thanks to Jeff Rogers, Guy Norton, Sam Wasser, Tanzanian National Parks' leadership, and members of the Animal Behavior Research Unit at the Mikumi National Park (Tanzania); past and present managers of the Ethiopian Wildlife Conservation Organization and park wardens for permission to work at the Awash National Park; Jan Rafert and Sandra Vojik of the Brookfield Zoo (1985), and game scouts, assistants, volunteers, and graduate students who helped us in the gathering of these data. Our research has been supported by Earthwatch/Center for Field Research (Ethiopia), the Harry Frank Guggenheim Foundation (Ethiopia), the National Science Foundation (Ethiopia and Tanzania), New York University, and Washington University. The comments of anonymous reviewers were helpful.

## REFERENCES

Abegglen, J.-J., 1984, *On Socialization in Hamadryas Baboons*, Associated University Presses, Cranbury, NJ.

Alberts, S. C. and Altmann, J., 1995a, Preparation and Activation: Determinants of age at reproductive maturity in male baboons, *Behav. Ecol. Sociobiol.* **36**:397–406.

Alberts, S. C. and Altmann, J., 1995b, Balancing costs and opportunities: Dispersal in male baboons, *Am. Nat.* **145**:279–306.

Alberts, S. C. and Altmann, J., 2001, Immigration and hybridization patterns of yellow and olive baboons in and around Amboseli, Kenya, *Am. J. Primatol.* **53**:139–154.

Boese, G. K., 1975, Social behavior and ecological considerations of West African baboons (*Papio papio*), in: *Socioecology and Psychology of Primates*, R. H. Tuttle, ed., Mouton, The Hague, pp. 205–230.

Barton, R. A., 2000, Socioecology of baboons: The interaction of male and female strategies, in: *Primate Males: Causes and Consequences of Variation in Group Composition*, Kappeler, P. M., ed., Cambridge University Press, New York, PP.97–107.

Cracraft, J., 1989, Speciation and its ontology: The empirical consequences of alternative species concepts for understanding patterns and processes of differentiation, in: *Speciation and Its Consequences*, D. Otte, and J. Endler, eds., Sinauer Associates, Sunderland, MA, pp. 28–59.

Dunbar, R. I. and Nathan, M. F., 1972, Social organization of the Guinea baboon, *Papio papio, Folia Primatol.* **17**:321–334.

Glander, K. E., Wright P. C., Daniels, P. S., and Merenlender A. M., 1992, Morphometrics and testicle size of rain forest lemur species from southeastern Madagascar, *J. Hum. Evol.* **22**:1–18.

Groves, C., 2001, *Primate Taxonomy*, Smithsonian Institution Press, Washington.

Harcourt, A. H., 1997, Sperm competition in primates, *Am. Nat.* **149**:189–194.

Harcourt, A. H., Harvey P. H., Larson S. G., and Short R. V., 1981, Testis weight, body weight, and breeding system in primates, *Nature* **293**:55–57.

Harvey, P. H. and Harcourt, A. H., 1984, Sperm competition, testes size and breeding systems in primates, in: *Sperm Competition and the Evolution of Animal Mating Systems*, Smith, R. L., ed., Academic Press, Orlando, FL, pp. 589–600.

Henzi, S. P. and Barret, L., 2005, The historical socioecology of savanna baboons (*Papio hamadryas*) *J. Zool., Lond.* **265,** 215–226.

Jolly, C. J., 1964, Origins and specialisations of the long-faced cercopithecoidea. Ph.D. Dissertation, University of London.

Jolly, C. J., 1993, Species, subspecies, and baboon systematics, in: *Species, Species Concepts, and Primate Evolution*, W. H. Kimbel, and L. B. Martin, eds., Plenum Press, New York, pp. 67–107.

Jolly, C. J. and Phillips-Conroy, J.E., 2003, Testicular size, mating system and maturation schedules in wild olive baboon and hamadryas baboons, *Int. J. Primatol.* **24**:124–142.

Kingdon, J., 1997, *The Kingdon Field Guide to African Mammals*, Academic Press, San Francisco.

Kinsky, M., 1960, Quantitative Unterschungen an athiopischen Saugetieren: Absolute und relative Gewichte der Hoden athiopischer Affen, *Anat. Anzeig.* **108**:65–82.

Kummer, H., 1968, Two variations in the social organization of baboons, in: *Primates*, P. Jay, ed., Holt, Rinehart and Winston, New York, pp. 293–312.

Kummer, H., 1995, *In Quest of the Sacred Baboon*, Princeton University Press, Princeton.

Maestripieri, D., Leoni, M., Raza, S. S., Hirsch, E. J., and Whitham, J. C., 2005, Female copulation calls in Guinea baboons: Evidence for post-copulatory female choice? *Int. J. Primatol.* **26**:737–755.

Møller, A. P., 1988, Ejaculate quality, testes size and sperm competition in primates, *J. Hum. Evol.* **17**:479–488.

Newman, T. K., Jolly, C. J., and Rogers, J., 2004, Mitochondrial phylogeny and systematics of baboons (*Papio*), *Am. J. Phys. Anthropol.* **124**:17–27.

Phillips-Conroy, J. E. and Jolly, C. J., 1981, Sexual dimorphism in two subspecies of Ethiopian baboons (*Papio hamadryas*) and their hybrids, *Am. J. Phys. Anthropol* **56**:115–129.

Phillips-Conroy, J. E. and Jolly, C. J., 1986, Changes in the structure of the baboon hybrid zone in the Awash National Park, Ethiopia, *Am. J. Phys. Anthropol.* **71**:337–349.

Phillips-Conroy, J. E. and Jolly, C. J., 1988, Dental eruption schedules of wild and captive baboons, *Am. J. Primatol.* **15**:17–29.

Phillips-Conroy, J. E. and Jolly, C. J., 2004, Dispersal and philopatry in the Awash Baboon Hybrid Zone in Jones, C and Schwibbe, M. (eds) *Dispersal and Philopatry in Primates.* Primate Report (special Issue) **68**:27–52.

Phillips-Conroy, J. E., Jolly, C. J., and Rogers, J. A., 1987, Hematocrits of free-ranging baboons: Variation within and among populations, *J. Med. Primatol.* **16**:389–402.

Phillips-Conroy, J. E., Lambrecht, F. L., and Jolly, C. J., 1988, Incidence of *Hepatocystis* in free-ranging baboons (*Papio hamadryas* s.l.) of Ethiopia and Tanzania, *J. Med. Primatol.* **17**:145–152.

Phillips-Conroy, J. E., Jolly, C. J., and Brett, F. L., 1991, Characteristics of hamadryas-like male baboons living in olive baboon troops in the Awash Hybrid Zone, Ethiopia, *Am. J. Phys. Anthropol.* **86**:353–368.

Phillips-Conroy, J. E., Jolly, C. J., Nystrom, P. D., and Hemmalin, H. A., 1992, Migration of male hamadryas baboons into olive baboon groups in the Awash National Park, in: *Dispersal and Migration in Nonhuman Primates*, E. O. Smith, and D. Sprague, eds., *Int. J. Primatol.* **13**:455–476.

Pusey, A. E. and Packer, C., 1987, Dispersal and philopatry, in: *Primate Societies*, B. B. Smuts, D. L. Cheney, R. M. Seyfarth, R. W. Wrangham, and T. T. Struhsaker, eds., University of Chicago Press, Chicago, pp. 259–266.

Rhine, R. J., Norton, G. W., and Wasser, S. K., 2000, Lifetime reproductive success, longevity, and reproductive life history of female yellow baboons (*Papio cynocephalus*) of Mikumi National Park, Tanzania, *Am J Primatol.* **51**:229–241.

Rogers, J. and Kidd, K. K., 1993, Nuclear DNA polymorphisms in a wild population of yellow baboons (*Papio hamadryas cynocephalus*) from Mikumi National Park, Tanzania, *Am. J. Phys. Anthropol.* **90**:477–486.

Short, R. V., 1979, Sexual selection and its component parts, somatic and genital selection, as illustrated by man and the great apes, *Adv. Stud. Behav.* **9**:131–158.

Smuts, B. B., 1985, *Sex and Friendship in Baboons*, Aldine, New York.

Smuts, B. B., Cheney, D. L., Seyfarth, R. M., Wrangham, R. W., and Struhsaker, T. T., eds., 1987, *Primate Societies*, University of Chicago Press, Chicago.

Stockley P. and Purvis A., 1993, Sperm competition in mammals: A comparative study of male roles and relative investment in sperm production, *Funct. Ecol.* 7:560–570.

Wildman, D. E., Bergman, T. J., al-Aghbari, A., Sterner, K. N., Newman, T. K., Phillips-Conroy, J. E., Jolly, C. J., and Disotell, T. R., 2004, Mitochondrial evidence for the origin of hamadryas baboons, *Mol. Phylogenet. Evol.* **31**:287–296.

CHAPTER TWELVE

# The Evolutionary Past and the Research Future: Environmental Variation and Life History Flexibility in a Primate Lineage

*Susan C. Alberts and Jeanne Altmann*

## CHAPTER SUMMARY

The chapters in this volume demonstrate that baboons (genus *Papio*) are a diverse and flexible group that occupies a wide range of habitats. Further, the habitats occupied by baboons are characterized by frequent short-term environmental change (particularly seasonal change) and, for at least some populations, striking long-term change as well. Some of the environmental change is relatively predictable but some is highly unpredictable. What is the relationship between the diversity we observe in baboon life history and behavior and the variability the baboons experience in their physical and social

**Susan C. Alberts** • Department of Biology, Duke University, Durham, NC, USA **Jeanne Altmann** • Department of Ecology and Evolutionary Biology, Princeton University, Princeton, NJ, USA **Susan C. Alberts** • **Jeanne Altmann** • Institute for Primate Research, National Museums of Kenya, Nairobi, Kenya **Jeanne Altmann** • Department of Conservation Biology, Chicago Zoological Society, Brookfield, IL, USA

environments? We explore answers to this question in the context of a recent hypothesis for hominid evolution proposed by Potts (1996, 1998a,b), the variability selection hypothesis. This hypothesis proposes that the constantly changing environments that typified the East African paleoenvironment resulted in selection for flexible, generalist organisms (such as humans and baboons) that possess the ability to take advantage of the new ecological opportunities that emerged with habitat change. We propose that baboons are a particularly good lineage in which to explore the consequences of variability selection—selection in response to changing environments. We provide examples of baboons' adaptability, particularly in the context of environmental variability and change, and we suggest directions for future research that take advantage of the remarkable opportunity for comparative work in this species, given the relatively large number of baboon field studies. We emphasize that understanding the response of baboons to environmental variability and change is potentially important both because of the light it can shed on the role of changing environments in selecting for flexible species like baboons and humans, and because environmental change will increasingly characterize the global environment in which wild primates—most of them less flexible than baboons—will need to persist.

## 1. BABOON ADAPTABILITY AND LIFE IN CHANGING ENVIRONMENTS

The glaring heat of mid-day in the Amboseli basin, Kenya, is followed by a 20°C temperature drop during the cool night. A few months of lush grasses, hordes of insect larvae, and tree blossoms are quickly replaced by a long dry season of dust, bare earth, and grass stubble. Renewal occurs with the rains but the rains fail, unpredictably, approximately 1 year out of 5, and are highly variable even when they do not fail. These short-term changes, both predictable and unpredictable, occur against a backdrop of larger scale ecological changes that accumulate over decades—once thick woodland becomes open grassland, daily temperatures increase, the water table rises, and the ice caps shrink on Mt. Kilimanjaro, the mountain that dominates Amboseli's landscape (Struhsaker, 1973; Western and van Praet, 1973; Struhsaker, 1976; Hauser et al., 1986; Isbell et al., 1990, 1991; Behrensmeyer, 1993; Hastenrath and Greischar, 1997; Altmann, 1998; Altmann et al., 2002; Thompson et al., 2002).

Our work in the Amboseli basin indicates that short- and long-term habitat change has profound consequences for the behavior and life history of the baboons that occupy the basin. Foraging patterns change over seasons and across years as rainfall, temperature, and habitat productivity change. Home ranges shift in response to ecological changes, as tree die-offs in one part of the basin force the baboons to take advantage of new ecological opportunities—with attendant new risks—in other parts of the basin. Vital rates change, sometimes dramatically, as birth rates increase in rich habitats and decrease in poorer ones, and mortality rates rise and fall as well (Bronikowski and Altmann, 1996; Altmann and Alberts, 2003a,b; Alberts et al., 2005). Change in weather and habitat of the sort described for Amboseli—both short term and long term—is typical of the environments in which baboons live. The habitats of every population described in this volume show strong seasonal changes and most probably experience long-term change as well, although this is much more difficult to document and requires long-term ecological data.

Baboons (genus *Papio*) have achieved a nearly continental distribution in Africa and occupy habitats ranging from moist evergreen forests to deserts, from seashore to mountains, and from equatorial to subtropical regions (Estes, 1991; Jolly, 1993; Kingdon, 1997). They are increasingly treated as a single species, *P. hamadryas* (Jolly, 1993; Newman et al., 2004; but see Kingdon, 1997; Groves, 2001), which places them, among primates, second only to humans in geographical and environmental range. Thus, baboon populations experience very different environments and perhaps different degrees of variability and predictability. In addition, any single population of baboons experiences variability over time and space. The environmental variability typical of baboon habitats is accompanied by marked behavioral and life history flexibility. Probably the most striking and best-studied example is foraging behavior. Baboons are very flexible foragers, combining omnivory (plants, invertebrates, small vertebrates) with great selectivity on particular plant species and parts (Hamilton et al., 1978; Post, 1982; Norton et al., 1987; Whiten et al., 1991; Byrne et al., 1993; Altmann, 1998; Alberts et al., 2005). In other words, they are highly selective generalist foragers, and the dietary flexibility that results undoubtedly lies at the heart of baboons' evolutionary success. As this volume illustrates, however, baboons are strikingly diverse in a number of other areas of behavior and life history as well.

## 2. THE EVOLUTIONARY PAST: THE "VARIABILITY SELECTION" HYPOTHESIS FOR THE EVOLUTION OF ADAPTABILITY

What is the relationship between the variability we observe in baboon life history and behavior and the variability they experience in their environments? More specifically, can phenotypic flexibility be understood as a consequence of environmental variability? This is not a question about the socioecological model, which posits that different sets of environmental conditions require particular responses so that the dry habitat of hamadryas baboons selects for small foraging parties and eventually one-male social units. Instead, we are asking whether the flexibility that baboons exhibit within subspecies and populations can be understood as a common adaptive, evolved response to the fact that they occupy—and are successful in—constantly changing and often unpredictable environments. To explore this question we turn to a hypothesis for human evolution developed by Potts (1996, 1998a,b).

The emergence and spread of African savannas over the past 5 million years, such as Amboseli and others described in this volume, is a cornerstone for hypotheses of hominid evolution. According to these hypotheses, early human ancestors moved into savannas in response to increasing climatic cooling and drying. Under such conditions, the exigencies of the savanna environment were central selective pressures that shaped the hominid lineage (see reviews in Foley, 1987, 1993, 1994; Potts, 1998a,b; Klein, 1999). A common thread in a number of these hypotheses is that climatic change occurred at a single, critical point in human evolution.

However, in recent years a novel and more compelling scenario for the evolution of the human lineage has emerged: Potts (1996, 1998a,b) has proposed that it is not the savanna environment per se that represents the key selective force shaping human evolution. Rather, the key selective force shaping human evolution was the persistent pattern of environmental change over millennia, a pattern that has characterized the African environment both during and after the initial emergence of savannas.

Potts (1998a,b) first marshals evidence from marine and ice core oxygen isotope records, ocean dust records, sedimentary analysis, and a number of other sources to describe the East African paleoenvironment, particularly during the past 6 million years, as one of the dramatic environmental changes of increasing magnitude over time. This analysis is supported by a number of independent sources and seems uncontroversial. He then develops the hypothesis that, in the face of such environmental change, organisms will experience "multiple, substantial disparities in selective environment over time" resulting in selection for "complex mechanisms for dealing with unexpected, episodic change" (Potts, 1998a, p. 112). Specifically, variability selection will select for traits that result in highly flexible, generalist organisms that are able to cope not only with a range of habitat types, but with environmental change as well, particularly relatively unpredictable change. Potts describes several lineages—including papionins and hominids—which he says show evidence of variability selection (Potts, 1998b). He also describes several alternative "environmental hypotheses" for hominid evolution and describes specific tests of these hypotheses in comparison to the variability selection hypothesis (Potts, 1998a,b). Recently Bobe and Behrensmeyer (2004) conducted comparative tests of these hypotheses and found good support for the variability selection hypothesis.

Whether or not the variability selection hypothesis ultimately provides an explanatory framework for understanding human evolution, it underscores the fact that environmental variability and change, whether predictable or unpredictable, short term or long term, provide organisms with both challenges and opportunities. Understanding responses to environmental change will provide crucial insights into both current population processes and the history and future of species.

## 3. BABOON FLEXIBILITY: INSIGHTS INTO THE OUTCOME OF VARIABILITY SELECTION

For several reasons, baboons provide a good model for understanding how variability selection—selection for a flexible, generalist organism driven by environmental change—might shape life history and behavior in humans, nonhuman primates, and other taxa. First, baboons share important evolved characteristics with humans. Baboons have adapted to a very wide range of environments, and in most habitats they show little or no seasonality in reproduction (e.g., Melnick and Pearl, 1987; Bercovitch and Harding, 1993; Bentley-Condit and Smith, 1997; Alberts et al., 2005; Cheney et al., this volume). In these characteristics they are like humans and are unlike the large majority of other primate species. In other words, baboons have both adapted to diverse habitats and, in major aspects of their life histories, have largely broken free of the seasonal constraints in even highly seasonal habitats. This is an ability shared by very few other primates. In addition, baboons and humans share traits that Potts proposes evolved in response to variability selection, including a high encephalization quotient like other primates (Dunbar, 1998), a flexible locomotor system (baboons readily utilize both arboreal and terrestrial habitats; Estes, 1991; Fleagle, 1999), and a flexible mating and social system (e.g., Altmann and Altmann, 1970; Dunbar and Dunbar, 1977; Barton et al., 1996; Henzi et al., 1999; Henzi and Barrett, 2003, and chapters in this volume).

Second, *Papio* baboons survived Pleistocene environmental change while closely related taxa went extinct. Potts (1998a) cites *Papio* as an example of a taxon that survived periods of extensive Pleistocene environmental change by evolving ecological flexibility, possibly in response to variability selection. By contrast, the closely related taxon *Theropithecus oswaldi*, which presumably exhibited less ecological flexibility, went extinct 600,000–800,000 years ago.

Third, baboons in at least one well-studied habitat (Amboseli) have experienced environmental change of a type and magnitude typical of the changes that characterized East Africa paleoenvironments, and have persisted in the face of this environmental change. The Amboseli baboon population experienced a dramatic decline in the 1960s at the onset of woodland die-off (Altmann et al., 1985). However, the population recovered even as the woodland die-off continued, and population size has continued to increase somewhat over the past several decades (Altmann et al., 1985; Alberts and Altmann, 2003; Altmann and Alberts, 2003a), although not without fluctuations. The baboons' success is in striking contrast to the failure of Amboseli's vervet monkeys to adapt. Vervet monkeys, like baboons, are widespread savanna-dwelling monkeys, and the two species show considerable overlap in habitat and diet (Struhsaker, 1967; Wrangham and Waterman, 1981; Altmann et al., 1987; Altmann, 1998). However, Amboseli's vervet population has undergone dramatic decline, including local extinction in some locales, as a consequence of environmental change (Struhsaker, 1967, 1973, 1976; Hauser et al., 1986; Isbell et al., 1990).

If baboons represent a model for understanding the behavioral flexibility of early hominids and other species successful in the face of environmental change, what traits might such species have exhibited? A critically important trait, of course, would be dietary and foraging flexibility, the focus of the most diverse range of detailed studies within and between populations (Hamilton et al., 1978; Post, 1982; Norton et al., 1987; Whiten et al., 1991; Byrne et al., 1993; Altmann, 1998; Alberts et al., 2005). The Amboseli baboons have responded quite effectively to habitat change in this regard, and in contrast to chimpanzees and a number of other species (e.g., Wrangham et al., 1991; Malenky and Wrangham, 1994; Wrangham et al., 1998), the baboons have done this without relying heavily on a few "fallback foods" during difficult times (although fallback foods certainly play a role in their diet; Alberts et al., 2005). Instead, they engage in what we call "handoff foraging," in which temporal variability in food abundance is mitigated by careful tracking and exploitation of shifting food resources as they become available across different seasons and years (Alberts et al., 2005). Concomitant with this skill is an ability to find alternatives when important foods become scarce as the habitat changes.

The chapters in this book point to several other ways in which baboons effectively and adaptively respond to habitat change. One clearly important example is growth and development. Growth rates in young baboons vary substantially and are influenced not only by food availability (e.g., Eley et al., 1989; Strum, 1991) but also by maternal rank and parity (Johnson, 2003; Altmann and Alberts, 2005; Johnson, this volume). A significant result of variation in growth is variation in age at first reproduction; baboons are able to take advantage of temporary increases in food availability to accelerate age at first reproduction in a manner that has substantial implications for their fitness (Altmann et al., 1988; Altmann and Alberts, 2003a, 2005). This might be considered a shift on the "fast versus slow" life history continuum, but differs from interspecific shifts on that continuum in that baboons (and humans) generally do not face trade-offs in other life history parameters by accelerating first reproduction. In particular, contexts that allow earlier age at first reproduction typically also enhance a whole suite of life history components in concert—infant survival will typically increase, interbirth interval will often decrease, and adult body size and reproductive span may increase as well (e.g., Lyles and Dobson, 1988; Sade, 1990; Sterck, 1999).

Leigh and Bernstein's data (this volume) on brain growth also suggest that baboons are particularly well adapted to take advantage of these opportunities, because neonates have relatively large brains (compared to other primates) that may even fall within the adult size range. That is, brain growth in baboons occurs at an earlier ontogenetic stage than other similar-sized primates and is complete earlier. The result is, presumably, a youngster poised for rapid somatic growth and relatively free (compared to similar-sized primates) of brain growth requirements. This ability to accelerate age at first reproduction when conditions are good is surely a hallmark of a successful species. It depends crucially on the ability baboons have to break free of the seasonal constraints of their habitat and reproduce throughout the year, which in turn depends on flexible foraging.

The early completion of brain growth may not only leave a younger animal better poised to take advantage of subsequent challenges and opportunities, but it may also enhance the importance and lifetime consequences of fetal `imprinting' (Barker, 2001; Hales and Barker, 2001) and of the so-called maternal effects that occur while the brain growth is ongoing. In fact, maternal investment patterns, too, show illuminating variation. Interbirth intervals vary greatly within as well as between populations (e.g., Altmann et al., 1978; Strum and Western, 1982; Hill et al., 2000; Altmann and Alberts, 2003a; Barrett et al., this volume; Cheney et al., this volume), and the ability of baboons to modify the duration of maternal investment is perhaps one of the most flexible traits of their life history. It reflects both the general lack of birth seasonality in baboons (even populations with measurable birth seasonality generally show births throughout the year) and the ability of females and their maturing infants to respond rapidly to changes in environmental conditions (e.g., Dunbar et al., 2002). While it has long been recognized that interbirth intervals depend strongly on food availability (e.g., Strum and Western, 1982), other determinants are also in play—in particular how readily infants achieve independence in different environments, which depends on more than simply habitat productivity. The presence of particularly appropriate "weaning foods," the travel distance, and energetic expenditure required for foraging, and predation risk will all contribute to the ability of mothers to terminate intense investment in one offspring and initiate investment in the next without entailing an intolerable increase in mortality cost for the weanling (Altmann and Samuels, 1992; Barrett et al., this volume). In addition, variability in social and demographic environment as well as physical environment may be important, males may provide offspring care (e.g., Palombit et al., 1997; Buchan et al., 2003), and the mother's social and demographic environment may contribute to offspring survival (Palombit et al., 1997; Silk et al., 2003; Wasser et al., 2004; Cheney et al., this volume). Thus, maternal investment, facilitating social environments, and offspring development are seen to covary in significant ways.

A final way in which baboons exhibit important variability is in their reproductive behavior. A great deal of this variability occurs among subspecies.

Olive and yellow males—particularly middle-ranking males—engage in frequent coalitionary behavior that increases their mating access to reproductive females, whereas male chacmas apparently never do so (e.g., Packer, 1977; Noë, 1986, 1993; Bercovitch, 1988; Bulger ,1993; Henzi et al., 1999; Henzi and Barrett, 2003). The consequence is that high-ranking chacma males obtain nearly exclusive access to reproductive females during their tenure at high rank, while olive and yellow males are more limited in their ability to monopolize access to females (e.g., Packer, 1979; Bulger, 1993; Weingrill et al., 2000; Alberts et al., 2003). Concomitantly, olive and yellow baboons are the most polygynous and polyandrous of the subspecies on short time scales—females frequently change partners within sexual cycles, even within days, and males may move between partners from one day to the next. Female chacmas tend to be monandrous on short time scales, and hamadryas females are monandrous on both short and long time scales (see also discussion by Swedell and Saunders, this volume). And, as a consequence, olive and yellow males are relatively rarely infanticidal whereas male chacmas are commonly so (Janson and van Schaik, 2000; Palombit et al., 2000; Henzi and Barrett 2003; Cheney et al., this volume; data on infanticide in hamadryas are still difficult to characterize, see review in Swedell and Saunders, this volume). To some extent this fascinating variability among subspecies must be associated with the highly flexible nature of other baboon life history and behavioral traits. In that regard, it represents an important set of traits to consider when describing baboon flexibility. However, from another perspective it may reflect limits on baboon flexibility; within-population flexibility in patterns of polygyny, polyandry, and coalitionary behavior is relatively poorly described at the moment. Nonetheless, it is clear that the full range of reproductive behavior seen across subspecies is not exhibited within any subspecies. This suggests that different reproductive patterns have become fixed in different subspecies, and consequently that flexibility—otherwise a cornerstone of baboon biology—has been lost. We consider this and other limits to baboon flexibility in the next section.

## 4. LIMITS TO BABOON FLEXIBILITY

In spite of the adaptability that baboons exhibit in a number of dimensions, they are relatively invariant in others, and the limits to baboon flexibility may be as illuminating as the flexibility itself. Like most primate species, baboons have a litter size of one. Moreover, unlike many mammalian species, even if high food availability were to lead to higher twinning rates, twins or triplets would rarely if ever survive in natural environments. This is because of a number of invariant traits, including late development of locomotor independence, milk composition that requires very frequent suckling and almost continuous contact with the mother, long travel distances, and the very lack of seasonal breeding that leads to a low probability of communal suckling or creching. These traits constrain not only litter size but also offspring growth rates; growth rates that are too high would hinder a mother's ability to carry an infant and perhaps extract too high a cost on both mother and offspring (e.g., Altmann and Alberts, 1987, 2005; Altmann and Samuels, 1992).

The chapters in this book also reveal other constraints upon baboons' flexibility. One striking limit lies in the fact that female—and hence male—fertility is limited by the rate at which infants can develop, as illustrated nicely in the chapter by Barrett and colleagues (this volume). This limitation sets baboons at the lower end of the cooperative breeding continuum, distant from humans as well as from callitrichids and possibly siamangs, in which mothers can reproduce even while their current offspring is still highly dependent. Recently, Hrdy (1999, 2005) has developed the hypothesis that broad-sense cooperative breeding, allomaternal care or social facilitation in rearing of young, has been critical in the evolution of both human and nonhuman primate life histories. Cooperative care of offspring means that young can have long periods of maturation and yet mothers can reproduce repeatedly while their young are still relatively vulnerable. Further, infants in primate species that have higher levels of alloparental care experience more rapid growth (see review in Ross, 1998). The ways in which baboons are and are not able to share the burden of offspring care suggests an interesting comparison; evaluating baboons alongside species that allonurse (e.g., white-faced capuchins; Perry, 1998) or in which helpers other than the mother carry dependent young (e.g., callitrichids; Goldizen, 1987) might shed light on critical differences, as well as similarities, between species that do share offspring care, such as callitrichids and humans, and those that do not.

A related constraint, an ontogenetic and perhaps heritable one as well, is reflected in the surprising pervasiveness of maternal effects on offspring development. In spite of the fact that baboons are poised, ontogenetically, to take advantage of opportunities for rapid growth, their growth patterns are strikingly influenced by maternal effects. Baboon mothers influence the growth of their offspring through their own dominance rank and their own developmental status, age, and/or parity (Altmann and Alberts, 2005; Johnson, this volume), and these maternal effects persist well past weaning in both Okavango and in Amboseli, when immatures are completely independent of their mothers' direct care. This effect is less surprising for daughters than it is for sons, because the dominance rank of daughters is dependent upon the dominance rank of mothers, providing an avenue for the continued influence of dominance rank, if not parity, on daughters. However, this maternal effect in itself—the dependence of daughters' ranks upon their mothers' ranks—represents an interesting constraint upon female baboons, albeit one with clear benefits for at least some members of the social group (the high-ranking females), and one that is highly phylogenetically conserved (Melnick and Pearl, 1987).

The variability of baboon reproductive behaviors suggests another interesting constraint upon baboon flexibility, at least relative to humans. As noted in "Baboon Flexibility", female baboons exhibit a range of reproductive strategies, from a tendency to mate with a single male over several reproductive events (which results, among other things, in "paternity concentration"), to multiple mating (and consequent "paternity confusion," e.g., Swedell and Saunders, this volume). Male baboons, too, exhibit a range of reproductive strategies that fall along a continuum from permanent associations with females in hamadryas, which exclude other males and persist regardless of the female's reproductive state, to temporary mate-guarding episodes concentrated during the follicular phase of the female's cycle (e.g., Bergman, this volume).

Thus far, baboons are like humans, in that humans exhibit the same range of reproductive behaviors. However, unlike humans, which appear to possess similar mating dispositions the world over in spite of great variation in the number of partners each sex has (e.g., Buss 1990; Schmitt 2003; Gottschall et al., 2004), the clear subspecific differentiation in baboon mating behavior appears to have a genetic basis. Bergman (this volume) and Beehner and Bergman (this volume) describe the behavior of hamadryas–olive hybrids, all living in the same environment (indeed the same social group) as representing a set of clear points along a continuum. Some hybrid individuals consistently exhibited olive-like behavior—same-sex bonding in the case of females and minimal tendencies to herd or lead nonestrous females in the case of males—while others consistently exhibited hamadryas-like behavior—cross-sex bonding in the case of females and strong tendencies to herd and lead non-estrous females in the case of males—and still others were intermediate in their behavior. This strongly suggests a lack of plasticity (in the classic sense of different phenotypes developing from the same genotype in response to different environments) in mating behavior in these subspecies. Interestingly, the behavior of hybrids was highly correlated with their morphological phenotypes, at least for females; females that appeared more hamadryas-like in their coloring and body shape were more hamadryas-like in their behavior and vice versa (Beehner and Bergman, this volume).

As we have noted, there appears to be clear subspecific differentiation in male–male coalitionary behavior as well; chacma males do not show coalitionary behavior while olive and yellow males—particularly middle-ranking males—do so frequently (Packer, 1977; Noë, 1986, 1993; Bercovitch, 1988; Bulger, 1993; Henzi et al., 1999; Henzi and Barrett, 2003). Unlike the case of mating behavior in hamadryas–olive hybrids, no "natural experiment" involving yellow–chacma hybrids allows us to rule out the possibility that coalitionary behavior is plastic. However, chacma baboons have been studied in many different habitats by many different researchers (although not in the area of overlap between chacmas and yellows, a useful gap to fill), and male coalitionary behavior has not been described (see reviews in Bulger, 1993; Henzi and Barrett, 2003). The consequences of no male coalitionary behavior in chacmas are profound, for both males and females. Chacma males live in a "winner takes all" world in which the highest-ranking male in the group obtains nearly all of the mating opportunities (Bulger, 1993; Palombit et al., 1997; Weingrill et al., 2000; Henzi and Barrett, 2003). In contrast, in olive and yellow groups, high-ranking males are sometimes able to enforce a strict queuing system based upon rank, and to obtain nearly all the mating opportunities, but their ability to do this is contingent upon both their own fighting ability and on the number, and the behavior (particularly coalitionary behavior), of other males (Packer, 1979; Bercovitch, 1988; Alberts et al., 2003). The consequences of these differences for female reproductive options and for infanticidal behavior are enormous, as noted in "Baboon Flexibility" (Palombit et al., 1997, 2000; Henzi et al., 1999; Weingrill et al., 2000; Henzi and Barrett 2003; Cheney et al., this volume; Swedell and Saunders, this volume).

The apparent fixation of different reproductive patterns in different subspecies—the apparent lack of flexibility within subspecies—suggests important areas of future study. It provides a good opportunity for understanding how different ecological conditions might select for different reproductive behaviors. It also allows us to pose the question: What differentiates a lineage such as humans, in which individuals apparently retain the potential to express a wide range of reproductive patterns in different ecological contexts, from a lineage such as baboons, in which individuals apparently lack this potential even though the entire range is expressed within the taxon? Does protracted gene flow in humans, in contrast with periods of allopatry for different baboon subspecies, explain the difference? Or are there additional traits that must be maintained over evolutionary time in order for that kind of flexibility to be exhibited by individuals?

## 5. THE RESEARCH FUTURE: PRIORITIES FOR BABOON RESEARCH IN THE COMING DECADES

The behavioral and life history flexibility of baboons makes them critical potential contributors to a number of key research problems. They certainly represent an important model for understanding the attributes of a successful species. However, because they occur in a wide range of habitats, and because some of their traits vary systematically across these habitats, they represent an important model for understanding how life history and behavioral traits are likely to evolve or vary in response to environmental variability and change. Crucially, this also means that they represent a model—probably a best-case scenario—for how primates may cope with global climate change.

The diversity and duration of field research on baboons, at sites from Ethiopia through South Africa, is unparalleled among mammals. This provides an opportunity for comparative work that is just beginning to be realized (e.g., Bulger, 1993; Bronikowski et al., 2002). The manifestations of baboon flexibility and adaptability described in this chapter and others in this volume, and the limits to baboon flexibility that we have identified, present a number of compelling opportunities for advancing our understanding of baboon ecology and evolution. In our view, these compelling opportunities for comparative work fall into five major categories: (1) understanding the impact of short- and long-term, predictable and unpredictable environmental change on baboon behavior, particularly foraging behavior and social behavior; (2) identifying the manner in which life history components (fertility and survival) vary and covary under different environmental conditions; (3)

describing the suite of covarying traits associated with subspecific differences in male and female reproductive behavior; (4) identifying the genetic, ontogenetic, and physiological bases of subspecific differences; and (5) facilitating comparative research across sites.

### 5.1. Understanding the impact of environmental change—short- and long-term, predictable and unpredictable—on baboon behavior

Although not discussed in depth in this volume, evidence suggests that an ability to respond to changing environments is of primary importance to the success of baboons. The wide geographical and environmental range that they occupy suggests that they are quite adaptable in this regard. Further, as we noted above, baboons have survived and adapted to a changing environment in Amboseli, while vervet monkeys, with whom they share many resources and behavioral strategies, have experienced local extinction (Struhsaker, 1973, 1976; Hauser et al., 1986; Isbell et al., 1990, 1991). Baboons have a number of behavioral traits that are likely to contribute to flexibility in changing environments.

The most urgent demands of a changing environment involve nutritional intake. The ability to track and adapt to changing food sources, and to actually alter their own environment by finding and moving to more suitable habitats (Altmann and Alberts, 2003a; Alberts et al., 2005), are characteristics of baboons that warrant more study. In addition, social bonds appear very important to baboons. For instance, females that are socially well-integrated experience enhanced infant survival relative to poorly integrated females (Silk et al., 2003) and females attend carefully, and appear to have a sophisticated understanding of, their own social relationships and those of others (Bergman et al., 2003). These and other data suggest that a well-buffered social organization in which individual relationships are carefully serviced may be a cornerstone of successful adaptation to environmental change.

Habitat change will influence many ancillary behaviors as well. For instance, it will affect the demographic environment by intensifying, or alleviating, the effects of density as conditions deteriorate, or improve, respectively. The demographic environment has profound consequences for both males and females because both sexes experience density dependence in some aspects of their behavior and life history (e.g., Bulger and Hamilton, 1987; Packer et al., 2000; Alberts et al., 2003; Wasser et al., 2004). As with the effects of habitat change on food availability, the effects of habitat change on the demographic environment may be mitigated to some extent by the ability of both sexes—males through dispersal and females through permanent social group fission—to modify their own group membership. For males, this is a frequently recurring but relatively high-risk opportunity; for females the opportunity is quite rare, probably only occurring about once in each female's lifetime, but is relatively low risk and garners many potential benefits—most notably a temporary return to density-independent survival and fertility.

It is important to evaluate all of these responses both in terms of long- versus short-term habitat change, and in terms of predictable versus unpredictable habitat change, as the two dimensions offer very different types of challenges and opportunities and are likely to result in different types of adaptations. Baboons in many parts of Africa will increasingly experience changing environments, and the manner in which they respond behaviorally to this change must become an important area of study as primates the world over face both shrinking and fragmented habitats, and local as well as global climate change.

### 5.2. Identifying how life history components vary—and co-vary—under different environmental conditions

The consequences of environmental variability for baboons are profound. Variation in the extent of birth seasonality among baboon populations can be directly attributed to variation in the patterns and extent of seasonality in habitats (e.g., compare Bercovitch and Harding, 1993; Bentley-Condit and Smith, 1997; Alberts et al., 2005; Barrett et al., this volume; Cheney et al., this volume). Variation in growth patterns and in mortality patterns, of both adults and immatures, are partly attributable to variable habitats (e.g., Altmann and Alberts, 2003a, 2005; Johnson 2003). Variation in infant development and survival, where lifetime fitness has its highest sensitivity, is certainly partly attributable to environmental variation (e.g., compare Bentley-Condit and Smith, 1997; Packer et al., 2000; Altmann and Alberts, 2003a; Wasser et al., 2004; Barrett et al., this volume; Cheney et al., this volume; Beehner et al. in press). These life history traits also covary, and life history correlations probably change as habitats change, in ways that are not as yet described. Variation in social organization—in mating patterns, in the extent and nature of social bonds between individuals—is often attributed to differences between habitats, but may also be attributable to variability within habitats. That is, the extent to which habitats are predictable in their productivity patterns may contribute to selection for large versus small group sizes, and for fission–fusion versus stable groupings. This may contribute not only to the variety of baboon social organizations across habitats but also to the sometimes highly flexible organizations within habitats (Galat-Luong et al., this volume).

Baboon studies also offer the opportunity to solve a long-standing puzzle: Why do the effects of dominance rank on female reproductive success vary across study populations (and also primate species)? Different baboon studies (and studies on other primates as well) show different effects of rank on age at first reproduction, fertility, offspring survival, offspring sex ratio and growth, to name some key traits (reviewed in Silk, 1987; Bercovitch and Harding, 1993; Packer et al., 2000; Altmann and Alberts, 2003a; Johnson, 2003; Wasser et al., 2004; Cheney et al., this volume; Johnson et al., this volume). One possible explanation for this is that the effects of rank on female reproductive success are density dependent as they are for male baboons (see e.g., Wasser et al., 2004). Alternatively or in addition, the effects of rank may vary depending upon the nature and quantity of food resources—possibly, rank "matters more" in some foraging contexts than in others. Similarly, the importance of kin may change as density or habitat changes. Baboon studies—especially long-term studies where such effects may vary over time in the same population—offer particularly good opportunities to examine these questions.

### 5.3. Describing the suite of covarying traits associated with subspecific differences in male and female reproductive behavior

The fascinating and well-documented variation across baboon subspecies in reproductive behavior has received increasing attention in recent years. The time is ripe for comparative work aimed at clearly describing the suite of traits that characterize each subspecies. One possibility is that the hamadryas suite of traits initially arose in response to habitats that selected for small foraging groups, with resulting high infanticide risks when foraging parties encountered each other or exchanged males (Dunbar, 1988; Anderson, 1990; Kummer, 1990, 1995; Henzi and Barrett, 2003; Swedell and Saunders, this volume); the behavioral precursor to male hamadryas behavior would have consisted of extended mate guarding (Bergman, this volume). In this view, the chacma pattern, with multimale groups, no male coalitions, a tendency of the highest-ranking male to monopolize mating, and high risk of infanticide after takeovers, represents the next point on the continuum; the lack of male coalitions results from an evolutionary history of small group sizes in which suitable coalition partners were rare (Henzi et al., 1999; Henzi and Barrett, 2003). Yellow and olive baboons fall further along the continuum and represent a response to habitats that allow and/or possibly select for larger foraging group sizes, resulting in opportunities for males to form coalitions and, therefore, a breakdown of the relatively strict rank-based queuing system observed in chacmas.

Some key and relatively easy to test predictions arise from this proposed scenario. We would expect male chacmas to have smaller testes than yellow or olive baboons, but larger than hamadryas (see Jolly and Phillips-Conroy, this volume). We might also expect male chacmas to have shorter maximum adult lifespans than yellow or olive males, because their reproductive lifespan is shorter and hence their reproductive value drops more rapidly after their prime years. The work of Bergman (this volume) predicts that high-ranking male chacmas should show interest in cycling females earlier in the follicular phases than their high-ranking olive or yellow counterparts, and by extension we would predict that high-ranking olive and yellow males would show greater selectivity among females in the follicular phases of their cycles, preferring more fertile females (multiparous females, and those in conceptive versus nonconceptive cycles) over less fertile ones (adolescent and other nonconceptive females).

With respect to female behavior, we might predict that the extent to which female preferences—either through overt choice or cryptic choice—influence mating behavior and mating outcomes will vary systematically as a function of the intensity of male–male competition over particular cycles. For instance, in chacmas, where high-ranking males generally monopolize mating opportunities and females are therefore relatively monandrous in the short term, overt female choice may be relatively minimal, but selection for cryptic female choice would certainly occur. In contrast, in contexts where even low-ranking males may achieve paternity, females may have more opportunity to play a role in determining the success of preferred males in maintaining mate-guarding episodes.

A related set of traits, which may either differentiate or unify baboon subspecies, concerns paternal behavior. It has long been known that male and female baboons sometimes form preferences for opposite-sex partners that persist beyond the follicular phase of the female's sexual cycle, and that these preferences, on the part of males, may be related to paternity (e.g., Altmann, 1980; Smuts, 1985; Palombit et al., 1997, 2000). What has recently emerged is that males play a more pronounced paternal role than had previously been suspected—they provide direct paternal care and differentiate their own offspring from those of other males (Palombit et al., 1997, 2000; Buchan et al., 2003). Many aspects of this paternal behavior remain unexplored. For instance, males may contribute to the survival of their offspring not only by protecting them against infanticide, but also by providing paternal care in noncrisis situations, carrying and grooming them, providing a buffer zone for uninterrupted foraging, and reducing common harassment. Males may facilitate the development of paternal sib relationships (Widdig et al., 2001; Smith et al., 2003) by providing a common relationship—a connection—through which paternal sibs become familiar with each other. And males may provide assistance during agonistic encounters that facilitates rank attainment—for both sons and daughters—in the same way that assistance from mothers facilitates rank attainment for daughters. Indeed, fathers may be even more effective allies than mothers in this regard because of their large size. They can assist sons, which mothers generally cannot when their sons' opponents surpass them in size, and they can assist daughters against higher-ranking females, which mothers are more limited in their ability to do. None of these possible male social functions has been tested, but all represent key aspects of baboon social and behavioral flexibility that need to be understood.

### 5.4. Identifying the genetic, ontogenetic, and physiological bases of subspecific and intrapopulational differences

As more data emerge from more sites, the hypothesis that baboon populations have undergone behavioral differentiation in allopatry—that subspecies are genetically differentiated with respect to behavior—is receiving greater support. Some differences are relatively well described, while others are still being explored. For instance, while the chacma–olive–yellow differences in male reproductive behavior have received a good deal of recent attention, some data suggest that behavioral and genetic differentiation may have occurred within the chacma lineage as well, with the more southern populations representing a morphological extreme relative to the more northern populations, and behavioral differentiation perhaps occurring on an even finer geographic scale (Anderson, 1990; Newman et al., 2004; Babb et al., 2005).

As the suites of traits that describe each subspecies become better defined, the opportunity for identifying the genetic bases of these differences becomes real. Candidate gene approaches will increasingly be an option, with new high-throughput genotyping technologies. Studies of quantitative trait loci (QTLs) may be realistic even sooner, if hybrid populations (captive or wild) can yield pedigrees of even moderate depth, with good phenotypic behavioral data and high-quality genetic samples to accompany them. QTL analysis in combination with candidate gene approaches will be the most powerful technique in the immediate future, as exemplified by work on other organisms (e.g., Geiger-Thornsberry and Mackay, 2004; Pasyukova et al., 2004; Wheeler et al., 2005).

Behavioral and life history differences arise at various demographic levels and under various environmental conditions. Another major research agenda should be to identify the processes—not only genetic, but also ontogenetic, and in terms of both organizing and activating physiological mechanisms—by which such differences arise. Several new technologies, such as the increasing study of steroid hormones through noninvasive sampling, make this feasible. Hypotheses can be developed and tested that relate behavioral experience, ontogenetic trajectories, and differences in steroid concentrations to behavioral differences, not only among juveniles and adults within the same population, but also among populations. This research agenda will require individual-based longitudinal life history studies.

As new technologies become available, both the puzzle of subspecific differentiation and the ontogeny of differences among individuals within populations will certainly be tackled. In this context, biologists with good phenotypic data on wild populations will hold an increasingly valuable resource.

## 5.5 Facilitating comparative research across sites

Finally, to accomplish the agenda that we outline as well as other priorities that emerge, a high priority will need to be given to methods of data collection, analysis, and presentation that facilitate comparison and more rigorous theory testing. For example, while it has emerged relatively clearly that chacma, yellow, and olive males differ in their tendencies to form male–male coalitions, the related question of whether females in the subspecies differ in coalition formation remains unanswered (Henzi and Barrett, 2003; Silk et al., 2004). This is simply because researchers at different sites have tended to use different measures of coalition frequency and formation, making it difficult to compare across sites (Silk et al., 2004). Additionally, efforts that are collaborative across study sites—using the same methodology to produce the similar data sets that can be directly compared (e.g., Bronikowski et al., 2002; Barrett et al., this volume)—are likely to have rich payoffs that will make the greater time investment and leaps of trust well worth the effort.

Indeed, this is true not only across baboon studies, but also across primate studies more generally. For instance, rhesus macaques, like baboons, appear highly adaptable and have a very large geographic range. This makes them a prime candidate for comparisons with baboons. Studies on the two species have generally differed markedly in that behavioral studies of macaques have involved primarily captive or provisioned populations, while studies on baboons have involved primarily wild populations, and this must be taken into consideration in comparative studies, as it constrains the kinds of comparisons that can be made. Nonetheless, there is interesting potential for identifying patterns of flexibility common across the two species.

## 6. CONCLUSIONS

Baboons show great flexibility within populations and also interesting differentiation among subspecies. This provides a remarkable opportunity for comparative work in this species, especially given the relatively large number of baboon field studies, several of them now with long-term data. We have identified key areas for future research—which we view as priorities—that represent just a subset of the opportunities that baboons provide. As the habitats of wild primates become increasingly fragmented, and as both global and local climate change proceed, the urgency of producing good comparative work within a single, well-studied species increases. The potential fruits include an understanding of behavioral and life history diversity as generated in response to environmental variability and change, and a comprehensive understanding of the limits to adaptability that primate species can be expected to exhibit.

## ACKNOWLEDGMENTS

This work was supported by NSF-IOB 0322613, NSF BCS-0323553, NSF IOB-0322781, NSF BCS-0323596, R03 MH65294, and the Chicago Zoological Society. Our long-term research in Amboseli is facilitated by cooperation from the Office of the President, Republic of Kenya, the Kenya Wildlife Services, its Amboseli staff and Wardens, the Institute of Primate Research, the National Museums of Kenya, and the members of the Amboseli-Longido pastoralist communities.

## REFERENCES

Alberts, S. C. and Altmann, J., 2003, Matrix models for primate life history analysis, in: *Primate Life Histories and Socioecology*, P. M. Kappeler, and M. E. Pereira, eds., University of Chicago Press, Chicago.

Alberts, S. C., Watts, H. E., and Altmann, J., 2003, Queuing and queue-jumping: Long-term patterns of reproductive skew in male savannah baboons, *Papio cynocephalus, Anim. Behav.* **65**:821–840.

Alberts, S. C., Hollister-Smith, J. A., Mututua, R. S., Sayialel, S. N., Muruthi, P. M., Warutere, J. K., and Altmann, J., 2005, Seasonality and long-term change in a savannah environment, in: *Seasonality in Primates: Studies of Living and Extinct Human and Non-human Primates*, D. K. Brockman, and C. P. van Schaik, eds., Cambridge University Press, Cambridge, pp. 157–196.

Altmann, J., 1980, *Baboon Mothers and Infants*, Harvard University Press, Cambridge, MA.

Altmann, S. A., 1998, *Foraging for Survival*, University of Chicago Press, Chicago.

Altmann, J. and Alberts, S., 1987, Body mass and growth rates in a wild primate population, *Oecologia* **72**:15–20.

Altmann, J. and Alberts, S. C., 2003a, Intraspecific variability in fertility and offspring survival in a nonhuman primate: Behavioral control of ecological and social sources, in: *Offspring: Human Fertility Behavior in Biodemographic perspective*, K. W. Wachter, and R. A. Bulatao, eds., National Academies Press, Washington, DC.

Altmann, J. and Alberts, S. C., 2003b, Variability in reproductive success viewed from a life-history perspective in baboons, *Am. J. Hum. Biol.* **15**:401–409.

Altmann, J. and Alberts, S. C., 2005, Growth rates in a wild primate population: ecological influences and maternal effects, *Behav. Ecol. Sociobiol.* **57**:490–501.

Altmann, S. A. and Altmann, J., 1970, *Baboon Ecology*. University of Chicago Press, Chicago.

Altmann, J. and Samuels, A., 1992, Costs of maternal care: Infant carrying in baboons, *Behav. Ecol. Sociobiol.* **29**:391–398.

Altmann, J., Altmann, S. A., and Hausfater, G., 1978, Primate infant's effects on mother's future reproduction, *Science* **201**:1028–1030.

Altmann, J., Hausfater, G., and Altmann, S., 1985, Demography of Amboseli baboons, 1963–1983, *Am. J. Primatol.* **8**:113–125.

Altmann, S. A., Post, D. G., and Klein, D. F., 1987, Nutrients and toxins of plants in Amboseli, Kenya, *Afr. J. Ecol.* **25**:279–293.

Altmann, J., Altmann, S., and Hausfater, G., 1988, Determinants of reproductive success in savannah baboons (*Papio cynocephalus*), in: *Reproductive Success*, T. H. Clutton-Brock, ed., University of Chicago Press, Chicago.

Altmann, J., Alberts, S. C., Altmann, S. A., and Roy, S. B., 2002, Dramatic change in local climate patterns in the Amboseli basin, Kenya, *Afr. J. Ecol.* **40**:248–251.

Anderson, C. M., 1990, Desert, mountain and savanna baboons: A comparison with special reference to the Suikerborsrand population, in: *Baboons, Behaviour and Ecology, Use and Care: Selected Proceedings of the 12th Congress of the International Primatological Society*, M. Thiego de Mello, A. Whiten, and R. W. Byrne, eds., Brasilia, Brazil.

Babb, P. L., Sithaldeen, R., Ackermann, R. R., and Newman, T. K., 2005, Mitochondrial DNA sequence evidence for a deep phylogenetic split in chacma baboons (*Papio hamadryas ursinus*) and the phylogeographic implications for Papio systematics, *Am. J. Phys. Anthropol.* 67–68.

Barker, D. J. P., 2001, Preface: Type 2 diabetes: The thrifty phenotype, *Br. Med. Bull.* **60**:1–3.

Barton, R. A., Byrne, R. W., and Whiten, A., 1996, Ecology, feeding competition and social structure in baboons. *Behav. Ecol. Sociobiol* **38**:321–329.

Beehner, J. C., Onderdonk, D. A., Alberts, S. C., and Altmann, J., 2006, The ecology of conception and pregnancy failure in wild baboons. *Behav. Ecol.* in press.

Behrensmeyer, A. K., 1993, The bones of Amboseli: The taphonomic record of ecological change in Amboseli Park, Kenya, *Nat. Geogr. Res. Explor.* **9**:402–421.

Bentley-Condit, V. K. and Smith, E. O., 1997, Female reproductive parameters of Tana River yellow baboons, *Int. J. Primatol.* **18**:581–596.

Bercovitch, F. B., 1988, Coalitions, cooperation and reproductive tactics among adult male baboons, *Anim. Behav.* **36**:1198–1209.

Bercovitch, F. B. and Harding, R. S. O., 1993, Annual birth patterns of savanna baboons (*Papio cynocephalus anubis*) over a ten-year period at Gilgil, Kenya, *Folia Primatologica* **61**:115–122.

Bergman, T. J., Beehner, J. C., Cheney, D. L., and Seyfarth, R. M., 2003, Hierarchial classification by rank and kinship in baboons, *Science* **302**:1234–1236.

Bobe, R. and Behrensmeyer, A. K., 2004, The expansion of grassland ecosystems in African in relation to mammalian evolution and the origin of the genus Homo, *Palaeogeogr. Palaeoclimatol. Palaeoecol.* **207**:399–420.

Bronikowski, A. and Altmann, J., 1996, Foraging in a variable environment: Weather patterns and the behavioral ecology of baboons, *Behav. Ecol. Sociobiol.* **39**:11–25.

Bronikowski, A. M., Alberts, S. C., Altmann, J., Packer, C., Carey, K. D., and Tatar, M., 2002, The aging baboon: Comparative demography in a non-human primate, *Proc. Natl Acad. Sci. USA* **99**:9591–9595.

Buchan, J. C., Alberts, S. C., Silk, J. B., and Altmann, J., 2003, True paternal care in a multi-male primate society, *Nature* **425**:179–181.

Bulger, J. B., 1993, Dominance rank and access to estrous females in male savanna baboons, *Behaviour* **127**:67–103.

Bulger, J. and Hamilton, W. J., III, 1987, Rank and density correlates of inclusive fitness measures in a natural chacma baboon (*Papio ursinus*) troop, *Int. J. Primatol.* **8**:635–650.

Buss, D. M., 1990, International preferences in selection mates: A study of 37 cultures, *J. Cross-cult. Psychol.* **21**:5–47.

Byrne, R. W., Whiten, A., Henzi, S. P., and McCulloch, F. M., 1993, Nutritional constraints on mountain baboons (*Papio ursinus*): Implications for baboon socioecology, *Behav. Ecol. Sociobiol.* **33**:233–246.

Dunbar, R. I. M., 1988, *Primate Social Systems*, Chapman and Hall, London.

Dunbar, R. I. M., 1998, The social brain hypothesis, *Evol. Anthropol.* **6**:178–190.

Dunbar, R. I. M. and Dunbar, E. P., 1977, Dominance and reproductive success among female gelada baboons. *Nature* **266**:351–352.

Dunbar, R. I. M., Hannah-Stewart, L., and Dunbar, P., 2002, Forage quality and the costs of lactation for female gelada baboons, *Anim. Behav.* **64**:801–805.

Eley, R. M., Strum, S. C., Muchemi, G., and Reid, G. D. F., 1989, Nutrition, body condition, activity patterns, and parasitism of free-ranging troops of olive baboons (*Papio anubis*) in Kenya, *Am. J. Primatol.* **18**:209–219.

Estes, R. D., 1991, *The Behavior Guide to African Mammals*, University of California Press, Berkeley.

Fleagle, J. G., 1999, *Primate Adaptation and Evolution*. Academic press, San Diego.

Foley, R., 1987, *Another Unique Species: Patterns in Human Evolutionary Ecology*, Wiley, New York.

Foley, R. A., 1993, The influence of seasonality on hominid evolution, in: *Seasonality and Human Ecology: 35th Symposium Volume of the Society for the Study of Human Biology*, S. J. Ulijaszek and S. S. Strickland, eds., Cambridge University Press, Cambridge, UK.

Foley, R. A., 1994, Speciation, extinction and climatic change in hominid evolution. *J. Hum. Evol.* **26**:275–289.

Geiger-Thornsberry, G. L. and Mackay, T. F. C., 2004, Quantitative trait loci affecting natural variation in Drosophila longevity, *Mech. Ageing Devel.* **125**(3):179–189.

Goldizen, A. W., 1987, Tarmins and marmosets: communal care of offspring, in: *Primate Societies*, B. B. Smuts, D. L. Cheney, R. Seyfarth, R. W. Wrangham, and T. T. Struhsaker, eds., University of Chicago Press, Chicago.

Gottschall, J., Martin, J., Quish, H., and Rea, J., 2004, Sex differences in mate choice criteria are reflected in folktales from around the world and in historical European literature, *Evol. Hum. Behav.* **25**(2):102–112.

Groves, C., 2001, *Primate Taxonomy*. Smithsonian Institution press. Washington and London.

Hales, C. N. and Barker, D. J. P., 2001, The thrifty phenotype hypothesis: Type 2 diabetes, *Br. Med. Bull.* **60**:5–20.

Hamilton, W. J. I., Buskirk, R. E., and Buskirk, W. H., 1978, Omnivory and utilization of food resources by chacma baboons, *Papio ursinus, Am. Nat.* **112**:911–924.

Hastenrath, S. and Greischar, L., 1997, Glacier recession on Kilimanjaro, East Africa, 1912–89, *J. Glaciol.* **43**(145):455–459.

Hauser, M. D., Cheney, D. L., and Seyfarth, R. M., 1986, Group extinction and fusion in free-ranging vervet monkeys, *Am. J. Primatol.* **11**:63–77.

Henzi, P. and Barrett, L., 2003, Evolutionary ecology, sexual conflict, and behavioral differentiation among baboon populations, *Evol. Anthropol.* **12**:217–230.

Henzi, S. P., Weingrill, T., and Barrett, L., 1999, Male behaviour and the evolutionary ecology of chacma baboons, *S. Afr. J. Sci.* **95**:240–242.

Hill, R. A., Lycett, J. E., and Dunbar, R. I. M., 2000, Ecological and social determinants of birth intervals in baboons, *Behav. Ecol.* **11**:560–564.

Hrdy, S. B., 1999, *Mother Nature: A History of Mothers, Infants, and Natural Selection*, Pantheon Books, New York.

Hrdy, S. B., 2005, Evolutionary context of human development, in: *Attachment and Bonding: A New Synthesis*, C. S. Carter, K. E. Ahnert, S. B. Grossman, S. B. Hrdy, M. E. Lamb, S. W. Porges, and N. Sachser, eds., MIT Press, Cambridge, MA.

Isbell, L. A., Cheney, D. L., and Seyfarth, R. M., 1990, Costs and benefits of home range shifts among vervet monkeys (*Cercopithecus aethiops*) in Amboseli National Park, Kenya, *Behav. Ecol. Sociobiol.* **27**:351–358.

Isbell, L. A., Cheney, D. L., and Seyfarth, R. M., 1991, Group fusions and minimum group sizes in vervet monkeys (*Cercopithecus aethiops*), *Am. J. Primatol.* **25**:57–65.

Janson, C. H. and van Schaik, C. P., 2000, The behavioral ecology or infanticide by males, in: *Infanticide by Males and Its Implications*, C. P. van Schaik and C. P. Janson, eds., Cambridge University Press, Cambridge, UK.

Johnson, S. E., 2003, Life history and the competitive environment: Trajectories of growth, maturation, and reproductive output among Chacma baboons, *Am. J. Phys. Anthropol.* **120**:83–98.

Jolly, C. J., 1993, Species, subspecies, and baboon systematics, in: *Species, Species Concepts, and Primate Evolution*, W. H. Kimbel, and L. B. Martin, eds., Plenum Press, New York.

Kingdon, J., 1997, *The Kingdon Field Guide to African Mammals*, Academic Press, San Diego.

Klein, R. G., 1999, *The Human Career: Human Biological and Cultural Origins*, second Edition, University of Chicago Press, Chicago.

Kummer, H., 1990, The social system of hamadryas baboons and its presumable evolution, in: *Baboons, Behaviour and Ecology, Use and Care: Selected Proceedings of the 12th Congress of the International Primatological Society*, M. Thiego de Mello, A. Whiten, and R. W. Byrne, eds., Brasilia, Brazil.

Kummer, H., 1995, *In Quest of the Sacred Baboon*, Princeton University Press, Princeton.

Lyles, A. M. and Dobson, A. P., 1988, Dynamics of provisioned and unprovisioned primate populations, in: *Ecology and Behavior of Food-enhanced Primate Groups*, J. E. Fa, and C. H. Southwick, eds., Liss, New York.

Malenky, R. K. and Wrangham, R. W., 1994, A quantitative comparison of terrestrial herbaceous food consumption by *Pan paniscus* in the Lomako Forest, Zaire, and *Pan troglodytes* in the Kibale Forest, Uganda, *Am. J. Primatol.* **32**:1–12.

Melnick, D. J. and Pearl, M. C., 1987, Cercopithecines in multimale groups: Genetic diversity and population structure, in: *Primate Societies*, B. B. Smuts, D. L. Cheney, R. Seyfarth, R. W. Wrangham, and T. T. Struhsaker, eds., University of Chicago Press, Chicago.

Newman, T. K., Jolly, C. J., and Rogers, J., 2004, Mitochondrial phylogeny and systematics of baboons (Papio), *Am. J. Phys. Anthropol.* **124**:17–27.

Noë, R., 1986, Lasting alliances among adult male savannah baboons, in: *Primate Ontogeny, Cognition, and Social Behaviour*, J. G. Else, and P. C. Lee, eds., Cambridge University Press, Cambridge.

Noë, R., 1993, Alliance formation among male baboons: shopping for profitable partners, in: *Cooperation and Conflict: Coalitions and Alliances in Animals and Humans*, A. H. Harcourt, and F. B. M. de Waal, eds., Oxford University Press, Oxford.

Norton, G. W., Rhine, R. J., Wynn, G. W., and Wynn, R. D., 1987, Baboon diet: A five-year study of stability and variability in the plant feeding and habitat of the yellow baboons (*Papio cynocephalus*) of Mikumi National Park, Tanzania, *Folia Primatol.* **48**:78–120.

Packer, C., 1977, Reciprocal altruism in *Papio anubis, Nature* **265**:441–443.

Packer, C., 1979, Male dominance and reproductive activity in *Papio anubis, Anim. Behav.* **27**:37–45.

Packer, C., Collins, D. A., and Eberly, L. E., 2000, Problems with primate sex ratios, *Phil. Trans. R. Soc.* **355**:1627–1635.

Palombit, R. A., Seyfarth, R. M., and Cheney, D. L., 1997, The adaptive value of friendships to female baboons: Experimental and observational evidence, *Anim. Behav.* **54**:599–614.

Palombit, R. A., Cheney, D. L., Fischer, J., Johnson, S., Rendall, D., Seyfarth, R. M., and Silk, J. B., 2000, Male infanticide and defense of infants in chacma baboons, in: *Infanticide by Males and Its Implications*, C. P. van Schaik, and C. P. Janson, eds., Cambridge University Press, Cambridge, UK.

Pasyukova, E. G., Roshina, N. V., and Mackay, T. F. C., 2004, Shuttle craft: A candidate quantitative trait gene for Drosophila lifespan, *Aging Cell* **3**(5):297–307.

Perry, S., 1998, Female–female social relationships in wild white-faced capuchins, *Cebus capucinus, Am. J. Primatol.* **40**:167–182.

Post, D. G., 1982, Feeding behavior of yellow baboons (*Papio cynocephalus*) in the Amboseli National Park, Kenya, *Int. J. Primatol.* **3**:403–430.

Potts, R., 1996, Evolution and climate variability, *Science* **273**:922–923.

Potts, R., 1998a, Variability selection in hominid evolution, *Evol. Anthropol.* 7:81–96.

Potts, R., 1998b, Environmental hypotheses of hominin evolution, *Yrbk Phys. Anthropol.* **41**:93–136.

Ross, C., 1998, Primate life histories, *Evol. Anthropol.* **6**(2):54–63.

Sade, D. S., 1990, Intrapopulation variation in life-history parameters, in: *Primate Life History and Evolution*, C. J. DeRousseau, ed., Wiley-Liss, New York.

Schmitt, D. P., 2003, Universal sex differences in the desire for sexual variety: Tests from 52 nations, 6 continents, and 13 islands, *J. Personality Social Psychol.* **85**(1):85–104.

Silk, J. B., 1987, Social behavior in evolutionary perspective, in: *Primate Societies*, B. B. Smuts, D. L. Cheney, R. M. Seyfarth, R. W. Wrangham, and T. T. Struhsaker, eds., University of Chicago Press, Chicago.

Silk, J. B., Alberts, S. C., and Altmann, J., 2003, Social bonds of female baboons enhance infant survival, *Science* **302**:1231–1234.

Silk, J. B., Alberts, S. C., and Altmann, J., 2004, Patterns of coalition formation by adult female baboons in Amboseli, Kenya, *Anim. Behav.* 67:573–582.

Smith, K., Alberts, S. C., and Altmann, J., 2003, Wild female baboons bias their social behaviour towards paternal half-sisters, *Proc. R. Soc. Lond. B* **270**(1514):503–510.

Smuts, B. B., 1985, *Sex and Friendship in Baboons*, Aldine, Hawthorn, NY.

Sterck, E. H. M., 1999, Variation in langur social organization in relation to the socioecological model, human habitat alternation, and phylogenetic constraints, *Primates* **40**:199–213.

Struhsaker, T. T., 1967, Ecology of vervet monkeys (*Cercopithecus aethiops*) in the Masai-Amboseli Game Reserve, Kenya, *Ecology* **48**:891–904.

Struhsaker, T. T., 1973, A recensus of vervet monkeys in the Masai-Amboseli Game Reserve, Kenya, *Ecology* **54**:930–932.

Struhsaker, T. T., 1976, A further decline in numbers of Amboseli vervet monkeys, *Biotropica* **8**:211–214.

Strum, S. C., 1991, Weight and age in wild olive baboons, *Am. J. Primatol.* **25**:219–237.

Strum, S. C. and Western, J. D., 1982, Variations in fecundity with age and environment in olive baboons (*Papio anubis*), *Am. J. Primatol.* **3**:61–76.

Thompson, L. G., Mosley-Thompson, E., Davis, M. E., Henderson, K. A., Brecher, H. H., Zagotodnov, V. S., Mashiotta, T. A., Lin, P.-N., Mikhalenko, V. N., Hardy, D. R., and Beer, J., 2002, Kilimanjaro ice core records: Evidence of holocene climate change in tropical Africa, *Science* **298**:589–593.

Wasser, S. K., Norton, G. W., Kleindorfer, S., and Rhine, R. J., 2004, Population trend alters the effects of maternal dominance rank on lifetime reproductive success in yellow baboons (*Papio cynocephalus*), *Behav. Ecol. Sociobiol.* **56**(4):338–345.

Weingrill, T., Lycett, J. E., and Henzi, S. P., 2000, Consortship and mating success in chacma baboons (*Papio cynocephalus ursinus*), *Ethology* **106**:1033–1044.

Western, D. and van Praet, C., 1973, Cyclical changes in the habitat and climate of an East African ecosystem, *Nature* **241**:104–106.

Wheeler, F. C., Fernandez, L., Carlson, K. M., Wolf, M. J., Rockman, H. A., and Marchuk, D. A., 2005, QTL mapping in a mouse model of cardiomyopathy reveals an ancestral modifier allele affecting heart function and survival, *Mammal. Genome* **16**(6):414–423.

Whiten, A., Byrne, R. W., Barton, R. A., Waterman, P. G., and Henzi, S. P., 1991, Dietary and foraging strategies of baboons, *Phil. Trans. R. Soc. Lond. B* **334**: 187–197.

Widdig, A., Nürnberg, P., Krawczak, M., and Bercovitch, F. B., 2001, Paternal relatedness and age proximity regulate social relationships among adult female rhesus macaques, *Proc. Natl Acad. Sci. USA* **98**:13769–13773.

Wrangham, R. W. and Waterman, P. G., 1981, Feeding behaviour of vervet monkeys on *Acacia tortilis* and *Acacia xanthophloea*: With special reference to reproductive strategies and tannin production, *J. Anim. Ecol.* **50**:715–731.

Wrangham, R. W., Conklin, N. L., Chapman, C. A., and Hunt, K. D., 1991, The significance of fibrous foods for Kibale Forest chimpanzees, *Phil. Trans. R. Soc. Lond. S. B* **334**:171–178.

Wrangham, R. W., Conklin-Brittain, N. L., and Hunt, K. D., 1998, Dietary response of chimpanzees and cercopithecines to seasonal variation in fruit abundance. I. Antifeedants, *Int. J. Primatol.* **19**:949–970.

# Index

Page numbers followed by f and t indicate figures and tables, respectively.